Identification Tables for Minerals in Thin Sections

Identification Tables
for Minerals in Thin Sections

E. P. Saggerson

Professor of Geology, University of Natal, Durban

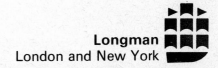

Longman
London and New York

Longman Group Limited London

*Associated companies, branches and
representatives throughout the world*

Published in the United States of America
by Longman Inc., New York

© Longman Group Limited 1975

First published 1975

ISBN 0 582 44343 1

Printed by J. W. Arrowsmith Ltd., Bristol

Contents

Introduction

This book arose from the need by students for a ready reference and rapid method of identification of the commoner rock-forming minerals. The student frequently has difficulty in identifying some of the lesser known but important minerals with any degree of certainty. Although many excellent texts containing numerous tables and methods of mineral identification in thin section already exist, it is felt that a summary of much of the published information would serve the needs of many students at undergraduate level using those reference works such as Deer, Howie and Zussman's *Rock-Forming Minerals*. The tables, however, are neither exhaustive nor do they afford a simple and infallible recipe for mineral recognition but are designed for convenience once the fundamentals of mineral optics have been mastered. It is hoped that as well as giving the student greater confidence in the microscopic determination of thin section material, it will also serve to draw attention to minerals of similar properties that might otherwise escape attention and possibly be omitted in the course of routine examinations.

Acknowledgements

The tables presented have resulted from the teaching of practical classes in mineralogy and I am, therefore, indebted to many students who by their many helpful suggestions persuaded me to make this compilation. In particular I would like to express my appreciation to Professor J. McIver, T. Mackie, A. Cain, and J. Guy for their discussions; to Professor E. W. Heinrich for his willingness in allowing me to use information from his book, *Microscopic Identification of Minerals*, and to Professors Deer, Howie and Zussman and their publisher Longman for their suggestions and permission to use the information given in their *Rock-Forming Minerals* on which the tables are mainly based. Further I would like to thank Messrs. J. Wiley and Sons Limited, for permission to quote additional information from *Feldspars* by T. F. W. Barth, and *Elements of Optical Mineralogy* by A. N. Winchell, and also to Messrs. McGraw-Hill Book Company to quote excerpts from *Microscopic Identification of Minerals* by E. W. Heinrich and *Optical Mineralogy* by P. F. Kerr. Finally but not least I would very gratefully like to acknowledge the willing help given by Mrs D. Lyall and Mrs D. J. Kirrane in typing a difficult manuscript.

The Tables and Their Use

The tables summarise the optical properties of rock-forming minerals as seen in thin sections and are designed principally for rapid identification of those minerals listed by Deer, Howie and Zussman in their reference work *Rock-Forming Minerals* (1962). The material has been collated using most of the data given in the reference work though supplemented in certain instances from other texts. It is not intended that the tables should be exhaustive, they are primarily for the use of the student: advanced workers will have recourse to the more detailed and exhaustive texts such as those by Dana, Winchell, Tröger and Larsen and Berman or to the more sophisticated methods of the X-ray diffractometer.

Determinative tables based on optical properties are numerous and can be combined in different ways. The present tables attempt to summarise the main properties of approximately 370 minerals in a discriminative manner. Students with a knowledge of the polarising microscope and good observational powers should have little trouble, therefore, in utilising the information provided. Two main tables are given using identical information, the only difference being in the order of the subdivision; in Table 1 birefringence has preference over relief whereas in Table 2 relief has preference over birefringence. The succeeding tables are the assembled data for the main specific mineral groups and include amphiboles, chlorites, epidotes, feldspars, feldspathoids, micas (including the brittle micas), olivines, pyroxenes and zeolites. In some instances, for example the plagioclase series, properties for individual members of the series as well as the group as a whole have been given.

Colour is not always a reliable means for identifying minerals but as Winchell (1957, p. 91)* has pointed out 'colour is so easily observed' that visual separation of minerals according to colour is automatic. The tables are therefore subdivided into minerals that are colourless, pink, red, brown, orange, yellow, green, blue, purple, black, grey and white. The intensity of colour is rarely, if ever, constant and many mineral species, especially those that are pleochroic, display different tints as well as differences in absorption on rotation. In these cases as well as in those with other variable optical properties, the minerals are shown in as many places as necessary to express all known variations in their optical properties. Pleochroism may be expressed in terms of variation in intensity of colour and this has been shown as follows:

Neutral N Weak W Moderate M Strong S Extreme E

The second subdivision is in order of increasing optical complexity and minerals are arranged according to the following groups:

Isometric Uniaxial positive Uniaxial negative
Biaxial positive Biaxial negative

The most common biaxial minerals with $2V \simeq 0°$ have been listed under uniaxial as well as biaxial characteristics.

The next subdivision in the discriminative system is that based on the birefringence of a mineral. If the *maximum* birefringence can be obtained for a mineral in random thin sections, then this is a property of real determinative value and may in some instances be of greater use as a means of identification than the qualitative method of relief. The method has the advantage in that it relies in the first place, on colour identification and in this respect has an immediate visual impact when a mineral is viewed between crossed polars. Experience with students has shown that whilst a large number use relief as an important diagnostic property in thin section examination, there are equally large numbers who prefer to use the property of birefringence for identification purposes. Table 1 is similar in all respects to Table 2 except that the minerals in the former have been placed in order of increasing birefringence whereas in the latter table

*Winchell, A. N. (1957), *Elements of Optical Mineralogy*, Part III, Determinative Tables, 2nd impression, Wiley.

the opposite holds. For thin sections with a thickness $t = 30m\mu$ the following designations apply:

Birefringence	Abbreviation
0·000—0·009	1
0·009—0·019	2
0·019—0·036	3
0·036—0·055	4
>0·055	5

In Table 1 the minerals are thereafter arranged in the order of their mean indices of refraction (ω or β). In mounted thin sections it is not possible to determine quantitatively the refractive indices of minerals though these may be compared with other known minerals or with that of the mounting medium, usually accepted as $n \simeq 1·54$. If they are lower than the medium the relief is said to be negative and if higher, the relief is positive. The author has followed Heinrich (1965, p. 389)* in using abbreviations as an expression of mineral relief viz:

Index	Relief	Abbreviation
<1·45	negative, moderate	—M
1·45—1·54	negative, low	—L
1·54	neutral	N
1·54—1·60	positive, low	+L
1·60—1·70	positive, moderate	+M
1·70—1·80	positive, high	H
1·80—2·00	positive, very high	VH
>2·00	positive, extreme	E

Thereafter the tables contain the remaining distinctive optical properties that are considered to have diagnostic value, minerals of similar classification under the initial three sub-

---

*Heinrich, E. W. (1965), *Microscopic Identification of Minerals*, McGraw-Hill.

divisions being placed in order of increasing optic axial angle (2V). In those instances where a specific property can be recognised in thin section this is documented, as for example, twinning, whereas properties not normally seen in thin section are omitted. The entry of a symbol or property does not necessarily imply that these are always seen in thin sections; those properties shown in brackets tend to be less common.

In the penultimate column the class of rocks with which the mineral is most closely associated is given. The rock classes, Igneous, Sedimentary, Metamorphic and Ores are shown appropriately as I, S, M, O respectively, being arranged in the relative order of importance whenever a mineral occurs in more than one class.

The properties of direct determinative value as given on the left-hand page are in most instances sufficient to identify the minerals by a process of elimination. It is logical to assess the properties, thereby coming to a conclusion with regard to the mineral name and it is for this reason that name, chemical composition and crystal system are shown separating quantitative properties from those of confirmatory and qualitative value which appear on the right-hand page.

Crystal systems are indicated by appropriate abbreviations:

Cubic Cub. Tetragonal Tet. Hexagonal Hex.
Trigonal Trig. Orthorhombic Orth. Monoclinic Mon.
Triclinic Tr. Mineraloid or Cryptocrystalline M'loid.

A number of substances such as volcanic glass and mineraloids do not qualify as minerals but their identification nevertheless may be important and therefore, are included for completeness.

In Tables 3—11 is assembled the optical data for the principal mineral groups. These have been arranged in alphabetical order and though the arrangement in each group is largely similar to that in Tables 1 and 2, the crystal system has additionally been used as a means of subdivision.

Other abbreviations and symbols

(001), (010)	crystal faces
mμ	millimicron
n	refractive index
O.A.P.	optic axial plane
r < v, (or r > v)	the optic axial angle in red light is less than (or greater than) that in violet light = dispersion
2V	optic axial angle
x, y, z	the crystal axes
α, β, γ	the vibration directions of the fast, intermediate and slow rays; the angles between the positive directions of the y and x, x and z, and x and y crystal axes
δ	birefringence
ε	extraordinary ray, refractive index
ω	ordinary ray, refractive index
Z	zoning
//	parallel to
$\perp$	perpendicular to
$\simeq$	equal to or nearly equal to
An	anorthite
anom.	anomalous
cf.	compare
incl.	including
elong. sect.	elongate section(s)
long. sect.	longitudinal section(s)
rect. sect.	rectangular sections
x-sect.	cross section(s)
x-dispersion	crossed dispersion
(weak), (strong), etc.	may be weak, may be strong, etc.

Table 1

Colour, optical group and birefringence

COLOURLESS MINERALS	Pleochroism	δ	Relief	2V Dispersion	Extinction	Orientation	Mineral	Composition
Isometric								
			−M				FLUORITE	CaF_2
			−L				LECHATELIERITE	SiO_2
			N				HALITE	NaCl
			+L				CLIACHITE	$Al_2O_3(H_2O)x$
			+M				PLAZOLITE	$3CaO \cdot Al_2O_3 \cdot 2SiO_2 \cdot 2H_2O$
	(aniso-tropic)		+M			length slow	MELILITE	$(Ca,Na)_2\{(Mg,Fe^{+2},Al,Si)_3O_7\}$
			H				HELVITE	$Mn_4\{Be_3Si_3O_{12}\}S$
			H				DANALITE	$Fe_4\{Be_3Si_3O_{12}\}S$
			H				GENTHELVITE	$Zn_4\{Be_3Si_3O_{12}\}S$
			H				PERICLASE	MgO
			H				PYROPE	$Mg_3Al_2Si_3O_{12}$
			H − E				SPINEL	(Mg,Fe^{+2},Zn,Mn,Ni) $(Al_2,Fe_2^{+3},Cr_2,Ti)O_4$
			VH				ALMANDINE	$Fe_3^{+2}Al_2Si_3O_{12}$

System	Form	Cleavage	Twinning	Zoning	Alteration	Occurrence	Remarks
Cub.	anhedral, hexagonal x-sect.	two or three perfect	interpenetrant	Z		I, S, O, M	colour spots cf. cryolite, halite
M'loid.	amorphous, vesicular					fulgurites	cf. opal
Cub.	anhedral	perfect cubic				S, I	inclusions common cf. sylvite
M'loid.	pisolitic, massive	contraction cracks				S	often with gibbsite and siderite
Cub.	dodecahedra	parting, irregular fractures				M	cf. grossularite, hydrogrossular
Tet.	tabular, peg structure	moderate, single crack		Z	zeolites, carbonate	I	cf. zoisite, vesuvianite, apatite, nepheline
Cub.	tetrahedral, triangular x-sect., granular	one poor	simple		ochre, manganese oxide	I, M	cf. garnet, danalite
Cub.	tetrahedral, triangular x-sect., granular	one poor				I, M	cf. garnet, helvite
Cub.	tetrahedral, triangular x-sect., granular	one poor				I, M	cf. garnet
Cub.	small crystals, aggregates	perfect cubic			brucite	M	cf. garnet, spinel
Cub.	four, six, eight-sided, polygonal x-sect., aggregates	parting, irregular fractures				M, I	cf. garnet group
Cub.	small grains, cubes, octahedra, rhombic x-sect.	parting		Z		M, I	colour variable according to composition cf. periclase, garnet
Cub.	four, six, eight-sided, polygonal x-sect., aggregates	parting, irregular fractures			chlorite	M, I	inclusions cf. garnet group

COLOURLESS MINERALS	Pleochroism	δ	Relief	2V Dispersion	Extinction	Orientation	Mineral	Composition
Isometric								
	(weak)	−M					OPAL	SiO_2
	(weak)	−L					FAUJASITE	$(Na_2,Ca)\{Al_2Si_4O_{12}\}\cdot 8H_2O$
	(weak)	−L					SODALITE (incl. HACKMANITE)	$Na_8\{Al_6Si_6O_{24}\}Cl_2$
	1	−L				length fast/slow	CRISTOBALITE	SiO_2
	1 (weak)	−L					ANALCITE	$Na\{AlSi_2O_6\}\cdot H_2O$
	(weak)	−L					NOSEAN	$Na_8\{Al_6Si_6O_{24}\}SO_4$
	(weak)	−L					HAÜYNE	$(Na,Ca)_{4-8}\{Al_6Si_6O_{24}\}(SO_4,S)_{1-2}$
	(aniso-tropic)	−L			wavy		LEUCITE	$K\{AlSi_2O_6\}$
	(weak)	−L − +M					VOLCANIC GLASS	
	(weak)	−L − +M					PALAGONITE	Altered glass
	1	+L			small	length slow	HALLOYSITE	$Al_4\{Si_4O_{10}\}(OH)_8\cdot 8H_2O$
	1	+L				length fast/slow	COLLOPHANE	$Ca_5(PO_4)_3(OH,F,Cl)$

System	Form	Cleavage	Twinning	Zoning	Alteration	Occurrence	Remarks
M'loid.	cryptocrystalline, colloform, veinlets, cavity fillings	irregular fractures				I, S	cf. lechatelierite
Cub.	octahedral, rounded	one distinct				I	cf. zeolites
Cub.	hexagonal x-sect., anhedral aggregates	poor	simple	Z	zeolites, diaspore, gibbsite	I	cf. fluorite, leucite
Tet.	minute square crystals, aggregates, spherulitic	curved fractures	polysynthetic, interpenetrant			I	cf. tridymite
Cub.	trapezohedral, rounded, radiating, irregular	poor	polysynthetic, complex, interpenetrant			I, S, M	cf. leucite, sodalite, wairakite
Cub.	hexagonal x-sect., anhedral aggregates	imperfect	simple	Z	zeolites, diaspore, gibbsite, limonite	I	clouded with inclusions
Cub.	hexagonal x-sect., anhedral aggregates	imperfect	simple	Z	zeolites, diaspore, gibbsite	I	
Tet. (Pseudo-Cub.)	always euhedral, octagonal	poor	polysynthetic, complex	Z		I	inclusions common cf. analcite, microcline
M'loid.	amorphous, massive	perlitic parting			frequent devit-rification	I	often with crystallites and phenocrysts cf. tachylyte, lechatelierite
M'loid.	amorphous, oolitic				chlorite	I	cf. volcanic glass, opal, collophane
Tr. Mon(?)	fine aggregates, colloform, clay, fibrous	shatter cracks			sericite	S, I	cf. hydrohalloysite (endellite)
M'loid.	amorphous, pisolitic, massive, cryptocrystalline	irregular fractures				S, I	contraction cracks cf. opal

COLOURLESS MINERALS	Pleochroism	δ	Relief	2V Dispersion	Extinction	Orientation	Mineral	Composition
Isometric								
	W $\varepsilon > \omega$	1 (iso-tropic)	+L – +M				EUDIALYTE-EUCOLITE	$(Na,Ca,Fe)_6Zr\{(Si_3O_9)_2\}$ (OH,F,Cl)
		anom.	+M				HIBSCHITE	$3CaO \cdot Al_2O_3 \cdot 2SiO_2 \cdot 2H_2O$
		(anom.)	H				GROSSULARITE	$Ca_3Al_2Si_3O_{12}$
		(anom.)	H – VH				GARNET	$(Mg,Fe^{+2},Mn,Ca)_3$ $(Fe^{+3},Ti,Cr,Al)_2\{Si_3O_{12}\}$
		(weak)	E				SPHALERITE	ZnS
	W $\gamma > \alpha$	(weak)	E				PEROVSKITE (incl. KNOPITE-LOPARITE-DYSANALYTE)	$(Ca,Na,Fe^{+2},Ce)(Ti,Nb)O_3$
Uniaxial +								
		1	–L		wavy		LEUCITE	$K\{AlSi_2O_6\}$
		1	–L		parallel	length slow	ERIONITE	$(Na_2,K_2,Ca,Mg)_{4.5}$ $\{Al_9Si_{27}O_{72}\} \cdot 27H_2O$
		1	–L		parallel		DAVYNE-NATRODAVYNE	K-cancrinite
		1	–L	$0° - 32°$	symmetrical		CHABAZITE	$Ca\{Al_2Si_4O_{12}\} \cdot 6H_2O$

System	Form	Cleavage	Twinning	Zoning	Alteration	Occurrence	Remarks
Trig.	rhombohedral aggregates	one good, one poor		Z		I	normally uniaxial, some zones isotropic cf. catapleite, låvenite, rosenbuschite, garnet
Cub.	octahedra	parting, irregular fractures				M	cf. grossularite
Cub.	four, six, eight-sided, polygonal x-sect., aggregates	parting, irregular fractures		Z		M	cf. garnet group, periclase, vesuvianite
Cub.	four, six, eight-sided, polygonal x-sect., aggregates	parting, irregular fractures	complex, sector	Z	chlorite	M, I, S	inclusions cf. spinel
Cub.	irregular, anhedral, curved surfaces	six perfect	polysynthetic, lamellar intergrowths	Z		O	colour variable, (uniaxial) cf. cassiterite
Mon? Pseudo-Cub.	small cubes, skeletal	poor to distinct	polysynthetic, complex, interpenetrant	Z	leucoxene	I, M	cf. melanite, picotite, ilmenite
Tet. (Pseudo-Cub.)	always euhedral, octagonal	poor	polysynthetic, complex	Z		I	inclusions common cf. analcite, microcline
Hex.	fibrous, radiating					I	cf. zeolites
Hex.	anhedral	two perfect				I	cf. cancrinite, microsommite
Trig.	rhombohedral, approaching cube, anhedral, granular	one poor	interpenetrant	Z		I	basal section in six segments cf. gmelinite, analcite

COLOURLESS MINERALS	Pleochroism	δ	Relief	2V Dispersion	Extinction	Orientation	Mineral	Composition
Uniaxial +								
		1 anom.	-L – N		parallel	length slow	APOPHYLLITE	$KFCa_4\{Si_8O_{20}\}\cdot 8H_2O$
		1	-L – N		parallel	length slow	ASHCROFTINE	$KNaCa\{Al_4Si_5O_{18}\}\cdot 8H_2O$
		1	N		parallel, symmetrical	length slow	QUARTZ	SiO_2
	W $\alpha=\beta>\gamma$	1 anom.	+L	0° – 20° r<v	$\simeq$ parallel	cleavage length fast	PENNINITE	$(Mg,Al,Fe)_{12}\{(Si,Al)_8O_{20}\}(OH)_{16}$
	W $\alpha=\beta>\gamma$	1	+L	$\simeq 20^o$ r<v	$\simeq$ parallel	cleavage length fast	SHERIDANITE	$(Mg,Al,Fe)_{12}\{(Si,Al)_8O_{20}\}(OH)_{16}$
	W $\varepsilon>\omega$	1 (isotropic)	+L – +M				EUDIALYTE	$(Na,Ca,Fe)_6Zr\{(Si_3O_9)_2\}(OH,F,Cl)$
	W $\alpha=\beta>\gamma$	1 anom.	+L – +M	0° – 30° r<v	$\simeq$ parallel	cleavage length fast	RIPIDOLITE- (PROCHLORITE) KLEMENTITE	$(Mg,Al,Fe)_{12}\{(Si,Al)_8O_{20}\}(OH)_{16}$
	W $\omega>\varepsilon$	1 anom. (isotropic)	+M		parallel	length slow	MELILITE	$(Ca,Na)_2\{(Mg,Fe^{+2},Al,Si)_3O_7\}$
		1	+M		parallel	length fast	ÅKERMANITE	$Ca_2\{MgSi_2O_7\}$
		1 anom.	H		parallel	length fast	VESUVIANITE (IDOCRASE) (incl. WILUITE)	$Ca_{10}(Mg,Fe)_2Al_4\{Si_2O_7\}_2\{SiO_4\}_5(OH,F)_4$
	W $\alpha=\beta>\gamma$	1 – 2	+L	0° – 40° r<v	0° – 9°	cleavage length fast	CLINOCHLORE	$(Mg,Al,Fe)_{12}\{(Si,Al)_8O_{20}\}(OH)_{16}$

System	Form	Cleavage	Twinning	Zoning	Alteration	Occurrence	Remarks
Tet.	prismatic, granular, anhedral - euhedral	two, one perfect				I, O	cf. zeolites
Tet.	prismatic, granular anhedral - euhedral, needles	two at 90°, one perfect				I	cf. zeolites
Trig.	variable, anhedral, flamboyant, intergrowths		rare in thin sections			S, I, M, O	inclusions, Boehm lamellae cf. cordierite, beryl, scapolite, feldspar
Mon. (Pseudo-Hex.)	tabular, vermicular, radiating, pseudomorphs	perfect basal	simple, pennine law	Z		I, M	pleochroic haloes cf. clinochlore, prochlorite
Mon. (Pseudo-Hex.)	tabular, spherulites	perfect basal	simple			I	
Trig.	rhombohedral, aggregates	one good, one poor		Z		I	some zones isotropic cf. catapleite, låvenite, rosenbuschite, garnet, eucolite
Mon. (Psuedo-Hex.)	tabular, scaly, vermicular, fan-shaped aggregates	perfect basal				M, I, O	cf. clinochlore, penninite
Tet.	tabular, peg structure	moderate, single crack		Z	zeolites, carbonate	I	cf. zoisite, vesuvianite, apatite, nepheline
Tet.	tabular, peg structure	moderate, single crack		Z	zeolites, carbonate	I	cf. zoisite, vesuvianite, apatite, nepheline
Tet.	variable, prismatic, fibrous, granular, radial	imperfect	sector	Z		M, I, S	cf. zoisite, clinozoisite, apatite, grossularite, melilite, andalusite
Mon.	tabular, fibrous	perfect basal	polysynthetic			M	pleochroic haloes cf. penninite, prochlorite, leuchtenbergite, katschubeite

COLOURLESS MINERALS	Pleochroism	δ	Relief	2V Dispersion	Extinction	Orientation	Mineral	Composition
Uniaxial +								
	W – S $\alpha=\beta>\gamma$	1 – 2 (anom.)	+L – +M	$0°$ – $60°$ r<v	$0°$ – $9°$	cleavage length fast	CHLORITE (Unoxidised)	$(Mg,Al,Fe)_{12}\{(Si,Al)_8O_{20}\}(OH)_{16}$
		1 – 3 (iso-tropic)	−L	anom.	parallel	length slow	MICROSOMMITE	$K,NaAlSiO_4 \cdot Ca(Cl_2,SO_4)$
		2 anom.	+L		parallel	x-sect. of plates length fast	BRUCITE	$Mg(OH)_2$
	W $\alpha=\beta>\gamma$	2	+L – +M	$31°$ r<v	$8°$ – $10°$	cleavage length fast	CORUNDOPHILITE	$(Mg,Al,Fe)_{12}\{(Si,Al)_8O_{20}\}(OH)_{16}$
		2 – 3	+L	$0°$ – $30°$ r<v	$3°$	cleavage fast	CATAPLEITE	$(Na,Ca)_2Zr\{Si_3O_9\} \cdot 2H_2O$
	N – M $\gamma=\alpha>\beta$ $\beta>\alpha=\beta$	3	+M – H	$0°$ – $30°$ r$\lessgtr$v	$37°$ – $44°$	extinction nearest cleavage slow	PIGEONITE	$(Mg,Fe^{+2},Ca)(Mg,Fe^{+2})\{Si_2O_6\}$
		4	VH				STISHOVITE	SiO_2
	W $\epsilon>\omega$	4 – 5	VH		parallel	length slow	ZIRCON	$Zr\{SiO_4\}$
	W	5	H		straight		XENOTIME	YPO_4
	W – S $\epsilon>\omega$	5	VH – E	$0°$ – $38°$ anom. strong	parallel, oblique to twin plane	length slow	CASSITERITE	SnO_2

System	Form	Cleavage	Twinning	Zoning	Alteration	Occurrence	Remarks
Mon. (Pseudo-Hex.)	tabular, scaly, radiating pseudomorphs	perfect basal	simple, polysynthetic	Z		I, M, S, O	pleochroic haloes
Hex.	anhedral, prismatic	one perfect, one poor	polysynthetic, rare	Z		I	some zones isotropic cf. nepheline, quartz, cancrinite, vishnevite
Trig.	fibrous, plates, whorls, scaly aggregates	one perfect			hydro-magnesite	M	alteration of periclase cf. alunite, muscovite, talc
Mon. (Pseudo-Hex.)	tabular, radiating	perfect basal				M	
Mon. (Pseudo-Hex.)	tabular	one perfect, one poor				I	cf. eudialyte
Mon.	prismatic, anhedral, overgrowths	two at 87°, parting	simple, polysynthetic	Z		I	exsolution lamellae cf. augite, olivine
Tet.	fine-grained					M	impact metamorphism
Tet.	minute prisms	poor, absent		Z	metamict	I, S, M	cf. apatite
Tet.	small, prismatic					I, M, S	inclusions, pleochroic haloes around grains cf. zircon, sphene, monazite
Tet.	subhedral, veinlets, diamond-shaped x-sect.	prismatic	geniculate, cyclic, common	Z		O, I, S	cf. sphalerite, rutile, sphene

COLOURLESS MINERALS	Pleochroism	δ	Relief	2V Dispersion	Extinction	Orientation	Mineral	Composition
Uniaxial –								
		1	–L		parallel	length fast/slow	CRISTOBALITE	SiO_2
		1	–L		parallel		LEVYNE	$Ca\{Al_2Si_4O_{12}\}\cdot6H_2O$
		1	–L		parallel		GARRONITE	$NaCa_{2.5}\{Al_6Si_{10}O_{32}\}\cdot13\cdot5H_2O$
		1	–L		parallel		KALSILITE-KALIOPHILITE	$K\{AlSiO_4\}$
		1	–L		parallel		DAVYNE	K-cancrinite
		1	–L	anom.	parallel	length fast	VISHNEVITE	$(Na,Ca,K)_{6-8}\{Al_6Si_6O_{24}\}$ $(CO_3,SO_4,Cl)_{1-2}\cdot1-5H_2O$
		1	–L	0° – moderate	symmetrical		GMELINITE	$(Na_2,Ca)\{Al_2Si_4O_{12}\}\cdot6H_2O$
	1 (anom.)		–L	0° – 25° r>v	α:(001) 5° – 8°	optic plane $\perp$(010)	SANIDINE	$(K,Na)\{AlSi_3O_8\}$
		1	–L	0° – 32°	symmetrical		CHABAZITE	$Ca\{Al_2Si_4O_{12}\}\cdot6H_2O$
		1	–L – N		parallel	rect. sect. length fast	NEPHELINE	$Na_3(Na,K)\{Al_4Si_4O_{16}\}$
		1 anom.	–L – N		parallel	length fast	APOPHYLLITE	$KFCa_4\{Si_8O_{20}\}\cdot8H_2O$
		1	N		parallel		EUCRYPTITE	$LiAlSiO_4$
	W $\gamma=\beta>\alpha$	1 anom.	+L	0° – 40° r>v	$\simeq$ parallel	cleavage length slow	PENNINITE	$(Mg,Al,Fe)_{12}\{(Si,Al)_8O_{20}\}(OH)_{16}$

System	Form	Cleavage	Twinning	Zoning	Alteration	Occurrence	Remarks
Tet. (Pseudo-Cub.)	minute, square crystals, aggregates, spherulitic	curved fractures	polysynthetic, interpenetrant			I	cf. tridymite
Trig.	tabular, sheaf-like aggregates, rhombohedral	indistinct, rhombohedral	interpenetrant			I	
Tet.							
Hex.	prismatic, hexagonal	two poor	rare	Z		I	cf. nepheline
Hex.	anhedral	two perfect					cf. natrodavyne, cancrinite, microsommite
Hex.	anhedral	one perfect, one poor	polysynthetic, rare			I	cf. cancrinite, microsommite, nepheline
Trig.	tabular, prismatic, rhombohedral, approaching cube, radiating	one good, one imperfect, parting	interpenetrant			I	cf. chabazite
Mon.	clear, distinct crystals, tabular, microlites	two, parting	Carlsbad, Baveno, Manebach	Z		I, M	perthitic cf. orthoclase, nepheline
Trig.	rhombohedral, approaching cube, anhedral, granular	one poor	interpenetrant	Z		I	basal section in six segments cf. gmelinite, analcite
Hex.	prismatic, hexagonal	two poor	rare	Z	zeolites, cancrinite, muscovite	I	inclusions cf. alkali feldspars, analcite, sodalite, leucite, scapolite
Tet.	prismatic, granular, anhedral – euhedral	two, one perfect				I, O	cf. zeolites
Hex.	anhedral	one distinct				I	cf. spodumene
Mon. (Pseudo-Hex.)	tabular, vermicular, radiating, pseudomorphs	perfect basal	simple, pennine law	Z		I, M	pleochroic haloes cf. clinochlore, prochlorite

COLOURLESS MINERALS	Pleochroism	δ	Relief	2V Dispersion	Extinction	Orientation	Mineral	Composition
Uniaxial –								
	W	1	+L		parallel	long. sect. length fast, x-sect. length slow	BERYL	$Be_3Al_2\{Si_6O_{18}\}$
		1	+L		parallel	cleavage length fast	MARIALITE	$Na_4\{Al_3Si_9O_{24}\}Cl$
	W ε>ω	1 (isotropic)	+L – +M				EUCOLITE	$(Na,Ca,Fe)_6Zr\{(Si_3O_9)_2\}(OH,F,Cl)$
	W ε>ω	1	+M		parallel	length fast, tabular length slow	APATITE (incl. DAHLLITE, FRANCOLITE)	$Ca_5(PO_4)_3(OH,F,Cl)$
	W ω>ε	1 anom. (isotropic)	+M		parallel	length slow	MELILITE	$(Ca,Na)_2\{(Mg,Fe^{+2},Al,Si)_3O_7\}$
	W γ=β>α	1 anom.	+M	0° – small r>v	≃ parallel	cleavage length slow	RIPIDOLITE (PROCHLORITE)	$(Mg,Al,Fe)_{12}\{(Si,Al)_8O_{20}\}(OH)_{16}$
	W γ=β>α	1 anom.	+M	0° – 20° small		cleavage length slow	DAPHNITE	$(Mg,Al,Fe)_{12}\{(Si,Al)_8O_{20}\}(OH)_2$
		1 anom.	H̄		parallel	length fast	VESUVIANITE (IDOCRASE)	$Ca_{10}(Mg,Fe)_2Al_4\{Si_2O_7\}_2\{SiO_4\}_5(OH,F)_4$
	W ω>ε	1	H	0° – 30°	parallel	tabular length slow, prismatic length fast	CORUNDUM	$\alpha\text{-}Al_2O_3$
	W – S γ=β>α	1 – 2 (anom.)	+L – +M r>v	0° – 20° 0° – small		length slow	CHLORITE (Oxidised)	$(Mg,Al,Fe)_{12}\{(Si,Al)_8O_{20}\}(OH)_{16}$

System	Form	Cleavage	Twinning	Zoning	Alteration	Occurrence	Remarks
Hex.	prismatic, inclusions zoned	imperfect, rare		Z	kaolin	I, M	liquid inclusions cf. apatite, quartz
Tet.	columnar, aggregates	two good			calcite, zeolites, muscovite	M	cf. feldspar, quartz, cancrinite
Trig.	rhombohedral, aggregates	one good, one poor		Z		I	some zones isotropic cf. catapleite, låvenite, rosenbuschite, garnet, eudialyte
Hex.	small, prismatic, hexagonal	poor basal				I, S, M, O	cf. beryl, topaz, dahllite
Tet.	tabular, peg structure	moderate, central crack		Z	zeolites, carbonate	I	cf. zoisite, vesuvianite, apatite, nepheline
Mon. (Pseudo-Hex.)	tabular, scaly, vermicular, fan-shaped aggregates	perfect basal				M, I, O	cf. clinochlore, penninite
Mon. (Pseudo-Hex.)	concentric aggregates, fibrous plates	perfect basal				O	
Tet.	variable, prismatic, fibrous, granular, radial	imperfect	sector	Z		M, I, S	cf. zoisite, clinozoisite, apatite, grossularite, melilite, andalusite
Trig.	tabular, prismatic, six-sided x-sect.	parting	simple, lamellar seams	Z colour banding		M, I	inclusions cf. sapphirine
Mon. (Pseudo-Hex.)	tabular, scaly, radiating, pseudomorphs	perfect basal	simple, polysynthetic	Z		I, M, S, O	pleochroic haloes

COLOURLESS MINERALS	Pleochroism	δ	Relief	2V Dispersion	Extinction	Orientation	Mineral	Composition
Uniaxial –								
	W – S $\gamma=\beta>\alpha$	1 – 2 (anom.)	+L – +M	0° – 20° r>v	0° – small	cleavage length slow	CHLORITE (Unoxidised)	$(Mg,Al,Fe)_{12}\{(Si,Al)_8O_{20}\}(OH)_{16}$
		1 – 3	–L	anom.	parallel	length fast	CANCRINITE-VISHNEVITE	$(Na,Ca,K)_{6-8}\{Al_6Si_6O_{24}\}$ $(CO_3,SO_4,Cl)_{1-2}\cdot1\text{-}5H_2O$
		1 – 4	+L		parallel	cleavage length fast	SCAPOLITE (MARIALITE-DIPYRE-MIZZONITE-MEIONITE-WERNERITE)	$(Na,Ca,K)_4\{Al_3(Al,Si)_3Si_6O_{24}\}$ (Cl,CO_3,SO_4,OH)
		2	+M		parallel	length slow	GEHLENITE (MELILITE)	$Ca_2\{Al_2SiO_7\}$
	W	2	+M	0° – 20°	small	cleavage length slow	THURINGITE	$(Mg,Al,Fe)_{12}\{(Si,Al)_8O_{20}\}(OH)_{16}$
	S $\omega>\varepsilon$	2 – 3	+M		parallel	length fast	ELBAITE (TOURMALINE)	$Na(Li,Al)_3Al_6B_3Si_6O_{27}(OH,F)_4$
		3	–L	anom.	parallel	length fast	CANCRINITE	$(Na,Ca,K)_{6-8}\{Al_6Si_6O_{24}\}$ $(CO_3,SO_4,Cl)_{1-2}\cdot1\text{-}5H_2O$
	S $\omega>\varepsilon$	3	+M		parallel	length fast	DRAVITE (TOURMALINE)	$NaMg_3Al_6B_3Si_6O_{27}(OH,F)_4$
	W	3	+M		parallel		ZUSSMANITE	$K(Fe,Mg,Mn)_{13}(Si,Al)_{18}O_{42}(OH)_{14}$
		3 – 4	+L		parallel	cleavage length fast	MEIONITE (SCAPOLITE)	$Ca_4\{Al_6Si_6O_{24}\}CO_3$
	M $\gamma\geqslant\beta>\alpha$ $\alpha>\beta=\gamma$	3 – 4	+L – +M	0° – 15° r<v	0° – 5°	cleavage length slow	PHLOGOPITE	$K_2(Mg,Fe^{+2})_6\{Si_6Al_2O_{20}\}(OH,F)_4$
		5	–L – +M		symmetrical to cleavage		CALCITE	$CaCO_3$

System	Form	Cleavage	Twinning	Zoning	Alteration	Occurrence	Remarks
Mon. (Pseudo-Hex.)	tabular, scaly, radiating, pseudomorphs	perfect basal	simple, polysynthetic	Z		I, M, S, O	pleochroic haloes
Hex.	anhedral	one perfect, one poor	polysynthetic, rare			I	cf. microsommite, muscovite
Tet.	columnar, aggregates	two good			calcite, zeolites, muscovite	M	cf. feldspar, quartz, cancrinite
Tet.	tabular, peg structure	moderate, single crack		Z	zeolites, carbonate	I	cf. zoisite, vesuvianite, apatite, nepheline
Mon. (Pseudo-Hex.)	tabular, radiating	perfect basal				O	
Trig.	prismatic, radiating	fractures at 90°	rare	Z		I, M	cf. apatite
Hex.	anhedral	one perfect, one poor	polysynthetic, rare			I	cf. vishnevite, microsommite, muscovite
Trig.	prismatic	fractures at 90°	rare	Z		M	cf. chondrodite
Hex.	tabular	one perfect				M	
Tet.	columnar, aggregates	two good			muscovite, ill-defined fibrous aggregates	M	cf. scapolite, feldspar, quartz, cancrinite
Mon.	tabular, flakes, plates, pseudo-hex.	perfect basal	inconspicuous	Z colour zoning		I, M	inclusions common, birds-eye maple structure cf. muscovite, lepidolite
Trig.	anhedral, oolitic, spherulitic	rhombohedral	polysynthetic, // long diagonal			S, M, I, O	twinkling cf. rhombohedral carbonates

COLOURLESS MINERALS	Pleochroism	δ	Relief	2V Dispersion	Extinction	Orientation	Mineral	Composition
Uniaxial –								
		5	–L – +M		symmetrical to cleavage		DOLOMITE	$CaMg(CO_3)_2$
	W $\omega>\epsilon$ rare	5	–L – H		symmetrical to cleavage		MAGNESITE	$MgCO_3$
	W $\omega>\epsilon$	5	N – VH		symmetrical to cleavage		RHODOCHROSITE	$MnCO_3$
		5	+L – VH		symmetrical to cleavage		SIDERITE	$FeCO_3$
		5	+M – H		symmetrical to cleavage		ANKERITE	$Ca(Mg,Fe^{+2},Mn)(CO_3)_2$
	very strong		+M				HUNTITE	$Mg_3Ca(CO_3)_4$
Biaxial +								
		1	–L	$0°$ – moderate	symmetrical		GMELINITE	$(Na_2,Ca)\{Al_2Si_4O_{12}\}\cdot 6H_2O$
		1	–L	$0°$ – $32°$	symmetrical		CHABAZITE	$Ca\{Al_2Si_4O_{12}\}\cdot 6H_2O$
		1	–L	$0°$ – $48°$ variable r>v	$6°$ variable	cleavage length fast	HEULANDITE	$(Ca,Na_2)\{Al_2Si_7O_{18}\}\cdot 6H_2O$
		1	–L	$0°$ – $63°$			ISOSANIDINE	$(K,Na)\{AlSi_3O_8\}$

System	Form	Cleavage	Twinning	Zoning	Alteration	Occurrence	Remarks
Trig.	rhombohedral	rhombohedral	polysynthetic, // long and short diagonals	Z	huntite	S, M, I	twinkling cf. rhombohedral carbonates
Trig.	microcrystalline, subhedral	rhombohedral			huntite	S, I, M	twinkling cf. rhombohedral carbonates
Trig.	rhombohedral, aggregates, bands	rhombohedral	polysynthetic, rare	Z	manganese oxide	O, I, M	twinkling cf. rhombohedral carbonates
Trig.	anhedral, aggregates, oolitic, spherulitic, colloform	rhombohedral	polysynthetic, uncommon, // long diagonal		brown spots	S, O, I, M	twinkling, brown stain around borders and along cleavage cracks cf. rhombohedral carbonates
Trig.	rhombohedral	rhombohedral	polysynthetic	Z		S, O, M	twinkling cf. rhombohedral carbonates
Trig.	compact, porous					S	alteration of dolomite and magnesite
Trig.	tabular, prismatic, rhombohedral, approaching cube, radiating	one good, one imperfect, parting	interpenetrant			I	cf. chabazite
Trig.	rhombohedral, approaching cube, anhedral, granular	one poor	interpenetrant	Z		I	basal section in six segments cf. gmelinite, analcite
Pseudo-Mon.	tabular, aggregates	one perfect				I, M	cf. stilbite, clinoptilite, epistilbite, mordenite, brewsterite
Mon.	clear, distinct crystals, tabular	two	Carlsbad			I	cf. sanidine

COLOURLESS MINERALS	Pleochroism	δ	Relief	2V Dispersion	Extinction	Orientation	Mineral	Composition
Biaxial +								
		1	-L	small, anom.	wavy		LEUCITE	$K\{AlSi_2O_6\}$
		1	-L	$40^{o} - 90^{o}$			TRIDYMITE	SiO_2
		1	-L	50°	parallel	length slow	FERRIERITE	$(Na,K)_4Mg_2\{Al_6Si_{30}O_{72}\}$ $(OH)_2 \cdot 18H_2O$
		1	-L	$65^{o} - 73^{o}$	38°		DACHIARDITE	$(\frac{1}{2}Ca,Na,K)_5\{Al_5Si_{19}O_{48}\} \cdot 12H_2O$
		1	-L	70°			YUGAWARALITE	$Ca\{Al_2Si_5O_{14}\} \cdot 4H_2O$
		1	-L	$76^{o} - 90^{o}$	parallel	length slow	MORDENITE	$(Na_2,K_2,Ca)\{Al_2Si_{10}O_{24}\} \cdot 7H_2O$
	1 nearly isotropic		-L	80° r>v	8°	length fast/slow	MESOLITE	$Na_2Ca_2\{Al_2Si_3O_{10}\}_3 \cdot 8H_2O$
		1	-L	80° weak x-dispersion	$63^{o} - 67^{o}$		HARMOTOME	$Ba\{Al_2Si_6O_{16}\} \cdot 6H_2O$
		1	-L	$\simeq 90^{o}$ variable			ISOORTHOCLASE	$(K,Na)\{AlSi_3O_8\}$
		1	-L - N	$82^{o} - 90^{o}$ r>v	$0^{o} - 12^{o}$ (in albite twins)		OLIGOCLASE	$Na\{AlSi_3O_8\}-Ca\{Al_2Si_2O_8\}$
		1	-L - +L	$76^{o} - 90^{o}$ $r \lessgtr v$	$0^{o} - 70^{o}$ (in albite twins)		PLAGIOCLASE	$Na\{AlSi_3O_8\}-Ca\{Al_2Si_2O_8\}$
		1	N	$0^{o} - 10^{o}$ (uniaxial)	parallel	length slow	QUARTZ	SiO_2

System	Form	Cleavage	Twinning	Zoning	Alteration	Occurrence	Remarks
Tet. (Pseudo-Cub.)	always euhedral, octagonal	poor	polysynthetic, complex	Z		I	inclusions common cf. analcite, microcline
Orth.	tabular, radiating, aggregates	poor	wedge-shaped			I, M	cf. cristobalite
Orth.	tabular, laths, radiating	one perfect				I	cf. zeolites
Mon.	prismatic	two perfect	cyclic, sector			I	
Mon.						I	
Orth.	tabular, acicular, fibrous	one perfect				I	
Mon. (Pseudo-Orth.)	fibrous, aggregates, needles	two perfect at 90°	universal but inconspicuous			I	cf. natrolite, thomsonite, scolecite
Mon. (Orth.)	groups, radiating	two, one good	interpenetrant, complex, sector, cruciform			I, O	
Mon.	anhedral - subhedral, phenocrysts	three	Carlsbad			I	cf. orthoclase
Tr.	anhedral - euhedral, laths, perthite	three	albite, Carlsbad, pericline, complex	Z	sericite, calcite, kaolinite, zeolites	I, M	peristerite, antiperthitic cf. quartz, cordierite
Tr.	anhedral - euhedral, laths	three	albite, Carlsbad, pericline, complex	Z	sericite, calcite, kaolinite, zeolites, saussurite	I, M, S	peristerite, schiller, antiperthitic cf. cordierite
Trig.	variable, anhedral, flamboyant, intergrowths		rare in thin sections			S, I, M, O	inclusions, Boehm lamellae cf. cordierite, beryl, scapolite, feldspar

COLOURLESS MINERALS	Pleochroism	δ	Relief	2V Dispersion	Extinction	Orientation	Mineral	Composition
Biaxial +								
	W $\alpha=\beta>\gamma$	1 anom.	+L	0° - 20° r<v	≈ parallel	cleavage length fast	PENNINITE	$(Mg,Al,Fe)_{12}\{(Si,Al)_8O_{20}\}(OH)_{16}$
	W $\alpha=\beta>\gamma$	1	+L	≈ 20° r<v	≈ parallel	cleavage length fast	SHERIDANITE	$(Mg,Al,Fe)_{12}\{(Si,Al)_8O_{20}\}(OH)_{16}$
		1	+L	52° - 80° r<v	14° - 20° undulatory	length slow	DICKITE	$Al_4\{Si_4O_{10}\}(OH)_8$
		1	+L	76° - 86° r<v	$27\frac{1}{2}^{o}$ - 39° (in albite twins)		LABRADORITE	$Na\{AlSi_3O_8\}-Ca\{Al_2Si_2O_8\}$
		1	+L - +M	54° - 64°	unsymmetrical		COESITE	SiO_2
	W $\alpha=\beta>\gamma$	1 anom.	+L - +M	0° - 30° r<v	≈ parallel	cleavage length fast	RIPIDOLITE (PROCHLORITE) - KLEMENTITE	$(Mg,Al,Fe)_{12}\{(Si,Al)_8O_{20}\}(OH)_{16}$
	W $\gamma>\beta>\alpha$	1	+M	50° r<v	parallel	length slow	CELESTINE	$SrSO_4$
		1	+M	63° - 64°	15°		RANKINITE	$Ca_3\{Si_2O_7\}$
	M	1 anom.	+M - H	0° - 30° r>v	parallel	length fast	α - ZOISITE	$Ca_2Al\cdot Al_2O\cdot OH\{Si_2O_7\}\{SiO_4\}$
	M	1 normal	+M - H	0° - 60° r<v	parallel	length fast/slow	β - ZOISITE (Fe-rich)	$Ca_2Al\cdot Al_2O\cdot OH\{Si_2O_7\}\{SiO_4\}$
		1 anom.	H	5° - 65° strong		length fast	VESUVIANITE (IDOCRASE) (incl. WILUITE)	$Ca_{10}(Mg,Fe)_2Al_4\{Si_2O_7\}_2 \{SiO_4\}_5(OH,F)_4$
	W $\beta>\gamma>\alpha$	1	H	50° - 66° r<v	6° - 9°	length slow	SAPPHIRINE	$(Mg,Fe)_2Al_4O_6\{SiO_4\}$

System	Form	Cleavage	Twinning	Zoning	Alteration	Occurrence	Remarks
Mon.	tabular, vermicular, radiating, pseudomorphs	perfect basal	simple, pennine law	Z		I, M	pleochroic haloes cf. clinochlore, prochlorite
Mon. (Pseudo-Hex.)	tabular, spherulites	perfect basal	simple			I	
Mon.	flakes, pseudo-hex., radial, aggregates	perfect basal				O, I, S	cf. kaolinite, nacrite
Tr.	anhedral - euhedral, laths	three	albite, Carlsbad, pericline, complex	Z	sericite, calcite, kaolinite, zeolites, saussurite	I, M	schiller, antiperthitic cf. cordierite
Mon. (Pseudo-Hex.)	pseudo-hex. plates, aggregates					M	impact metamorphism
Mon. (Pseudo-Hex.)	tabular, scaly, vermicular, fan-shaped aggregates	perfect basal				M, I, O	cf. clinochlore, penninite
Orth.	tabular, fibrous	three				S	cf. barytes
Mon.	granular					M	cf. tilleyite, spurrite, larnite, merwinite
Orth.	columnar aggregates, euhedral	one perfect, one imperfect	polysynthetic, rare	Z		M	cf. clinozoisite, epidote
Orth.	columnar aggregates, euhedral	one perfect, one imperfect	polysynthetic, rare	Z		M	cf. clinozoisite, epidote, diopside, augite
Tet.	variable, prismatic, fibrous, granular, radial	imperfect	sector	Z		M, I, S	cf. zoisite, clinozoisite, apatite, grossularite, melilite, andalusite
Mon.	tabular	poor	polysynthetic, uncommon			M	cf. corundum, cordierite, kyanite, zoisite, Na-amphibole

COLOURLESS MINERALS	Pleochroism	δ	Relief	2V Dispersion	Extinction	Orientation	Mineral	Composition
Biaxial +								
	W $\gamma > \alpha$	1	E	90° r>v	$\simeq 45^\circ$		PEROVSKITE (incl. KNOPITE-LOPARITE-DYSANALYTE)	$(Ca,Na,Fe^{+2},Ce)(Ti,Nb)O_3$
		1 - 2	-L	$42^\circ - 75^\circ$ r>v	parallel	length fast/slow	THOMSONITE	$NaCa_2\{(Al,Si)_5O_{10}\}_2 \cdot 6H_2O$
		1 - 2	-L	$60^\circ - 80^\circ$ r<v	$46^\circ - 85^\circ$	length slow	PHILLIPSITE	$(\frac{1}{2}Ca,Na,K)_3\{Al_3Si_5O_{16}\} \cdot 6H_2O$
		1 - 2	-L	$64^\circ - 90^\circ$	$15^\circ - 20^\circ$	cleavage length fast	ISOMICROCLINE	$(K,Na)\{AlSi_3O_8\}$
		1 - 2	-L	$77^\circ - 82^\circ$ r<v	$12^\circ - 19^\circ$ (in albite twins)		ALBITE	$Na\{AlSi_3O_8\}$
	M	1 - 2	-L - +L	$65^\circ - 90^\circ$ r<v	parallel		CORDIERITE	$Al_3(Mg,Fe^{+2})_2\{Si_5AlO_{18}\}$
	W $\alpha = \beta > \gamma$	1 - 2	+L	$0^\circ - 40^\circ$ r<v	$0^\circ - 9^\circ$	cleavage length fast	CLINOCHLORE	$(Mg,Al,Fe)_{12}\{(Si,Al)_8O_{20}\}(OH)$
	W - S $\alpha = \beta > \gamma$	1 - 2 (anom.)	+L - +M	$0^\circ - 60^\circ$ r<v	$0^\circ - 9^\circ$	cleavage length fast	CHLORITE (Unoxidised)	$(Mg,Al,Fe)_{12}\{(Si,Al)_8O_{20}\}(OH)$
	N - W	1 - 2	+M	$48^\circ - 68^\circ$ r>v	parallel, symmetrical	cleavage fast	TOPAZ	$Al_2\{SiO_4\}(OH,F)_2$
		1 - 2 anom.	+M - H	$14^\circ - 90^\circ$ r$\lessgtr$v	$0^\circ - 7^\circ$	length fast/slow	CLINOZOISITE	$Ca_2Al \cdot Al_2O \cdot OH\{Si_2O_7\}\{SiO_4\}$
	W - M $\beta > \alpha > \gamma$	1 - 2 anom.	H	$45^\circ - 68^\circ$ r>v	$2^\circ - 30^\circ$	length fast	CHLORITOID (OTTRELITE)	$(Fe^{+2},Mg,Mn)_2(Al,Fe^{+3})Al_3O_2\{SiO_4\}_2(OH)_4$

System	Form	Cleavage	Twinning	Zoning	Alteration	Occurrence	Remarks
Mon? Pseudo-Cub.	small cubes, skeletal	poor to distinct	polysynthetic, complex, interpenetrant	Z	leucoxene	I, M	cf. melanite, picotite, ilmenite
Orth. (Pseudo-Tet.)	fibrous, columnar, radiating	two at 90°				I	cf. natrolite, scolecite, mesolite, cancrinite
Mon. (Orth.)	groups, radiating	two at 90°	complex, sector, interpenetrant			I, S	
Tr.	anhedral – subhedral	two, parting	albite, Carlsbad, pericline, tartan	Z	cloudy, sericite, kaolinite	I, M, S	cf. microcline, orthoclase, albite, plagioclase
Tr.	anhedral – euhedral, laths	three	albite, Carlsbad, pericline, complex	Z	sericite, calcite, kaolinite, zeolites	I, S, M	peristerite, perthitic, antiperthitic cf. cordierite
Orth. (Pseudo-Hex.)	pseudo-hex., anhedral	moderate to absent	simple, polysynthetic, cyclic, sector, interpenetrant		pinite, talc	M, I	pleochroic haloes, inclusions (including opaque dust) common cf. quartz, plagioclase
Mon.	tabular, fibrous pseudo-hex.	perfect basal	polysynthetic			M	pleochroic haloes cf. penninite, prochlorite, leuchtenbergite, katschubeite
Mon. (Pseudo-Hex.)	tabular, scaly, radiating pseudomorphs	perfect basal	simple, polysynthetic	Z		I, M, S, O	pleochroic haloes
Orth.	prismatic, aggregates	one perfect				I, M, S	cf. quartz, andalusite
Mon.	columnar, six-sided x-sect.	one perfect	polysynthetic, uncommon	Z		M	cf. epidote, zoisite, diopside, augite
Mon., Tr.	tabular	one perfect, one imperfect, parting	polysynthetic	Z hour-glass		M, I	inclusions common cf. chlorite, clintonite, biotite, stilpnomelane

COLOURLESS MINERALS	Pleochroism	δ	Relief	2V Dispersion	Extinction	Orientation	Mineral	Composition
Biaxial +								
		2	-L	47° r>v	22°	length slow	BREWSTERITE	$(Sr,Ba,Ca)\{Al_2Si_6O_{16}\}\cdot 5H_2O$
		2	-L	58° r>v	52°	cleavage length fast/slow	GYPSUM (SELENITE)	$CaSO_4\cdot 2H_2O$
		2	-L	$58^o - 64^o$ r<v	parallel, symmetrical	length slow	NATROLITE	$Na_2\{Al_2Si_3O_{10}\}\cdot 2H_2O$
		2	-L	$58^o - 64^o$	$7\frac{1}{2}^o$	length slow	METANATROLITE	$Na_2\{Al_2Si_3O_{10}\}$
		2	-L	$82^o - 84^o$ r>v	$2^o - 8^o$ $24^o - 30^o$	length fast	PETALITE	$Li\{AlSi_4O_{10}\}$
		2	+L	small			AMESITE	$(Mg_4Al_2)(Si_2Al_2)O_{10}(OH)_8$
		2	+L	$83^o - 90^o$	$\alpha : z$ $3^o - 5^o$		CELSIAN	$Ba\{Al_2Si_2O_8\}$
	W $\alpha=\beta>\gamma$	2	+L - +M	31° r<v	$8^o - 10^o$	cleavage length fast	CORUNDOPHILITE	$(Mg,Al,Fe)_{12}\{(Si,Al)_8O_{20}\}(OH)_{16}$
	W $\gamma>\beta>\alpha$	2	+M	37° r<v	parallel	best cleavage slow	BARYTES	$BaSO_4$
	W	2	+M	$67^o - 70^o$ r>v	$33^o - 40^o$	extinction nearest cleavage slow	JADEITE	$NaAl\{Si_2O_6\}$
		2	+M	$\simeq 80^o$	parallel	length fast	BOEHMITE	$\gamma-AlO(OH)$
	W $\beta>\alpha=\gamma$	2	+M - H	$26^o - 85^o$ r<v	$4^o - 32^o$	length slow	PUMPELLYITE	$Ca_4(Mg,Fe^{+2})(Al,Fe^{+3})_5O(OH)_3$ $\{Si_2O_7\}_2\{SiO_4\}_2\cdot 2H_2O$

System	Form	Cleavage	Twinning	Zoning	Alteration	Occurrence	Remarks
Mon.	prismatic	one perfect		Z		I	cf. heulandite, epistilbite
Mon.	aggregates, fibrous, acicular, lozenge-shaped	three perfect	polysynthetic			S	cf. celestine, anhydrite
Orth. (Pseudo-Tet.)	prismatic, fibrous, needles, radiating	// length				I	cf. scolecite, thomsonite
Orth. (Pseudo-Tet.)	prismatic, fibrous, needles, radiating	// length	sector			I	cf. natrolite
Mon.	prismatic	two good	polysynthetic			I	cf. quartz, feldspar
Mon. (Pseudo-Hex.)	tabular	perfect basal				M	
Mon.	prismatic	three	Carlsbad, Baveno, Manebach			O, I	cf. orthoclase, hyalophane
Mon. (Pseudo-Hex.)	tabular, radiating	perfect basal				M	
Orth.	granular aggregates	three	polysynthetic			I, S	cf. celestine
Mon.	prismatic	two at 87°	simple, polysynthetic	Z	tremolite-actinolite	M	cf. nephrite, diopside, omphacite, fassaite
Orth.	tabular, fine-grained, oolitic	one good				S	cf. diaspore, gibbsite
Mon.	fibrous, needles, bladed, radial aggregates	two moderate	common	Z		M	cf. clinozoisite, zoisite, epidote, lawsonite

Biaxial +

Pleochroism	δ	Relief	2V Dispersion	Extinction	Orientation	Mineral	Composition
	2	+M – H	$55°$ – $90°$ $r<v$	parallel	length slow	ENSTATITE	$(Mg,Fe^{+2})\{SiO_3\}$
	2	H	$35°$ – $46°$ $r>v$	$64°$		PYROXMANGITE	$(Mn,Fe)\{SiO_3\}$
W $\alpha>\gamma$	2	H	$61°$ – $76°$ $r<v$	$5°$ – $25°$		RHODONITE	$(Mn,Ca,Fe)\{SiO_3\}$
M $\gamma>\beta>\alpha$	2	H	$82°$ – $90°$ $r>v$	parallel, symmetrical	length slow	STAUROLITE	$(Fe^{+2},Mg)_2(Al,Fe^{+3})_9O_6$ $\{SiO_4\}_4(O,OH)_2$
	2 – 3	+L	$0°$ – $30°$ $r<v$	$3°$	cleavage fast	CATAPLEITE	$(Na,Ca)_2Zr\{Si_3O_9\}\cdot2H_2O$
W $\gamma>\beta=\alpha$	2 – 3	+M	$45°$ – $61°$ $r>v$	parallel, symmetrical	length slow	MULLITE	$3Al_2O_3\cdot2SiO_2$
W	2 – 3	+M	$58°$ – $68°$ $r<v$	$22°$ – $26°$	extinction nearest cleavage slow	SPODUMENE (HIDDENITE, KUNZITE)	$LiAl\{Si_2O_6\}$
W $\gamma=\beta>\alpha$ $\gamma>\beta=\alpha$	2 – 3	+M	$68°$ – $90°$ $r\lessgtr v$	parallel, symmetrical	length slow	ANTHOPHYLLITE (Fe-rich)	$(Mg,Fe^{+2})_7\{Si_8O_{22}\}(OH,F)_2$
N – W	2 – 3	+M – H	$25°$ – $83°$ $r>v$	$35°$ – $48°$	extinction nearest cleavage fast	AUGITE-FERROAUGITE (incl. SALITE)	$(Ca,Na,Mg,Fe^{+2},Mn,Fe^{+3},Al,Ti)_2$ $\{(Si,Al)_2O_6\}$
W	2 – 3	+M – H	$51°$ – $62°$	$35°$ – $48°$	extinction nearest cleavage fast	FASSAITE	$(Ca,Na,Mg,Fe^{+2},Mn,Fe^{+3},Al,Ti)_2$ $\{(Si,Al)_2O_6\}$
W	2 – 3	+M – H	$60°$ – $67°$	$39°$ – $41°$	extinction nearest cleavage fast	OMPHACITE	$(Ca,Na,Mg,Fe^{+2},Mn,Fe^{+3},Al,Ti)_2$ $\{(Si,Al)_2O_6\}$

System	Form	Cleavage	Twinning	Zoning	Alteration	Occurrence	Remarks
Orth.	prismatic, anhedral - euhedral, kelyphitic borders	two at 88°, parting	rare			I, M	often schillerized, Fe-enstatite = bronzite
Tr.	prismatic	four, two at 92°	simple, polysynthetic			O, M	cf. rhodonite, bustamite
Tr.	prismatic, square x-sect., anhedral	three, two at 92$\frac{1}{2}^{\circ}$	polysynthetic	Z	pyrolusite, rhodochrosite	O, M	rare, inclusions, exsolution lamellae cf. bustamite, pyroxmangite
Mon. (Pseudo-Orth.)	short, prismatic, six-sided x-sect., sieve texture	inconspicuous	interpenetrant	Z		M, S	inclusions cf. chondrodite, vesuvianite, melanite, Fe-olivine
Mon. (Pseudo-Hex.)	tabular	one perfect, one poor				I	cf. eudialyte
Orth.	prismatic, needles, fibres, square x-sect.	one distinct				M	rare cf. sillimanite
Mon.	tabular	two at 87°, parting	simple		muscovite, cymatolite, kaolinite, eucryptite	I	cf. aegirine-augite, diopside
Orth.	bladed, prismatic, fibrous, asbestiform	two at 54$\frac{1}{2}^{\circ}$			talc	M	cf. gedrite, tremolite, cummingtonite, holmquistite
Mon.	prismatic	two at 87°, parting	simple, polysynthetic, twin seams	Z	amphibole	I, M	herringbone structure, exsolution lamellae cf. pigeonite, diopside, epidote
Mon.	prismatic	two at 87°	simple, polysynthetic	Z	amphibole	I, M	cf. augite, diopside, omphacite, jadeite
Mon.	prismatic	two at 87°	simple, polysynthetic			M	inclusions cf. fassaite, diopside, jadeite

COLOURLESS MINERALS	Pleochroism	δ	Relief	2V Dispersion	Extinction	Orientation	Mineral	Composition
Biaxial +								
	W $\gamma=\beta>\alpha$ $\gamma>\beta=\alpha$	2 - 3	+M - H	$68^o - 90^o$ $r \lessgtr v$	parallel, symmetrical	length slow	GEDRITE	$(Mg,Fe^{+2})_5Al_2\{Si_6Al_2O_{22}\}(OH,F)$
		2 - 3	H	$52^o - 76^o$ $r>v$	36°		MERWINITE	$Ca_3Mg\{Si_2O_8\}$
		3	+L	$0^o - 40^o$ $r>v$	21°	elongate twinned crystals, length slow	GIBBSITE	$Al(OH)_3$
	W $\alpha=\beta>\gamma$	3	+L	$44^o - 50^o$ $r>v$	parallel		NORBERGITE	$Mg(OH,F)_2 \cdot Mg_2SiO_4$
	W $\gamma>\alpha$	3	+M	$21^o - 30^o$ $r>v$	parallel	length slow	SILLIMANITE	Al_2SiO_5
		3	+M		parallel	length slow	FIBROLITE (SILLIMANITE)	Al_2SiO_5
	W	3	+M	$50^o - 62^o$ $r>v$	$38^o - 48^o$	extinction nearest cleavage slow	DIOPSIDE (SALITE)	$Ca(Mg,Fe)\{Si_2O_6\}$
		3 (anom.)	+M	$65^o - 69^o$ $r>v$	parallel, often wavy	cleavage length fast	PREHNITE	$Ca_2Al\{AlSi_3O_{10}\}(OH)_2$
	W $\alpha>\beta=\gamma$	3	+M	$65^o - 84^o$ $r>v$	parallel		HUMITE	$Mg(OH,F) \cdot 3Mg_2SiO_4$
	M - S $\gamma\gtrless\beta>\alpha$	3	+M	$67^o - 90^o$ $r>v$	26°	length slow	PARGASITE	$NaCa_2Mg_4Al_3Si_6O_{22}(OH)_2$
	W $\gamma>\beta=\alpha$	3	+M	$68^o - 78^o$ $r>v$	28°	length slow	ROSENBUSCHITE	$(Ca,Na,Mn)_3(Zr,Ti,Fe^{+3})\{SiO_4\}_2(F,OH)$
	W $\alpha>\beta=\gamma$	3	+M	$71^o - 85^o$ $r>v$	$22^o - 31^o$	extinction nearest twin planes fast	CHONDRODITE	$Mg(OH,F)_2 \cdot 2Mg_2SiO_4$

System	Form	Cleavage	Twinning	Zoning	Alteration	Occurrence	Remarks
Orth.	bladed, prismatic, fibrous, asbestiform	two at 54½°			talc	M	cf. Fe-anthophyllite, tremolite, cummingtonite, grunerite, zoisite
Mon.	granular	one perfect	polysynthetic			M	cf. tilleyite, spurrite, larnite
Mon.	aggregates, tabular, very small, stalactitic	one perfect	polysynthetic			S, M, I	cf. chalcedony, dahlite, muscovite, kaolinite, boehmite, diaspore
Orth.	massive					M, O	cf. humite group
Orth.	needles, slender prisms, fibrous, rhombic x-sect., faserkiesel	one, // long diagonal of x-sect.				M	cf. andalusite, mullite
Orth.	fibrous, felted, faserkiesel					M	stained brown in fibrous mats, variety of sillimanite
Mon.	short prismatic	two at 87°	simple, polysynthetic	Z	tremolite-actinolite	M, I	cf. hedenbergite, tremolite, omphacite, wollastonite, epidote
Orth.	tabular, prismatic, sheaf-like aggregates, bow-tie structure	two	fine polysynthetic, in two directions at right angles	optical sectors, hour-glass		I, M	cf. lawsonite, topaz, wollastonite, datolite
Orth.	tabular, rounded	poor				M, O	cf. humite group, olivine
Mon.	prismatic	two at 56°, partings	simple, polysynthetic			M, I	cf. hornblende, hastingsite, cummingtonite
Tr.	prismatic, fibres, needles	one perfect, two poor				I	cf. låvenite
Mon.	rounded	poor	simple, polysynthetic, twin seams			M, O	cf. humite group, olivine, staurolite

COLOURLESS MINERALS	Pleochroism	δ	Relief	2V Dispersion	Extinction	Orientation	Mineral	Composition
Biaxial +								
	W $\alpha > \gamma$	3	+M	$76^{\circ} - 87^{\circ}$ r>v	parallel, symmetrical	length slow, long diagonal of rhombic sect. slow	LAWSONITE	$CaAl_2(OH)_2\{Si_2O_7\}H_2O$
	N - M $\gamma = \alpha > \beta$ $\beta > \alpha = \gamma$	3	+M - H	$0^{\circ} - 30^{\circ}$ r$\lessgtr$v	$37^{\circ} - 44^{\circ}$	extinction nearest cleavage slow	PIGEONITE	$(Mg,Fe^{+2},Ca)(Mg,Fe^{+2})\{Si_2O_6\}$
		3	H	moderate	$13^{\circ} - 14^{\circ}$		LARNITE	$Ca_2\{SiO_4\}$
		3	H	$55^{\circ} - 90^{\circ}$ r<v	parallel	length slow	ORTHOFERROSILITE	$Fe^{+2}\{SiO_3\}$
		3	H	$68^{\circ} - 70^{\circ}$ r>v	$46^{\circ} - 48^{\circ}$	extinction nearest cleavage slow	JOHANNSENITE	$Ca(Mn,Fe)\{Si_2O_6\}$
		3 - 4	+M	$35^{\circ} - 63^{\circ}$ r>v	$10^{\circ} - 19^{\circ}$	length slow	PECTOLITE (incl. SCHIZOLITE, SERANDITE)	$Ca_2NaH\{SiO_3\}_3$
	W $\gamma > \beta > \alpha$	3 - 4	+M	$65^{\circ} - 90^{\circ}$ Mg r<v Fe r>v	$15^{\circ} - 21^{\circ}$	length slow	CUMMINGTONITE	$(Mg,Fe^{+2})_7\{Si_8O_{22}\}(OH)_2$
		4	+L	$42^{\circ} - 44^{\circ}$	parallel		ANHYDRITE	$CaSO_4$
		4	+M	$0^{\circ} - 6^{\circ}$	$\alpha : z \ 9^{\circ}$		PSEUDOWOLLASTONITE	$\beta-CaSiO_3$
		4	+M	$82^{\circ} - 90^{\circ}$ r>v	parallel	cleavage length slow	FORSTERITE	Mg_2SiO_4
		4	+M	$85^{\circ} - 89^{\circ}$ r<v	$24^{\circ} - 26^{\circ}$		TILLEYITE	$Ca_3\{Si_2O_7\}\cdot2CaCO_3$

System	Form	Cleavage	Twinning	Zoning	Alteration	Occurrence	Remarks
Orth.	rectangular, rhombic	two perfect, one imperfect	simple, polysynthetic			M	cf. clinozoisite, zoisite, prehnite, pumpellyite, andalusite, scapolite
Mon.	prismatic, anhedral, overgrowths	two at 87°, parting	simple, polysynthetic	Z		I	exsolution lamellae cf. augite, olivine
Mon.	granular	one distinct, one imperfect at 90°	polysynthetic			M	cf. bredigite, merwinite, spurrite
Orth.	prismatic, anhedral	two at 88°	rare			M	
Mon.	prismatic	two at 87°	simple, polysynthetic			O	cf. pyroxene group
Tr.	elongate, stellate, needles	two at 85°	rare		stevensite	M, I	cf. wollastonite
Mon.	prismatic, subradiating, fibrous, asbestiform	two at 55°	simple, polysynthetic			M, I	cf. tremolite, actinolite, grunerite, anthophyllite
Orth.	aggregates, anhedral – subhedral	three, rectangular	polysynthetic, two at $83\frac{1}{2}^o$		gypsum	S	twinkling cf. gypsum, barytes
Tr.	prismatic, anhedral	one	polysynthetic			M	rare
Orth.	anhedral – euhedral, rounded	uncommon, irregular fractures	uncommon	Z	chlorite, antigorite, serpentine, iddingsite, bowlingite	I, M	deformation lamellae cf. diopside, augite, humite group, epidote
Mon.	granular	one perfect, two poor	polysynthetic			M	cf. merwinite, rankinite, larnite, spurrite

COLOURLESS MINERALS	Pleochroism	δ	Relief	2V Dispersion	Extinction	Orientation	Mineral	Composition
Biaxial +								
	W	4	+M – VH	48° – 90° $r \lesseqgtr v$	parallel	cleavage length slow	OLIVINE	$(Mg,Fe)_2\{SiO_4\}$
	W $\gamma > \beta > \alpha$	4	H	84° – 86° $r < v$	parallel	length fast	DIASPORE	α-AlO(OH)
	W $\beta > \alpha = \gamma$	4 – 5	H – VH	6° – 19° $r < v$	2° – 7°	length fast/slow	MONAZITE	$(Ce,La,Th)PO_4$
	W – S	5	VH – E	0° – 38° anom. strong	parallel, oblique to twin plane	length slow	CASSITERITE	SnO_2
	W – M $\alpha > \beta > \gamma$	5 (anom.)	VH – E	17° – 40° $r > v$	40° symmetrical		SPHENE	$CaTi\{SiO_4\}(O,OH,F)$
Biaxial –								
	1		–L	0° – 85°	parallel		ANALCITE	$Na\{AlSi_2O_6\} \cdot H_2O$
	1		–L	variable	variable	cleavage length fast	CLINOPTILITE	$(Na_2,K_2,Ca)\{Al_2Si_{10}O_{24}\} \cdot 7H_2O$
	1 (anom.)		–L	0° – 25° $r > v$	α : (001) 5° – 8°	optic plane $\perp$ (010)	SANIDINE	$(K,Na)\{AlSi_3O_8\}$
	1		–L	0° – 32°	symmetrical		CHABAZITE	$Ca\{Al_2Si_4O_{12}\} \cdot 6H_2O$
	1		–L	0° – 63° $r < v$	α : (001) 5° – 11°	optic plane // (010)	HIGH SANIDINE	$(K,Na)\{AlSi_3O_8\}$

System	Form	Cleavage	Twinning	Zoning	Alteration	Occurrence	Remarks
Orth.	anhedral - euhedral, rounded	uncommon, one moderate, irregular fractures	uncommon, vicinal	Z	chlorite, antigorite, serpentine, iddingsite, bowlingite	I, M	deformation lamellae cf. diopside, augite, humite group, epidote
Orth.	tabular, aggregates, scales, acicular, stalactitic	one perfect				S, M, I	cf. boehmite, gibbsite, anhydrite, corundum, sillimanite
Mon.	small, euhedral	parting	rare		metamict	I, S	cf. sphene, zircon
Tet.	subhedral, veinlets, diamond-shaped x-sect.	prismatic	geniculate, cyclic, common	Z		O, I, S	cf. sphalerite, rutile, sphene
Mon.	rhombs, irregular grains	parting	simple, polysynthetic		leucoxene	I, M, S	cf. monazite, rutile, cassiterite
Cub.	trapezohedral, rounded, radiating, irregular	poor	polysynthetic, complex, interpenetrant			I, S, M	cf. leucite, sodalite, wairakite
Pseudo-Mon.	tabular	one perfect				I	cf. heulandite
Mon.	clear, distinct crystals, tabular, microlites	two, parting	Carlsbad, Baveno, Manebach	Z		I, M	perthitic cf. orthoclase, nepheline
Mon.	rhombohedral, approaching cube, anhedral, granular	one poor	interpenetrant	Z		I	basal section in six segments cf. gmelinite, analcite
Mon.	clear, distinct crystals, tabular, microlites	two, parting	Carlsbad, Baveno, Manebach	Z		I, M	perthitic, cf. orthoclase, nepheline

COLOURLESS MINERALS	Pleochroism	δ	Relief	2V Dispersion	Extinction	Orientation	Mineral	Composition
Biaxial $-$								
		1	$-$L	$0°$ $-$ moderate	symmetrical		GMELINITE	$(Na_2,Ca)\{Al_2Si_4O_{12}\}\cdot 6H_2O$
		1	$-$L	small-large	α : (001) $5°$ $-$ $6°$		ADULARIA	$(K,Na)\{AlSi_3O_8\}$
		1	$-$L	$33°$ $-$ $85°$ r>v	α : (001) $5°$ $-$ $19°$	(010) cleavage length fast	ORTHOCLASE	$(K,Na)\{AlSi_3O_8\}$
		1	$-$L	$36°$ $-$ $56°$ r<v	$18°$	length fast	SCOLECITE	$Ca\{Al_2Si_3O_{10}\}\cdot 3H_2O$
		1	$-$L	$43°$ $-$ $60°$ r>v	α : (001) $1°$ $-$ $6°$		ANORTHOCLASE	$(K,Na)\{AlSi_3O_8\}$
		1	$-$L	$45°$ r>v	inclined to twin lamellae		HIGH ALBITE	$Na\{AlSi_3O_8\}$
		1	$-$L	$50°$	parallel	length fast	GONNARDITE	$Na_2Ca\{(Al,Si)_5O_{10}\}_2\cdot 6H_2O$
		1	$-$L	$52°$ $-$ $73°$ r>v	$0°$ $-$ $12°$ (in albite twins)		HIGH OLIGOCLASE	$Na\{AlSi_3O_8\}-Ca\{Al_2Si_2O_8\}$
		1	$-$L	$70°$ $-$ $90°$	parallel		WAIRAKITE	$CaAl_2Si_4O_{12}\cdot 2H_2O$
		1	$-$L	$76°$ $-$ $90°$	parallel	length slow	MORDENITE	$(Na_2,K_2,Ca)\{Al_2Si_{10}O_{24}\}\cdot 7H_2O$

System	Form	Cleavage	Twinning	Zoning	Alteration	Occurrence	Remarks
Trig.	tabular, prismatic, rhombohedral, approaching cube, radiating	one good, one imperfect, parting	interpenetrant			I	cf. chabazite
Mon.	minute crystals, rhombic, x-sect.	three	albite, pericline	Z sector		O, S, M	cf. alkali feldspars
Mon.	phenocrysts, anhedral - subhedral, spherulitic aggregates	three	Carlsbad, Baveno, Manebach	Z	sericite, kaolinite	I, M	perthitic, inclusions cf. sanidine, nepheline
Mon. (Pseudo-Tet.)	columnar, fibrous	two at 88°	interpenetrant, common			I, M	cf. metascolecite, natrolite, zeolites
Mon.	phenocrysts, microlites	two	albite, Carlsbad, pericline, fine gridiron			I	perthitic cf. sanidine, orthoclase
Tr.	anhedral - euhedral, plates, laths	three	albite, Carlsbad, pericline, complex	Z	sericite, calcite, kaolinite, zeolites	I	perthitic
Orth. (Pseudo-Tet.)	fibrous, spherulitic					I	
Tr.	anhedral - euhedral, laths	three	albite, Carlsbad, pericline, complex	Z	sericite, calcite, kaolinite, albite	I	
Pseudo-Cub.	trapezohedral, anhedral.	imperfect	simple			I	cf. analcite
Orth.	tabular, acicular, fibrous	one perfect				I	

COLOURLESS MINERALS	Pleochroism	δ	Relief	2V Dispersion	Extinction	Orientation	Mineral	Composition
Biaxial −								
		1	−L − +L	76° − 90° $r \lessgtr v$	0° − 70° (in albite twins)		PLAGIOCLASE	$Na\{AlSi_3O_8\}$− $Ca\{Al_2Si_2O_8\}$
		1	N − +L	86° − 90° $r>v$	0° − 12° (in albite twins)		OLIGOCLASE	$Na\{AlSi_3O_8\}$−$Ca\{Al_2Si_2O_8\}$
		1	+L				LIZARDITE (SERPENTINE)	$Mg_3\{Si_2O_5\}(OH)_4$
		1	+L		small	length slow	HALLOYSITE	$Al_4\{Si_4O_{10}\}(OH)_8 \cdot 8H_2O$
	W	1	+L	0° − 17°	parallel	long. sect. length fast, x-sect. length slow	BERYL	$Be_3Al_2\{Si_6O_{18}\}$
	W $\gamma=\beta>\alpha$	1 anom.	+L	0° − 40° $r>v$	$\simeq$ parallel	cleavage length slow	PENNINITE	$(Mg,Al,Fe)_{12}\{(Si,Al)_8O_{20}\}(OH)_{16}$
	N − W $\gamma=\beta>\alpha$	1	+L	24° − 50° $r>v$	$1 - 3\frac{1}{2}^o$	length slow	KAOLINITE GROUP (KANDITES) (KAOLINITE, NACRITE)	$Al_4\{Si_4O_{10}\}(OH)_8$
	$\gamma>\alpha$	1 anom.	+L	37° − 61° $r>v$	parallel	length slow	ANTIGORITE (SERPENTINE)	$Mg_3\{Si_2O_5\}(OH)_4$
		1	+L	76° − 90° $r<v$	13° − $27\frac{1}{2}^o$ (in albite twins)		ANDESINE	$Na\{AlSi_3O_8\}$−$Ca\{Al_2Si_2O_8\}$
	W $\gamma=\beta>\alpha$	1 anom.	+M	0° − small $r>v$	$\simeq$ parallel	cleavage length slow	RIPIDOLITE (PROCHLORITE)	$(Mg,Al,Fe)_{12}\{(Si,Al)_8O_{20}\}(OH)_{16}$

System	Form	Cleavage	Twinning	Zoning	Alteration	Occurrence	Remarks
Tr.	anhedral - euhedral, laths	three	albite, Carlsbad, pericline, complex	Z	sericite, calcite, kaolinite, zeolites, saussurite	I, M, S	peristerite, schiller, exsolution lamellae, antiperthitic cf. cordierite
Tr.	anhedral - euhedral, laths, perthite	three	albite, Carlsbad, pericline, complex	Z	sericite, calcite, kaolinite, zeolites	I, M	peristerite, antiperthitic cf. quartz
Mon.	tabular, fine grained aggregates	perfect basal				I, M	mesh, hour-glass structure cf. chlorite group
Tr. Mon.(?)	fine aggregates, colloform, clay, fibrous	shatter cracks			sericite	S, I	cf. hydrohalloysite (endellite)
Orth. (usually Hex.)	prismatic, inclusions zoned	imperfect, rare		Z	kaolin	I, M	liquid inclusions cf. apatite, quartz
Mon.	tabular, vermicular, radiating, pseudomorphs	perfect basal	simple, pennine law	Z		I, M	pleochroic haloes cf. clinochlore, prochlorite
Tr., Mon.	mosaic-like masses, veinlets, scales, fibrous, alteration product of feldspar, clay	perfect basal	rare			S, I	cf. dickite, nacrite, phengite, sericite, montmorillonite
Mon.	anhedral, fibrolamellar, flakes, laths, plates	perfect basal	occasional			I, M	mesh, hour-glass structure cf. chrysotile, serpophite, chlorite, amphibole
Tr.	anhedral - euhedral, laths	three	albite, Carlsbad, pericline, complex	Z	sericite, calcite, kaolinite, saussurite	I, M	exsolution lamellae, antiperthitic cf. cordierite
Mon.	tabular, scaly, vermicular, fan-shaped aggregates	perfect basal				M, I, O	cf. clinochlore, penninite

COLOURLESS MINERALS	Pleochroism	δ	Relief	2V Dispersion	Extinction	Orientation	Mineral	Composition
Biaxial –								
	W $\gamma=\beta>\alpha$	1 anom.	+M	$0^{\circ} - 20^{\circ}$	small	cleavage length slow	DAPHNITE	$(Mg,Al,Fe)_{12}\{(Si,Al)_8O_{20}\}(OH)_{16}$
	W $\epsilon>\omega$	1	+M	$0^{\circ} - 36^{\circ}$ (uniaxial)		length fast, tabular length slow	APATITE (incl. DAHLLITE, FRANCOLITE)	$Ca_5(PO_4)_3(OH,F,Cl)$
	W $\omega>\epsilon$	1	H	$0^{\circ} - 30^{\circ}$ (uniaxial) moderate	parallel	tabular length slow, prismatic length fast	CORUNDUM	$\alpha-Al_2O_3$
		1 anom.	H	$5^{\circ} - 65^{\circ}$ strong		length fast	VESUVIANITE (IDOCRASE)	$Ca_{10}(Mg,Fe)_2Al_4\{Si_2O_7\}_2\{SiO_4\}_5(OH,F)_4$
	W $\beta>\gamma>\alpha$	1	H	$50^{\circ} - 66^{\circ}$ r<v	$6^{\circ} - 9^{\circ}$	length slow	SAPPHIRINE	$(Mg,Fe)_2Al_4O_6\{SiO_4\}$
		1 - 2	-L	$66^{\circ} - 90^{\circ}$ r>v	α : (001) $15^{\circ} - 20^{\circ}$	cleavage length fast	MICROCLINE	$(K,Na)\{AlSi_3O_8\}$
		1 - 2	-L - N	$48^{\circ} - 79^{\circ}$ r>v	α : x $0^{\circ} - 20^{\circ}$		HYALOPHANE	$(K,Na,Ba)\{(Al,Si)_4O_8\}$
	M	1 - 2	-L - +L	$65^{\circ} - 90^{\circ}$ r<v	parallel		CORDIERITE	$Al_3(Mg,Fe^{+2})_2\{Si_5AlO_{18}\}$
		1 - 2	N	$15^{\circ} - 90^{\circ}$ r<v	small		GISMONDINE	$Ca\{Al_2Si_2O_8\}\cdot4H_2O$
	W - S $\gamma=\beta>\alpha$	1 - 2 (anom.)	+L - +M	$0^{\circ} - 20^{\circ}$ r>v	0° - small	cleavage length slow	CHLORITE (Oxidised)	$(Mg,Al,Fe)_{12}\{(Si,Al)_8O_{20}\}(OH)_{16}$
	W - S $\alpha=\beta>\gamma$	1 - 2 (anom.)	+L - M	$0^{\circ} - 60^{\circ}$ r<v	0° - small	cleavage length slow	CHLORITE (Unoxidised)	$(Mg,Al,Fe)_{12}\{(Si,Al)_8O_{20}\}(OH)_{16}$

System	Form	Cleavage	Twinning	Zoning	Alteration	Occurrence	Remarks
Mon.	concentric aggregates, fibrous plates	perfect basal				O	
Hex.	small, prismatic	poor basal				I, S, M, O	cf. beryl, topaz, dahllite
Trig.	tabular, prismatic, six-sided x-sect.	parting	simple, lamellar seams	Z colour banding		M, I	inclusions cf. sapphirine
Tet.	variable, prismatic, fibrous, granular, radial	imperfect	sector	Z		M, I, S	cf. zoisite, clinozoisite, apatite, grossularite, melilite, andalusite
Mon.	tabular	poor	polysynthetic, uncommon			M	cf. corundum, cordierite, kyanite, zoisite, Na-amphibole
Tr.	anhedral – subhedral, phenocrysts	two, parting	albite, Carlsbad, pericline, tartan	Z	cloudy, sericite, kaolinite	I, M, S	perthitic cf. isomicrocline, orthoclase, albite, plagioclase
Mon.	tabular, prismatic	two	Carlsbad, Baveno, Manebach			O, M	cf. orthoclase
Orth. (Pseudo-Hex.)	pseudo-hex., anhedral	moderate to absent	simple, polysynthetic, cyclic, sector, interpenetrant		pinite, talc	M, I	pleochroic haloes, inclusions (including opaque dust) common cf. quartz, plagioclase
Mon. (Pseudo-Tet.)	euhedral	distinct	complex, segmented			I, O	basal sections show four segments, opposite parts alike and extinction 5°
Mon. (Pseudo-Hex.)	tabular, scaly, radiating, pseudomorphs	perfect basal	simple, polysynthetic	Z		I, M, S, O	pleochroic haloes
Mon.	tabular, scaly, radiating, pseudomorphs	perfect basal	simple, polysynthetic	Z		I, M, S, O	pleochroic haloes

COLOURLESS MINERALS	Pleochroism	δ	Relief	2V Dispersion	Extinction	Orientation	Mineral	Composition
Biaxial –								
	S $\gamma>\beta>\alpha$ $\beta>\gamma>\alpha$	1 – 2 anom.	+M	$12°-65°$ r<v	$2°-30°$	length slow	CROSSITE	$Na_2(Mg_3,Fe_3{}^{+2},Fe_2{}^{+3},Al_2)\{Si_8O_{22}\}(OH)_2$
	N – S $\gamma>\beta>\alpha$	1 – 2	+M – H	$50°-90°$ r$\lessgtr$v	parallel	length slow	HYPERSTHENE (incl. BRONZITE-EULITE)	$(Mg,Fe^{+2})\{SiO_3\}$
	W – M $\beta>\alpha>\gamma$	1 – 2	H	$45°-68°$ r>v	$2°-30°$	length fast	CHLORITOID (OTTRELITE)	$(Fe^{+2},Mg,Mn)_2(Al,Fe^{+3})Al_3O_2\{SiO_4\}_2(OH)_4$
	S $\gamma>\beta>\alpha$	1 – 3 anom.	+M	$0°-50°$ r>v	$4°-14°$	length slow	GLAUCOPHANE	$Na_2Mg_3Al_2\{Si_8O_{22}\}(OH)_2$
		2	–L	$26°-44°$ r<v	$8°-33°$	length fast/slow	LEONHARDITE	$Ca\{Al_2Si_4O_{12}\}\cdot nH_2O$
		2	–L	$26°-47°$ r<v	$8°-11°$	length slow	LAUMONTITE	$Ca\{Al_2Si_4O_{12}\}\cdot 4H_2O$
		2	–L	$30°-49°$ r<v	$0°-5°$ wavy	cleavage length fast/slow	STILBITE	$(Ca,Na_2,K_2)\{Al_2Si_7O_{18}\}\cdot 7H_2O$
		2	–L	$44°$ r<v	$10°$	length slow	EPISTILBITE	$Ca\{Al_2Si_6O_{16}\}\cdot 5H_2O$
		2	+L	$30°-32°$	parallel	length slow	CHRYSOTILE (SERPENTINE)	$Mg_3\{Si_2O_5\}(OH)_4$
		2	+L	$54°$ r<v	parallel	length fast	EDINGTONITE	$Ba\{Al_2Si_3O_{10}\}\cdot 4H_2O$
		2	+L	$77°-79°$ r>v	$51°-70°$ (in albite twins)		ANORTHITE	$Ca\{Al_2Si_2O_8\}$

System	Form	Cleavage	Twinning	Zoning	Alteration	Occurrence	Remarks
Mon.	prismatic, columnar, aggregates	two at 58°	simple, polysynthetic	Z		M	cf. glaucophane, riebeckite
Orth.	prismatic, anhedral – euhedral	two at 88°	polysynthetic, twin seams	Z		I, M	exsolution lamellae, schiller inclusions cf. andalusite
Mon., Tr.	tabular	one perfect, one imperfect, parting	polysynthetic	Z hour-glass		M, I	inclusions common cf. chlorite, clintonite, biotite, stilpnomelane
Mon.	prismatic, columnar, aggregates	two at 58°	simple, polysynthetic	Z		M	resembles holmquistite in pleochroism cf. arfvedsonite, eckermannite, riebeckite
Mon.	prismatic, fibrous	two good				I	cf. laumontite
Mon.	prismatic, fibrous	two good			leonhardite	I, M	cf. leonhardite, phillipsite
Mon. (Pseudo-Hex.)	sheaf-like aggregates, spherulitic	one good	sector, interpenetrant, cruciform			I	cf. heulandite, phillipsite
Mon.	prismatic, sheaf-like aggregates, spherulitic	one good	sector, interpenetrant, cruciform			I	
Mon.	fibrous, cross-fibre veinlets	fibrous				I, M	mesh, hour-glass structure cf. antigorite, serpophite, chlorite, amphibole
Orth.	minute, fibrous	two at 90°				I	
Tr.	anhedral – euhedral, laths	three	albite, Carlsbad, pericline, complex	Z	calcite, albite, saussurite	I, M	antiperthitic cf. cordierite

COLOURLESS MINERALS	Pleochroism	δ	Relief	2V Dispersion	Extinction	Orientation	Mineral	Composition
Biaxial −								
		2	+L	$79^{\circ} - 88^{\circ}$ r>v	$39^{\circ} - 51^{\circ}$ (in albite twins)		BYTOWNITE	$Na\{AlSi_3O_8\}-Ca\{Al_2Si_2O_8\}$
		2	+L	$83^{\circ} - 90^{\circ}$	$\alpha:z\ 3^{\circ} - 5^{\circ}$		CELSIAN	$Ba\{Al_2Si_2O_8\}$
	W	2	+M	$0^{\circ} - 20^{\circ}$	small	cleavage length slow	THURINGITE	$(Mg,Al,Fe)_{12}\{(Si,Al)_8O_{20}\}(OH)_1$
	M $\gamma=\beta>\alpha$	2	+M	$0^{\circ} - 23^{\circ}$ r<v	parallel	cleavage length slow	XANTHOPHYLLITE	$Ca_2(Mg,Fe)_{4.6}Al_{1.4}$ $\{Si_{2.5}Al_{5.5}O_{20}\}(OH)_4$
	S patchy $\alpha>\beta\gtrless\gamma$	2	+M	$20^{\circ} - 40^{\circ}$ r$\lessgtr$v	parallel	length fast	DUMORTIERITE	$HBAl_8Si_3O_{20}$
	M $\gamma=\beta>\alpha$	2	+M	32° r<v	parallel	cleavage length slow	CLINTONITE	$Ca_2(Mg,Fe)_{4.6}Al_{1.4}$ $\{Si_{2.5}Al_{5.5}O_{20}\}(OH)_4$
		2	+M	$38^{\circ} - 60^{\circ}$ r>v	$\alpha:z\ 30^{\circ}-44^{\circ}$ $\simeq$ parallel	length fast/slow	WOLLASTONITE	$Ca\{SiO_3\}$
	N − W	2	+M	$40^{\circ} - 67^{\circ}$ r<v	$6^{\circ} - 8^{\circ}$	cleavage length slow	MARGARITE	$Ca_2Al_4\{Si_4Al_4O_{20}\}(OH)_4$
		2	+M	44°	$\alpha:z\ 38^{\circ}$ $\beta:\gamma\ 0^{\circ}$	length fast/slow	PARAWOLLASTONITE	$Ca\{SiO_3\}$
	W	2	+M	$63^{\circ} - 80^{\circ}$ r<v	inclined		AXINITE	$(Ca,Mn,Fe^{+2})_3Al_2BO_3\{Si_4O_{12}\}OH$
		2	+M	$72^{\circ} - 82^{\circ}$ r>v	parallel	cleavage length slow	MONTICELLITE	$CaMg\{SiO_4\}$
	W patchy $\alpha>\beta=\gamma$	2	+M	$73^{\circ} - 86^{\circ}$ r<v	parallel	length fast	ANDALUSITE (incl. CHIASTOLITE)	Al_2SiO_5

System	Form	Cleavage	Twinning	Zoning	Alteration	Occurrence	Remarks
Tr.	anhedral – euhedral, laths	three	albite, Carlsbad, pericline, complex	Z	calcite, albite, saussurite	I	antiperthitic cf. cordierite
Mon.	prismatic	three	Carlsbad, Baveno, Manebach			O, I	cf. orthoclase, hyalophane
Mon. (Pseudo-Hex.)	tabular, radiating	perfect basal				O	
Mon.	tabular, short prismatic	perfect basal	simple			M	cf. clintonite, chlorite, chloritoid
Orth.	prismatic, acicular	imperfect, cross fractures	interpenetrant, trillings		sericite	M, I	cf. tourmaline, sillimanite
Mon.	tabular, short prismatic	perfect basal	simple			M	cf. xanthophyllite, chlorite, chloritoid
Tr.	subhedral – euhedral, columnar, fibrous	three	polysynthetic	Z	pectolite, calcite	M, I	cf. tremolite, pectolite
Mon.	tabular, flakes, pseudo-hex.	perfect basal	polysynthetic		vermiculite	M	cf. muscovite, talc, chlorite, chloritoid
Tr.	columnar, fibrous aggregates	three	polysynthetic			M	cf. wollastonite
Tr.	bladed, wedge-shaped, clusters	imperfect in several directions		Z		I, M	inclusions common cf. quartz
Orth.	aggregates, euhedral, rounded	poor			idocrase, fassaite, serpentine	M, I	cf. olivine
Orth.	columnar, square x-sect., inclusions in shape of a cross (chiastolite)	two at 89°			sericite, kyanite, sillimanite	M, S	variety chiastolite cf. sillimanite, hypersthene, viridine

COLOURLESS MINERALS	Pleochroism	δ	Relief	2V Dispersion	Extinction	Orientation	Mineral	Composition
Biaxial –								
	W	2	+M – H	$30^\circ - 44^\circ$ r<v	$15^\circ - 35^\circ$		BUSTAMITE	$(Mn,Ca,Fe)\{SiO_3\}$
		2	H	$35^\circ - 46^\circ$ r>v	64°		PYROXMANGITE	$(Mn,Fe)\{SiO_3\}$
	W $\gamma>\beta>\alpha$	2	H	$82^\circ - 83^\circ$ r>v	$0^\circ - 32^\circ$	length slow	KYANITE	Al_2SiO_5
		2 – 3	–L	$40^\circ - 60^\circ$	$\simeq$ parallel	length slow	SEPIOLITE	$H_4Mg_2\{Si_3O_{10}\}$
		2 – 3	–L – +L	moderate			SAPONITE (SMECTITE)	$(\tfrac{1}{2}Ca,Na)_{0.7}(Al,Mg,Fe)_4\{(Si,Al)_8O_{20}\}(OH)_4 \cdot nH_2O$
	M $\beta>\gamma>\alpha$	2 – 3 anom.	+M	$66^\circ - 87^\circ$ r<v	$15^\circ - 40^\circ$	length slow	RICHTERITE-FERRORICHTERITE	$Na_2Ca(Mg,Fe^{+3},Fe^{+2},Mn)_5\{Si_8O_{22}\}(OH,F)_2$
	W $\gamma>\beta>\alpha$	2 – 3	+M	$68^\circ - 90^\circ$ r$\lesssim$v	parallel, symmetrical	length slow	ANTHOPHYLLITE	$(Mg,Fe^{+2})_7\{Si_8O_{22}\}(OH,F)_2$
	W $\gamma>\beta>\alpha$	2 – 3	+M	$73^\circ - 86^\circ$ r<v	$10^\circ - 17^\circ$	length slow	ACTINOLITE-FERROACTINOLITE	$Ca_2(Mg,Fe^{+2})_5\{Si_8O_{22}\}(OH,F)_2$
	W $\gamma=\beta>\alpha$ $\gamma>\beta=\alpha$	2 – 3	+M – H	$68^\circ - 90^\circ$ r$\lesssim$v	parallel, symmetrical	length slow	FERROGEDRITE	$Fe_5Al_4Si_6O_{22}(OH)_2$
	W $\gamma=\beta>\alpha$	2 – 4	+L	$0^\circ - 58^\circ$ $(30^\circ - 50^\circ)$ r>v	$0^\circ - 7^\circ$	cleavage length slow	LEPIDOLITE	$K_2(Li,Al)_{5-6}\{Si_{6-7}Al_{2-1}O_{20}\}(OH,F)_4$
		3	–L		mass extinction		PALYGORSKITE (ATTAPULGITE)	$2MgO \cdot 3SiO_2 \cdot 4H_2O$ $-Al_2O_3 \cdot 5SiO_2 \cdot 6H_2O$
		3	–L	small			HECTORITE (SMECTITE)	$2MgO \cdot 3SiO_2 \cdot nH_2O$
	W $\gamma>\beta>\alpha$	3	–L – +M	$0^\circ - 30^\circ$	small		MONTMORILLONITE-BEIDELLITE (SMECTITES)	$(\tfrac{1}{2}Ca,Na)_{0.7}(Al,Mg,Fe)_4\{(Si,Al)_8O_{20}\}(OH)_4 \cdot nH_2O$

System	Form	Cleavage	Twinning	Zoning	Alteration	Occurrence	Remarks
Tr.	prismatic, fibrous	three, two at 95°	simple			M, O	cf. rhodonite, pyroxmangite
Tr.	prismatic	four, two at 92°	simple, polysynthetic			O, M	cf. rhodonite, bustamite
Tr.	bladed, prismatic	two, parting	simple, polysynthetic			M	cf. sillimanite, pyroxene
Mon. (Orth.)	fibrous, aggregates, curved, matted					I, S	
Mon.	shards, massive, scales, lamellar, clay	perfect basal				I	cf. pyrophyllite, talc
Mon.	long prismatic, fibrous	two at 56°, parting	simple, polysynthetic	Z		M, I	
Orth.	bladed, prismatic, fibrous, asbestiform	two at 54$\frac{1}{2}^{o}$			talc	M	cf. gedrite, tremolite, cummingtonite, holmquistite
Mon.	long prismatic, fibrous	two at 56°, parting	simple, polysynthetic	Z		M, I	cf. tremolite, orthoamphibole, hornblende
Orth.	prismatic, asbestiform	two at 54$\frac{1}{2}^{o}$				M	cf. anthophyllite, grunerite, cummingtonite, zoisite
Mon. (Trig.)	tabular, short prismatic, flakes, pseudo-hex.	perfect basal	simple			I	pleochroic haloes, inclusions cf. muscovite, zinnwaldite, phlogopite
Mon(?)	fine aggregates, clay					S	cf. montmorillonite
Mon.	shards, massive, scales, clay	perfect basal				I	
Mon.	shards, scales, massive, microcrystalline	perfect basal				I, S	cf. nontronite

COLOURLESS MINERALS	Pleochroism	δ	Relief	2V Dispersion	Extinction	Orientation	Mineral	Composition
Biaxial −								
	W $\gamma=\beta>\alpha$	3	+L	$0^o - 8^o$ r≪v	$1^o - 2^o$	cleavage length slow	VERMICULITE	$(Mg,Ca)_{0.7}(Mg,Fe^{+3},Al)_{6.0}$ $\{(Al,Si)_8O_{20}\}(OH)_4 \cdot 8H_2O$
	W $\gamma>\beta>\alpha$	3	+L	$0^o - 40^o$ r>v	$0^o - 2^o$	cleavage length slow	ZINNWALDITE	$K_2(Fe^{+2}_{2-1},Li_{2-3},Al_2)$ $\{Si_{6-7}Al_{2-1}O_{20}\}(F,OH)_4$
	W	3	+L	$< 10^o$	small		ILLITE GROUP (incl. ILLITE, BRAMMALLITE, HYDROMUSCOVITE)	$K_{1-1.5}Al_4\{Si_{7-6.5}Al_{1-1.5}O_{20}\}$ $(OH)_4$
		3	+M	$65^o - 86^o$ r<v	$15^o - 21^o$	length slow	TREMOLITE	$Ca_2(Mg,Fe^{+2})_5\{Si_8O_{22}\}(OH,F)_2$
	W $\gamma=\beta>\alpha$	3 - 4	+L	small	≃ parallel	cleavage length slow	MINNESOTAITE	$(Fe^{+2},Mg)_3\{(Si,Al)_4O_{10}\}OH_2$
	M $\gamma\geqslant\beta>\alpha$ $\alpha>\beta=\gamma$	3 - 4	+L - +M	$0^o - 15^o$ r<v	$0^o - 5^o$	cleavage length slow	PHLOGOPITE	$K_2(Mg,Fe^{+2})_6\{Si_6Al_2O_{20}\}(OH,F)_4$
		3 - 4	+L - +M	$0^o - 40^o$ r>v	≃ parallel	cleavage length slow	PARAGONITE	$Na_2Al_4\{Si_6Al_2O_{20}\}(OH)_4$
		3 - 4	+L - +M	small			SAUCONITE (SMECTITE)	$(\frac{1}{2}Ca,Na)_{0.7}Zn_{4-5}(Mg,Al,Fe^{+3})_{2-1}$ $\{(Si,Al)_8O_{20}\}(OH)_4 \cdot nH_2O$
	W $\gamma=\beta>\alpha$	3 - 4	+L - +M	$25^o - 70^o$	indistinct	length slow	NONTRONITE (SMECTITE)	$(\frac{1}{2}Ca,Na)_{0.7}(Al,Mg,Fe)_4$ $\{(Si,Al)_8O_{20}\}(OH)_4 \cdot nH_2O$
	W $\gamma>\beta=\alpha$	3 - 4	+M - H	$84^o - 90^o$ r>v	$10^o - 15^o$	length slow	GRUNERITE	$(Fe^{+2},Mg)_7\{Si_8O_{22}\}(OH)_2$
	W $\beta>\gamma>\alpha$	3 - 4 anom.	H	$74^o - 90^o$ r>v	$0^o - 15^o$ parallel in elong. sect.	length fast/slow	EPIDOTE	$Ca_2Fe^{+3}Al_2O \cdot OH\{Si_2O_7\}\{SiO_4\}$
	W $\gamma=\beta>\alpha$	4	+L	$0^o - 30^o$ r>v	≃ parallel	cleavage length slow	TALC	$Mg_6\{Si_8O_{20}\}(OH)_4$
	M $\gamma\geqslant\beta>\alpha$	4	+L	$32^o - 46^o$ r>v	$1^o - 3^o$	cleavage length slow	FUCHSITE	Cr- muscovite

System	Form	Cleavage	Twinning	Zoning	Alteration	Occurrence	Remarks
Mon.	minute particles, plates	perfect basal				S, I, M	cf. biotite, smectites
Mon.	tabular, short prismatic, flakes, pseudo-hex.	perfect basal	simple			I	cf. lepidolite, biotite
Mon.	plates, minute flakes, clay aggregates	perfect basal, parting				S, I	cf. montmorillonite, kaolinite, muscovite
Mon.	long prismatic, fibrous	two at 56°, partings	simple, polysynthetic			M, I	cf. actinolite, orthoamphibole, cummingtonite, wollastonite
Mon.	minute plates, needles, radiating aggregates	perfect basal				M	cf. talc
Mon.	tabular, flakes, plates, pseudo-hex.	perfect basal	inconspicuous	Z colour zoning		I, M	inclusions common, birds-eye maple structure cf. biotite, muscovite, lepidolite, rutile, tourmaline
Mon.	scaly aggregates	perfect basal				M, S	cf. muscovite
Mon.	shards, massive, scales	perfect basal				I, S	
Mon.	shards, scales, fibrous, massive	perfect basal				I, M	cf. montmorillonite, kaolinite
Mon.	prismatic, subradiating, fibrous, asbestiform	two at 55°, cross fractures	simple, polysynthetic		limonite	M	cf. tremolite, actinolite, cummingtonite anthophyllite
Mon.	distinct crystals, columnar, aggregates, six-sided x-sect.	one perfect	uncommon	Z		M, I, S	cf. zoisite, clinozoisite, diopside, augite, sillimanite
Mon.	platy, tabular, fibrous, aggregates	perfect basal				M, I	cf. muscovite, pyrophyllite, brucite
Mon.	thin tablets, shreds	perfect basal	simple			M	birds-eye maple structure cf. muscovite

COLOURLESS MINERALS	Pleochroism	δ	Relief	2V Dispersion	Extinction	Orientation	Mineral	Composition
Biaxial –								
	W	4	+L	$53^{\circ}-62^{\circ}$ $r>v$	$\simeq$ parallel	cleavage length slow, elongate crystals length fast	PYROPHYLLITE	$Al_4\{Si_8O_{20}\}(OH)_4$
	W	4	+L – +M	$30^{\circ}-47^{\circ}$ $r>v$	$1^{\circ}-3^{\circ}$	cleavage length slow	MUSCOVITE (incl. PHENGITE, SERICITE)	$K_2Al_4\{Si_6Al_2O_{20}\}(OH,F)_4$
		4	+M	$35^{\circ}-41^{\circ}$ $r>v$	$0^{\circ}-33^{\circ}$		SPURRITE	$2Ca_2\{SiO_4\}\cdot CaCO_3$
		4	+M	$72^{\circ}-75^{\circ}$ $r>v$	$1^{\circ}-4^{\circ}$		DATOLITE	$CaB\{SiO_4\}(OH)$
	W	4	+M – VH	$48^{\circ}-90^{\circ}$ $r\lessgtr v$	parallel	cleavage length slow	OLIVINE	$(Mg,Fe)_2\{SiO_4\}$
		4	M – VH	$52^{\circ}-90^{\circ}$ $r>v$	parallel	cleavage length slow	CHRYSOLITE- HYALOSIDERITE- HORTONOLITE- FERROHORTONOLITE	$(Mg,Fe)_2\{SiO_4\}$
	W $\gamma>\beta>\alpha$	4	H	$73^{\circ}-85^{\circ}$ $r<v$	$40^{\circ}-41^{\circ}$	length fast	LÅVENITE	$(Na,Ca,Mn,Fe^{+2})_3(Zr,Nb,Ti)\{Si_2O_7\}(OH,F)$
	W $\gamma>\beta>\alpha$	4	VH	$44^{\circ}-70^{\circ}$ $r>v$	parallel	length fast/slow	KNEBELITE	$(Mn,Fe)_2\{SiO_4\}$
	W $\gamma>\beta>\alpha$	4	VH	$44^{\circ}-70^{\circ}$ $r>v$	parallel	length fast/slow	TEPHROITE	$Mn_2\{SiO_4\}$
	W	4	VH	$48^{\circ}-52^{\circ}$ $r>v$	parallel	cleavage length slow	FAYALITE	Fe_2SiO_4

System	Form	Cleavage	Twinning	Zoning	Alteration	Occurrence	Remarks
Mon.	fine-grained, foliated lamellae, radiating, granular, spherulitic aggregates	perfect basal	ill-defined			M, O	rutile inclusions cf. talc, muscovite
Mon.	tabular, flakes, scales, aggregates	perfect basal	simple			M, I, S	birds-eye maple structure cf. talc, pyrophyllite, mica group
Mon.	granular masses	one distinct, one poor	polysynthetic, in two directions at 57°			M	cf. merwinite, rankinite, larnite
Mon.	aggregates, glassy					I, M	cf. danbourite, topaz
Orth.	anhedral - euhedral, rounded	uncommon, one moderate, irregular fractures	uncommon, vicinal	Z	chlorite, antigorite, serpentine, iddingsite, bowlingite	I, M	deformation lamellae cf. diopside, augite, pigeonite, humite group, epidote
Orth.	anhedral	moderate, irregular fractures		Z	chlorite, antigorite, serpentine, iddingsite, bowlingite	I, M	cf. olivine
Mon.	prismatic	one good	polysynthetic			I	
Orth.	anhedral - euhedral, rounded	two moderate, imperfect	uncommon			O, M	cf. tephroite, olivine
Orth.	anhedral - euhedral rounded	two moderate, imperfect	uncommon			O, M	cf. knebelite, olivine
Orth.	anhedral-euhedral, rounded	one moderate, irregular fractures	uncommon, vicinal		grunerite, serpentine	I, M	cf. olivine, knebelite, pyroxene

COLOURLESS MINERALS	Pleochroism	δ	Relief	2V Dispersion	Extinction	Orientation	Mineral	Composition
Biaxial –								
		5	–L – +M	$4^{\circ} – 14^{\circ}$ (uniaxial)	symmetrical to cleavage		CALCITE	$CaCO_3$
		5	+M	$7^{\circ} – 10^{\circ}$ r<v	parallel		STRONTIANITE	$SrCO_3$
		5	+M	16° r>v	parallel		WITHERITE	$BaCO_3$
		5	+M	$18^{\circ} – 18\frac{1}{2}^{\circ}$ r<v	parallel	length fast	ARAGONITE	$CaCO_3$

System	Form	Cleavage	Twinning	Zoning	Alteration	Occurrence	Remarks
Trig.	anhedral	rhombohedral	polysynthetic, // long diagonal			M	twinkling cf. rhombohedral carbonates
Orth.	fibrous masses	one good	simple, repeated, polysynthetic			S, I, O	twinkling cf. aragonite, witherite
Orth.	pyramidal, veins, pseudo-hex.	one distinct, one poor	repeated, cyclic, always present			S, I	twinkling cf. aragonite, strontianite
Orth.	columnar, fibrous, acicular, pseudo-hex.	imperfect, // length	repeated, polysynthetic, cyclic, interpenetrant		calcite	S, I, M	twinkling cf. calcite

PINK MINERALS	Pleochroism	δ	Relief	2V Dispersion	Extinction	Orientation	Mineral	Composition
Isometric								
			H				**DANALITE**	$Fe_4\{Be_3Si_3O_{12}\}S$
			H				**GENTHELVITE**	$Zn_4\{Be_3Si_3O_{12}\}S$
			H				**PYROPE**	$Mg_3Al_2Si_3O_{12}$
			H				**ALMANDINE**	$Fe_3^{+2}Al_2Si_3O_{12}$
	(anom.)		H – VH				**GARNET**	$(Mg,Fe^{+2},Mn,Ca)_3$ $(Fe^{+3},Ti,Cr,Al)_2\{Si_3O_{12}\}$
	(anom.)		VH				**SPESSARTITE**	$Mn_3Al_2Si_3O_{12}$
	(weak)		–L				**SODALITE** (incl. HACKMANITE)	$Na_8\{Al_6Si_6O_{24}\}Cl_2$
	W $\varepsilon>\omega$	1 (aniso-tropic)	–L – +M				**EUDIALYTE-EUCOLITE**	$(Na,Ca,Fe)_6Zr\{(Si_3O_9)_2\}$ (OH,F,Cl)
Uniaxial +								
	W $\varepsilon>\omega$	1 (iso-tropic)	+L – +M				**EUDIALYTE**	$(Na,Ca,Fe)_6Zr\{(Si_3O_9)_2\}$ (OH,F,Cl)
	M – S $\alpha=\beta>\gamma$	1	+L – +M	0° – 60° $r<v$	0° – small	length fast	**Mn and Cr-CHLORITES**	$(Mg,Mn,Al,Cr,Fe)_{12}$ $\{(Si,Al)_8O_{20}\}(OH)_{16}$

System	Form	Cleavage	Twinning	Zoning	Alteration	Occurrence	Remarks
Cub.	tetrahedral, triangular x-sect., granular	one poor				I, M	cf. garnet, helvite
Cub.	tetrahedral, triangular x-sect., granular	one poor				I, M	cf. garnet
Cub.	polygonal x-sect., aggregates	parting, irregular fractures				M, I	inclusions
Cub.	polygonal x-sect., aggregates	parting, irregular fractures			chlorite	M, I	inclusions
Cub.	four, six, eight-sided, polygonal x-sect., aggregates	parting, irregular fractures	complex, sector	Z	chlorite	M, I, S	inclusions cf. spinel
Cub.	polygonal x-sect., aggregates	parting, irregular fractures				M	inclusions cf. almandine
Cub.	hexagonal x-sect., anhedral aggregates	poor	simple		zeolites, diaspore, gibbsite	I	cf. fluorite, leucite
Trig.	rhombohedral, aggregates	one good, one poor		Z		I	usually anisotropic cf. catapleite, låvenite, rosenbuschite, garnet
Trig.	rhombohedral, aggregates	one good, one poor		Z		I	cf. catapleite, låvenite, rosenbuschite, garnet, eucolite
Mon.	tabular, radiating	perfect basal				O, I, M	

PINK MINERALS	Pleochroism	δ	Relief	2V Dispersion	Extinction	Orientation	Mineral	Composition
Uniaxial +								
	N – M $\gamma=\alpha>\beta$ $\beta>\alpha=\gamma$	3	+M – H	0° – 30°	37° – 44°	extinction nearest cleavage slow	PIGEONITE	$(Mg,Fe^{+2},Ca)(Mg,Fe^{+2})\{Si_2O_6\}$
	W	5	H			straight	XENOTIME	YPO_4
Uniaxial –								
	W $\varepsilon>\omega$	1 (iso-tropic)	+L – +M				EUCOLITE	$(Na,Ca,Fe)_6Zr\{(Si_3O_9)_2\}$ (OH,F,Cl)
	M – S $\gamma=\beta>\alpha$	1	+L – +M	0° – small	0° – small	length fast	Mn and Cr-CHLORITES	$(Mg,Mn,Al,Cr,Fe)_{12}$ $\{(Si,Al)_8O_{20}\}(OH)_{16}$
	W $\omega>\varepsilon$	1	H	0° – 30°	parallel, symmetrical	tabular length slow, prismatic length fast	CORUNDUM	$\alpha-Al_2O_3$
	S $\omega>\varepsilon$	2 – 3	+M		parallel	length slow	ELBAITE (TOURMALINE)	$Na(Li,Al)_3Al_6B_3Si_6O_{27}(OH,F)_4$
	W $\omega>\varepsilon$	5	N – VH		symmetrical to cleavage		RHODOCHROSITE	$MnCO_3$
Biaxial +								
	M – S $\gamma=\beta>\alpha$	1	+L – +M	0° – small	0° – small	length fast	Mn and Cr-CHLORITES	$(Mg,Mn,Al,Cr,Fe)_{12}$ $\{(Si,Al)_8O_{20}\}(OH)_{16}$
	N – W	1 – 2	+M	48° – 68° r>v	parallel, symmetrical	cleavage fast	TOPAZ	$Al_2\{SiO_4\}(OH,F)_2$
	M	1 – 3	+M – H	0° – 60° r>v	parallel	length fast/slow	THULITE	$Ca_2Mn^{+3}Al_2O\cdot OH\{Si_2O_7\}\{SiO_4\}$

System	Form	Cleavage	Twinning	Zoning	Alteration	Occurrence	Remarks
Mon.	prismatic, anhedral, overgrowths	two at 87°, parting	simple, polysynthetic	Z		I	exsolution lamellae cf. augite, olivine
Tet.	small prismatic					I, M, S	inclusions, pleochroic haloes around grains cf. zircon, sphene, monazite
Trig.	aggregates	one good, one poor		Z		I	some zones isotropic cf. catapleite, låvenite, rosenbuschite, garnet, eudialyte
Mon.	tabular, radiating	perfect basal				O, I, M	
Trig.	tabular, prismatic, six-sided x-sect.	parting	simple, polysynthetic, twin seams	Z		M, I	inclusions cf. sapphirine
Trig.	prismatic, radiating	fractures at 90°	rare	Z		M	cf. tourmaline group
Trig.	rhombohedral, aggregates, bands	rhombohedral	polysynthetic, rare	Z	manganese oxide	O, I, M	twinkling cf. rhombohedral carbonates
Mon.	tabular, radiating	perfect basal				O, I, M	
Orth.	prismatic, aggregates	one perfect				I, M, S	usually colourless cf. quartz, andalusite
Orth.	columnar aggregates	one perfect, one imperfect	polysynthetic, rare			M	cf. zoisite, clinozoisite, vesuvianite, sillimanite

PINK MINERALS	Pleochroism	δ	Relief	2V Dispersion	Extinction	Orientation	Mineral	Composition
Biaxial +								
	W $\alpha>\gamma$	2	H	$61^{\circ} - 76^{\circ}$ $r<v$	$5^{\circ} - 25^{\circ}$		RHODONITE	$(Mn,Ca,Fe)\{SiO_3\}$
	W $\gamma>\beta=\alpha$	2 - 3	+M	$45^{\circ} - 61^{\circ}$ $r>v$	parallel, symmetrical	length slow	MULLITE	$3Al_2O_3 \cdot 2SiO_2$
	M - S	2 - 3 (anom.)	+M - H	$25^{\circ} - 50^{\circ}$ $r>v$	$35^{\circ} - 48^{\circ}$	extinction nearest cleavage fast	TITANAUGITE	$(Ca,Na,Mg,Fe^{+2},Mn,Fe^{+3},Al,Ti)_2\{(Si,Al)_2O_6\}$
	N - M $\gamma=\alpha>\beta$ $\beta>\alpha=\gamma$	3	+M - H	$0^{\circ} - 30^{\circ}$ $r\lessgtr v$	$37^{\circ} - 44^{\circ}$	extinction nearest cleavage slow	PIGEONITE	$(Mg,Fe^{+2},Ca)(Mg,Fe^{+2})\{Si_2O_6\}$
	S $\gamma>\alpha>\beta$ $\gamma>\beta>\alpha$	3 - 5	H - VH	$64^{\circ} - 85^{\circ}$ $r\lessgtr v$	$2^{\circ} - 9^{\circ}$	length fast/slow	PIEMONTITE	$Ca_2(Mn,Fe^{+3},Al)_2AlO \cdot OH \{Si_2O_7\}\{SiO_4\}$
	W $\gamma>\beta>\alpha$	4	H	$84^{\circ} - 86^{\circ}$ $r<v$	parallel	length fast	DIASPORE	$\alpha \cdot AlO(OH)$
Biaxial -								
	W $\gamma>\alpha$	1	+L	small			KÄMMERERITE	$(Mg,Al,Cr,Fe)_{12}\{(Si,Al)_8O_{20}\}(OH)_{16}$
	M - S $\gamma=\beta>\alpha$	1	+L - +M	$0^{\circ} - 20^{\circ}$ $r>v$	0° - small	length slow	Mn and Cr-CHLORITES	$(Mg,Mn,Al,Cr,Fe)_{12}\{(Si,Al)_8O_{20}\}(OH)_{16}$
	W $\omega>\epsilon$	1	H	$0^{\circ} - 30^{\circ}$ (uniaxial) moderate	parallel, symmetrical	tabular length slow, prismatic length fast	CORUNDUM	$\alpha-Al_2O_3$
	N - S $\gamma>\beta>\alpha$	1 - 2	+M - H	$50^{\circ} - 90^{\circ}$ $r\lessgtr v$	parallel	length slow	HYPERSTHENE (incl. BRONZITE-EULITE)	$(Mg,Fe^{+2})\{SiO_3\}$

System	Form	Cleavage	Twinning	Zoning	Alteration	Occurrence	Remarks
Tr.	prismatic, square x-sect., anhedral	three, two at $92\frac{1}{2}°$	polysynthetic	Z	pyrolusite, rhodochrosite	O, M	rare, inclusions, exsolution lamellae cf. bustamite, pyroxmangite
Orth.	prismatic, needles, fibres, square x-sect.	one distinct				M	rare cf. sillimanite
Mon.	prismatic	two at $93°$	polysynthetic, twin seams	Z hourglass		I	cf. hypersthene, piemontite
Mon.	prismatic, anhedral, overgrowths	two at $87°$, parting	simple, polysynthetic	Z		I	exsolution lamellae cf. augite, olivine
Mon.	columnar, six-sided x-sect.	one perfect	polysynthetic, uncommon			M, I	cf. thulite, titanaugite, dumortierite
Orth.	tabular aggregates, scales, acicular, stalactitic	one perfect				S, M, I	cf. anhydrite, boehmite, gibbsite, corundum, sillimanite
Mon.	tabular	perfect				O	cf. chlorite
Mon.	tabular, radiating	perfect basal				O, I, M	
Trig.	tabular, prismatic, six-sided x-sect.	parting	simple, polysynthetic, twin seams	Z		M, I	inclusions cf. sapphirine
Orth.	prismatic, anhedral - euhedral	two at $88°$	polysynthetic, twin seams	Z		I, M	exsolution lamellae, schiller inclusions, cf. andalusite, titanaugite

PINK MINERALS	Pleochroism	δ	Relief	2V Dispersion	Extinction	Orientation	Mineral	Composition
Biaxial –								
	S patchy $\alpha > \beta \geqslant \gamma$	2	+M	$20^{\circ} - 40^{\circ}$ $r \gtrless v$	parallel	length fast	DUMORTIERITE	$HBAl_8Si_3O_{20}$
	W patchy $\alpha > \gamma$	2	+M	$73^{\circ} - 86^{\circ}$ $r < v$	parallel	length fast	ANDALUSITE	Al_2SiO_5
	W	2	+M – H	$30^{\circ} - 44^{\circ}$ $r < v$	$15^{\circ} - 35^{\circ}$		BUSTAMITE	$(Mn,Ca,Fe)\{SiO_3\}$
	S $\omega > \epsilon$	2 – 3	+M		parallel	length fast	ELBAITE (TOURMALINE)	$Na(Li,Al)_3Al_6B_3Si_6O_{27}(OH,F)_4$

System	Form	Cleavage	Twinning	Zoning	Alteration	Occurrence	Remarks
Orth.	prismatic, acicular	imperfect, cross fractures	interpenetrant, trillings		sericite	M, I	cf. tourmaline, sillimanite
Orth.	columnar, square x-sect.	two at 89°	rare		sericite, kyanite, sillimanite	M	variety chiastolite, inclusions in shape of a cross cf. hypersthene, viridine
Tr.	prismatic, fibrous	three, two at 95°	simple			M, O	cf. rhodonite, pyroxmangite
Trig.	prismatic, radiating	fractures at 90°	rare	Z		M	cf. tourmaline group

RED MINERALS	Pleochroism δ	Relief	2V Dispersion	Extinction	Orientation	Mineral	Composition
Isometric							
		+L				CLIACHITE	$Al_2O_3(H_2O)x$
		H				PYROPE	$Mg_3Al_2Si_3O_{12}$
	W $\gamma > \alpha$	H – VH				ALLANITE (ORTHITE)	$(Ca,Ce)_2(Fe^{+2},Fe^{+3})Al_2O \cdot OH$ $\{Si_2O_7\}\{SiO_4\}$
		H – E				SPINEL (MAGNESIOFERRITE, GALAXITE, FRANKLINITE)	$(Mg,Mn,Zn)(Fe_2^{+3},Al_2)O_4$
		VH				ALMANDINE	$Fe_3^{+2}Al_2Si_3O_{12}$
	(weak)	–L				SODALITE (incl. HACKMANITE)	$Na_8\{Al_6Si_6O_{24}\}Cl_2$
	(anom.)	H				GROSSULARITE (HESSONITE)	$Ca_3Al_2Si_3O_{12}$
	(anom.)	H – VH				GARNET	$(Mg,Fe^{+2},Mn,Ca)_3$ $(Fe^{+3},Ti,Cr,Al)_2\{Si_3O_{12}\}$
	(anom.)	VH				SPESSARTITE	$Mn_3Al_2Si_3O_{12}$
	(anom.)	VH				MELANITE	$Ca_3(Fe^{+3},Ti)_2Si_3O_{12}$
	(anom.)	VH				SCHORLOMITE	$Ca_3(Fe^{+3},Ti)_2Si_3O_{12}$

System	Form	Cleavage	Twinning	Zoning	Alteration	Occurrence	Remarks
M'loid.	pisolitic, massive	contraction cracks				S	often with gibbsite and siderite
Cub.	four, six, eight-sided, polygonal x-sect., aggregates	parting, irregular fractures				M, I	inclusions cf. garnet group
Mon.	distinct crystals, columnar, six-sided x-sect., irregular	two imperfect	uncommon	Z	metamict	I, M	anastomosing cracks, isotropic in metamict state cf. epidote, melanite
Cub.	small grains, cubes, octahedra, rhombic x-sect.	parting		Z		M, I, O	
Cub.	four, six, eight-sided, polygonal x-sect., aggregates	parting, irregular fractures			chlorite	M, I	inclusions cf. garnet group
Cub.	hexagonal x-sect., anhedral aggregates	poor	simple		zeolites, diaspore, gibbsite	I	cf. fluorite, leucite
Cub.	four, six, eight-sided, polygonal x-sect., aggregates	parting, irregular fractures	sector	Z		M	cf. garnet group, periclase, vesuvianite
Cub.	six or eight-sided, polygonal x-sect., aggregates	parting, irregular fractures	complex, sector	Z	chlorite	M, I	inclusions cf. spinel
Cub.	four, six, eight-sided, polygonal x-sect., aggregates	parting, irregular fractures				M	inclusions cf. almandine
Cub.	four, six, eight-sided, polygonal x-sect., aggregates	parting, irregular fractures	complex, sector	Z		I, M	cf. garnet group
Cub.	four, six, eight-sided, polygonal x-sect., aggregates	parting, irregular fractures	complex, sector	Z		I, M	cf. garnet group

RED MINERALS	Pleochroism	δ	Relief	2V Dispersion	Extinction	Orientation	Mineral	Composition
Isometric								
	(anom.)		VH	90° rare			ANDRADITE	$Ca_3(Fe^{+3},Ti)_2Si_3O_{12}$
	W $\gamma>\alpha$	(weak)	E				PEROVSKITE (KNOPITE-LOPARITE-DYSANALYTE)	$(Ca,Na,Fe^{+2},Ce)(Ti,Nb)O_3$
		(weak)	E				LIMONITE	$FeO \cdot OH \cdot nH_2O$
Uniaxial +								
	M – S $\alpha=\beta>\gamma$	1	+L – +M	$0^{\circ} - 60^{\circ}$ r<v	0° – small	cleavage length fast	Mn and Cr-CHLORITES	$(Mg,Mn,Al,Cr,Fe)_{12}$ $\{(Si,Al)_8O_{20}\}(OH)_{16}$
	W – S $\varepsilon>\omega$	5	VH – E		parallel, oblique to twin plane	length slow	CASSITERITE	SnO_2
	W – M $\varepsilon>\omega$	5	E		parallel		RUTILE	TiO_2
Uniaxial –								
	M – S $\gamma=\beta>\alpha$	1	+L – +M	$0^{\circ} - 20^{\circ}$ r>v	0° – small	cleavage length slow	Mn and Cr-CHLORITES (incl. DELESSITE)	$(Mg,Mn,Al,Cr,Fe)_{12}$ $\{(Si,Al)_8O_{20}\}(OH)_{16}$
	W $\omega>\varepsilon$	1	H	$0^{\circ} - 30^{\circ}$	parallel, symmetrical	tabular length slow, prismatic length fast	CORUNDUM	$\alpha-Al_2O_3$

System	Form	Cleavage	Twinning	Zoning	Alteration	Occurrence	Remarks
Cub.	four, six, eight-sided, polygonal x-sect., aggregates	parting, irregular fractures	complex, sector	Z		M	cf. garnet group
Mon? Pseudo-Cub.	small cubes, skeletal	poor to distinct	polysynthetic, complex, interpenetrant	Z	leucoxene	I, M	cf. melanite, picotite, ilmenite
M'loid.	stain or border to other minerals, pseudomorphs					I, M, S, O	opaque to translucent cf. goethite
Mon.	tabular, radiating	perfect basal				O, I, M	
Tet.	subhedral veinlets, diamond-shaped x-sect.	prismatic	geniculate, cyclic, common	Z		O, S	cf. sphalerite, rutile
Tet.	prismatic, acicular, reticulate network, inclusions	one good	geniculate			M, I, S	often opaque cf. brookite, anatase, sphene, baddeleyite, cassiterite,
Mon.	tabular, radiating	perfect basal				O, I, M	
Trig.	tabular, prismatic, six-sided x-sect.	parting	simple, lamellar seams	Z colour banding		M, I	inclusions

RED MINERALS	Pleochroism	δ	Relief	2V Dispersion	Extinction	Orientation	Mineral	Composition
Uniaxial −								
	S $\gamma \geqslant \beta > \alpha$ $\beta > \gamma > \alpha$	4 – 5	+M	$0^o - 25^o$ Mg r$\lessgtr$v Fe r$\gtrless$v	$0^o - 9^o$ wavy	cleavage length slow	BIOTITE	$K_2(Mg,Fe^{+2})_{6-4}(Fe^{+3},Al,Ti)_{0-2}$ $\{Si_{6-5}Al_{2-3}O_{20}\}(OH,F)_4$
	W	5	E				HAEMATITE	$\alpha\text{-}Fe_2O_3$
Biaxial +								
	M – S $\alpha = \beta > \gamma$	1	+L – +M	$0^o - 60^o$ r<v	0° – small	cleavage length fast	Mn and Cr- CHLORITES	$(Mg,Mn,Al,Cr,Fe)_{12}$ $\{(Si,Al)_8O_{20}\}(OH)_{16}$
	W $\gamma > \alpha$	1	E	$\simeq 90^o$ r>v	$\simeq 45^o$		PEROVSKITE (incl. KNOPITE- LOPARITE-DYSANALYTE)	$(Ca,Na,Fe^{+2},Ce)(Ti,Nb)O_3$
	W $\alpha > \gamma$	2	H	$61^o - 76^o$ r<v	$5^o - 25^o$		RHODONITE	$(Mn,Ca,Fe)\{SiO_3\}$
	M – S	2 – 3 (anom.)	+M – H	$25^o - 50^o$ r>v	$35^o - 48^o$	extinction nearest cleavage fast	TITANAUGITE	$(Ca,Na,Mg,Fe^{+2},Mn,Fe^{+3},Al,Ti)_2$ $\{(Si,Al)_2O_6\}$
	W $\gamma > \alpha$	2 – 3	H – VH	$40^o - 90^o$ r$\lessgtr$v	$1^o - 42^o$ straight, // elongation	length fast/slow	ALLANITE (ORTHITE)	$(Ca,Ce)_2(Fe^{+2},Fe^{+3})Al_2O \cdot OH$ $\{Si_2O_7\}\{SiO_4\}$
	S $\gamma > \alpha > \beta$ $\gamma > \beta > \alpha$	3 – 5	H – VH	$64^o - 85^o$ r$\lessgtr$v	$2^o - 9^o$	length fast/slow	PIEMONTITE	$Ca_2(Mn,Fe^{+3},Al)_2AlO \cdot OH$ $\{Si_2O_7\}\{SiO_4\}$
	W – M $\gamma > \beta > \alpha$	4	H	$25^o - 75^o$ r<v	parallel		IDDINGSITE	$Mg,Fe_2^{+3}\{Si_3O_{10}\}4H_2O$
	W – S	5	VH – E	$0^o - 38^o$ anom. strong	parallel, oblique to twin plane	length slow	CASSITERITE	SnO_2

System	Form	Cleavage	Twinning	Zoning	Alteration	Occurrence	Remarks
Mon. (Pseudo-Hex.)	tabular, flakes, plates, pseudo-hex.	perfect basal	simple	Z	chlorite, vermiculite, prehnite	M, I, S	pleochroic haloes, inclusions common, birds-eye maple structure cf. stilpnomelane, phlogopite
Trig.	scales, flakes, grains, irregular masses	parting	polysynthetic			M, I, O, S	opaque to translucent red cf. goethite, limonite
Mon.	tabular, radiating	perfect basal				O, I, M	
Mon? Pseudo-Cub.	small cubes, skeletal	poor to distinct	polysynthetic, complex, interpenetrant	Z	leucoxene	I, M	cf. melanite, picotite, ilmenite
Tr.	prismatic, square x-sect., anhedral	three, two at $92\frac{1}{2}^{\circ}$	polysynthetic	Z	pyrolusite, rhodochrosite	O, M	usually pink, rare, inclusions, exsolution lamellae cf. bustamite, pyroxmangite
Mon.	prismatic	two at 93°	polysynthetic, twin seams	Z hour-glass		I	cf. hypersthene, piemontite
Mon.	distinct crystals, columnar, six-sided x-sect., irregular	two imperfect	uncommon	Z	metamict	I, M	anastomosing cracks, isotropic in metamict state cf. epidote, melanite
Mon.	columnar, six-sided x-sect.	one perfect	polysynthetic, uncommon			M, I	cf. thulite, titanaugite, dumortierite
Orth.	tabular, pseudomorphs	reported			limonite	I, M	alteration of olivine
Tet.	subhedral, veinlets, diamond-shaped x-sect.	prismatic	geniculate, cyclic, common	Z		O, I, S	cf. sphalerite, rutile, sphene,

RED MINERALS	Pleochroism	δ	Relief	2V Dispersion	Extinction	Orientation	Mineral	Composition
Biaxial +								
	W – M $\epsilon > \omega$	5	E	anom. (uniaxial)	parallel		RUTILE	TiO_2
Biaxial –								
	M – S $\gamma = \beta > \alpha$	1	+L – +M	$0° - 20°$ r>v	$0°$ – small	cleavage length slow	Mn and Cr- CHLORITES (incl. DELESSITE)	$(Mg,Mn,Al,Cr,Fe)_{12}$ $\{(Si,Al)_8O_{20}\}(OH)_{16}$
	W $\omega > \epsilon$	1	H	$0° - 30°$ (uniaxial) moderate	parallel, symmetrical	tabular length slow, prismatic length fast	CORUNDUM	$\alpha-Al_2O_3$
	N – S $\gamma > \beta > \alpha$	1 – 2	+M – H	$50° - 90°$ r$\lessgtr$v	parallel	length slow	HYPERSTHENE (incl. BRONZITE, EULITE)	$(Mg,Fe^{+2})\{SiO_3\}$
	M $\gamma = \beta > \alpha$	2	+M	$0° - 23°$ r<v	parallel	cleavage length slow	XANTHOPHYLLITE	$Ca_2(Mg,Fe)_{4.6}Al_{1.4}$ $\{Si_{2.5}Al_{5.5}O_{20}\}(OH)_4$
	S patchy $\alpha > \beta \geqslant \gamma$	2	+M	$20° - 40°$ r$\lessgtr$v	parallel	length fast	DUMORTIERITE	$HBAl_8Si_3O_{20}$
	W patchy $\alpha > \gamma$	2	+M	$73° - 86°$ r<v	parallel	length fast	ANDALUSITE	Al_2SiO_5
	M $\beta > \gamma > \alpha$	2 – 3 anom.	+M	$66° - 87°$ r<v	$15° - 40°$	length slow	RICHTERITE- FERRORICHTERITE	$Na\ Ca(Mg,Fe^{+3},Fe^{+2},Mn)_5$ $\{Si_8O_{22}\}(OH,F)_2$
	W $\gamma > \alpha$	2 – 3	H – VH	$40° - 90°$ r$\lessgtr$v	$1° - 42°$ straight, // elongation	length fast/slow	ALLANITE (ORTHITE)	$(Ca,Ce)_2(Fe^{+2},Fe^{+3})Al_2O \cdot OH$ $\{Si_2O_7\}\{SiO_4\}$
	S $\gamma > \beta > \alpha$	2 – 5	+M – H	$60° - 82°$ r<v	$0° - 18°$	length slow	BASALTIC HORNBLENDE (LAMPROBOLITE, OXYHORNBLENDE)	$(Ca,Na)_{2-3}(Mg,Fe^{+2})_{3-2}$ $(Fe^{+3},Al)_{2-3}O_2\{Si_6Al_2O_{22}\}$

System	Form	Cleavage	Twinning	Zoning	Alteration	Occurrence	Remarks
Tet.	prismatic, acicular, reticulate network, inclusions	one good	geniculate			M, I, S	often opaque cf. brookite, anatase, sphene, baddeleyite, cassiterite,
Mon.	tabular, radiating	perfect basal				O, I, M	
Trig.	tabular, prismatic, six-sided x-sect.	parting	simple, lamellar seams	Z colour banding		M, I	inclusions
Orth.	prismatic, anhedral - euhedral	two at 88°	polysynthetic, twin seams	Z		I, M	exsolution lamellae, schiller inclusions cf. andalusite, titanaugite
Mon.	tabular, short prismatic	perfect basal	simple			M	cf. clintonite, chlorite, chloritoid
Orth.	prismatic, acicular	imperfect, cross fractures	interpenetrant, trillings		sericite	M, I	cf. tourmaline, sillimanite
Orth.	columnar, square x-sect.	distinct, two at 89°	rare		sericite, kyanite, sillimanite	M	variety chiastolite, inclusions in shape of a cross cf. hypersthene, viridine
Mon.	long prismatic, fibrous	two at 56°, partings	simple, polysynthetic	Z		M, I	
Mon.	distinct crystals, columnar, six-sided x-sect., irregular	two imperfect	uncommon	Z	metamict	I, M	anastomosing cracks, isotropic in metamict state cf. epidote group, melanite
Mon.	short prismatic	two at 56°	simple, polysynthetic	Z		I	cf. kaersutite, barkevikite, katophorite, cossyrite

RED MINERALS	Pleochroism	δ	Relief	2V Dispersion	Extinction	Orientation	Mineral	Composition
Biaxial -								
	M $\gamma>\beta>\alpha$ $\alpha>\beta=\gamma$	3 - 4	+L - +M	$0^\circ - 15^\circ$ r<v	$0^\circ - 5^\circ$	cleavage length slow	PHLOGOPITE	$K_2(Mg,Fe^{+2})_6\{Si_6Al_2O_{20}\}(OH,F)_4$
	W $\gamma>\beta>\alpha$	4	VH	$44^\circ - 70^\circ$ r>v	parallel	length slow	TEPHROITE (incl. ROEPPERITE)	$Mn_2\{SiO_4\}$
	S $\gamma\geq\beta>\alpha$ $\beta>\gamma>\alpha$	4 - 5	+M	$0^\circ - 25^\circ$ Mg r<v Fe r>v	$0^\circ - 9^\circ$ wavy	cleavage length slow	BIOTITE	$K_2(Mg,Fe^{+2})_{6-4}(Fe^{+3},Al,Ti)_{0-2}$ $\{Si_{6-5}Al_{2-3}O_{20}\}(OH,F)_4$
	M $\alpha>\beta>\gamma$	5	H	$70^\circ - 88^\circ$ r>v	13°	length slow	ASTROPHYLLITE (KUPLETSKITE)	$(K,Na)_3(Fe,Mn)_7Ti_2$ $\{Si_4O_{12}\}_2(O,OH,F)_7$
	M $\alpha>\beta>\gamma$	5	E	$0^\circ - 27^\circ$ extreme	parallel	length fast/slow	GOETHITE	$\alpha\cdot FeO\cdot OH$
	S $\gamma>\beta>\alpha$	5	E	83° slight	parallel		LEPIDOCROCITE	$\gamma\cdot FeO\cdot OH$

System	Form	Cleavage	Twinning	Zoning	Alteration	Occurrence	Remarks
Mon.	tabular, flakes, plates, pseudo-hex.	perfect basal	inconspicuous	Z colour banding		I, M	inclusions common, birds-eye maple structure cf. biotite, muscovite, lepidolite, rutile, tourmaline
Orth.		two moderate, imperfect				O, M	cf. olivine, knebelite
Mon. (Pseudo-Hex.)	tabular, flakes, plates, pseudo-hex.	perfect basal	simple	Z	chlorite, vermiculite, prehnite	M, I, S	pleochroic haloes, inclusions common, birds-eye maple structure cf. stilpnomelane, phlogopite
Tr.	tabular plates	perfect				I	cf. biotite, staurolite, ottrelite
Orth.	pseudomorphs, fibrous, oolitic, concretions	one perfect				O, S, I, M	weathering product cf. bowlingite, limonite
Orth.	tabular	two perfect, moderate				O, S	

BROWN MINERALS	Pleochroism	δ	Relief	2V Dispersion	Extinction	Orientation	Mineral	Composition
Isometric								
			+L				CLIACHITE	$Al_2O_3(H_2O)x$
			+M				GREENALITE	$Fe_6{}^{+2}Si_4O_{10}(OH)_8$
			H				HELVITE	$Mn_4\{Be_3Si_3O_{12}\}S$
			H				PYROPE	$Mg_3Al_2Si_3O_{12}$
			H – VH				ALLANITE (ORTHITE)	$(Ca,Ce)_2(Fe^{+2},Fe^{+3})Al_2O\cdot OH$ $\{Si_2O_7\}\{SiO_4\}$
			VH				ALMANDINE	$Fe_3{}^{+2}Al_2Si_3O_{12}$
			H – E				SPINEL (PICOTITE)	$(Mg,Fe)\{(Al,Cr)_2O_4\}$
	ir-regular		-M				OPAL	SiO_2
	(weak)		-L				NOSEAN	$Na_8\{Al_6Si_6O_{24}\}SO_4$
	l (weak)		-L				ANALCITE	$Na\{AlSi_2O_6\}\cdot H_2O$
	(weak)		-L – +M				PALAGONITE	Altered glass
	(weak)		-L – +M				VOLCANIC GLASS	

System	Form	Cleavage	Twinning	Zoning	Alteration	Occurrence	Remarks
M'loid.	pisolitic, massive	contraction cracks				S	often with gibbsite and siderite
M'loid.	massive					O	restricted to Iron Formation, Mesabi Range, Minnesota
Cub.	tetrahedral, triangular x-sect., granular	one poor	simple		ochre, manganese oxide	I, M	cf. garnet, danalite
Cub.	four, six, eight-sided, polygonal x-sect.	parting, irregular fractures				M, I	inclusions cf. garnet group
Mon.	distinct crystals, columnar, tabular, six-sided x-sect., irregular	two imperfect	uncommon	Z	metamict	I, M	anastomosing cracks, isotropic in metamict state cf. epidote, melanite
Cub.	four, six, eight-sided, polygonal x-sect.	parting, irregular fractures			chlorite	M, I	inclusions cf. garnet group
Cub.	small grains, cubes, rhombic x-sect.	parting				M, I, O	cf. garnet group
M'loid.	cryptocrystalline, colloform, veinlets, cavity fillings	irregular fractures				I, S	cf. lechatelierite
Cub.	hexagonal x-sect., anhedral aggregates	imperfect	simple	Z	zeolites, diaspore, gibbsite, limonite	I	clouded with inclusions
Cub.	trapezohedral, rounded, radiating, irregular	poor				I, S, M	cf. leucite, sodalite, wairakite
M'loid.	amorphous, oolitic				chlorite	I	cf. volcanic glass, opal, collophane
M'loid.	amorphous, massive	perlitic parting			frequent, devitrification	I	often with crystallites and phenocrysts cf. tachylyte, lechatelierite, palagonite

BROWN MINERALS	Pleochroism	δ	Relief	2V Dispersion	Extinction	Orientation	Mineral	Composition
Isometric								
		(aniso-tropic)	+M			length slow	MELILITE	$(Ca,Na)_2\{(Mg,Fe^{+2},Al,Si)_3O_7\}$
		(anom.)	H				GROSSULARITE (HESSONITE)	$Ca_3Al_2Si_3O_{12}$
		(anom.)	H – VH				GARNET	$(Mg,Fe^{+2},Mn,Ca)_3$ $(Fe^{+3},Ti,Cr,Al)_2\{Si_3O_{12}\}$
		(anom.)	VH	90° rare			ANDRADITE	$Ca_3(Fe^{+3},Ti)_2Si_3O_{12}$
		(anom.)	VH				MELANITE	$Ca_3(Fe^{+3},Ti)_2Si_3O_{12}$
		(anom.)	VH				SCHORLOMITE	$Ca_3(Fe^{+3},Ti)_2Si_3O_{12}$
		(weak)	E				SPHALERITE	ZnS
	W $\gamma>\alpha$	(weak)	E				PEROVSKITE (incl. KNOPITE-LOPARITE-DYSANALYTE)	$(Ca,Na,Fe^{+2},Ce)(Ti,Nb)O_3$
		(weak)	E				LIMONITE	$FeO \cdot OH \cdot nH_2O$

System	Form	Cleavage	Twinning	Zoning	Alteration	Occurrence	Remarks
Tet.	tabular, peg structure	moderate, single crack		Z	zeolites, carbonate	I	cf. zoisite, vesuvianite, apatite, nepheline
Cub.	four, six, eight-sided, polygonal x-sect.	parting, irregular fractures	sector	Z		M	cf. garnet group, periclase, vesuvianite
Cub.	four, six, eight-sided, polygonal x-sect., aggregates	parting, irregular fractures	complex, sector	Z	chlorite	M, I, S	inclusions cf. spinel
Cub.	four, six, eight-sided, polygonal x-sect.	parting, irregular fractures	complex, sector	Z		M	cf. garnet group
Cub.	four, six, eight-sided, polygonal x-sect.	parting, irregular fractures	complex, sector	Z		I, M	cf. garnet group
Cub.	four, six, eight-sided, polygonal x-sect.	parting, irregular fractures	complex, sector	Z		I, M	cf. garnet group
Cub.	irregular, anhedral, curved faces	six perfect	polysynthetic, lamellar intergrowths	Z		O	colour variable, (uniaxial) cf. cassiterite
Mon? Pseudo-Cub.	small cubes, skeletal	poor to distinct	polysynthetic, complex, interpenetrant	Z	leucoxene	I, M	cf. melanite, picotite, ilmenite
M'loid.	stain or border to other minerals, pseudomorphs					I, M, S, O	opaque to translucent cf. goethite

BROWN MINERALS	Pleochroism	δ	Relief	2V Dispersion	Extinction	Orientation	Mineral	Composition
Uniaxial +								
	W $\omega > \epsilon$	1 anom. (iso- tropic)	+M		parallel	length slow	**MELILITE**	$(Ca,Na)_2\{(Mg,Fe^{+2},Al,Si)_3O_7\}$
		1 anom.	+H		parallel	length fast	**VESUVIANITE (IDOCRASE) (incl. WILUITE)**	$Ca_{10}(Mg,Fe)_2Al_4\{Si_2O_7\}_2 \{SiO_4\}_5(OH,F)_4$
		2	E				**WURTZITE**	ZnS
	N - M $\gamma = \alpha > \beta$ $\beta > \alpha = \gamma$	3	+M - H	$0^\circ - 30^\circ$ $37^\circ - 44^\circ$	extinction nearest cleavage slow		**PIGEONITE**	$(Mg,Fe^{+2},Ca)(Mg,Fe^{+2})\{Si_2O_6\}$
	W $\epsilon > \omega$	4 - 5	VH		parallel	length slow	**ZIRCON**	$Zr\{SiO_4\}$
	W	5	H		straight		**XENOTIME**	YPO_4
	W - S $\epsilon > \omega$	5	VH - E	$0^\circ - 38^\circ$ anom. strong	parallel, oblique to twin plane	length slow	**CASSITERITE**	SnO_2
	W - M $\epsilon > \omega$	5	E		parallel		**RUTILE**	TiO_2

System	Form	Cleavage	Twinning	Zoning	Alteration	Occurrence	Remarks
Tet.	tabular, peg structure	moderate, single crack		Z	zeolites, carbonate	I	cf. zoisite, vesuvianite, apatite
Tet.	variable, prismatic, fibrous, granular, radial	imperfect		Z		M, I, S	cf. zoisite, clinozoisite, apatite, grossularite, melilite, andalusite
Hex.	irregular					O	
Mon.	prismatic, anhedral, overgrowths	two at 87°, parting	simple, polysynthetic	Z		I	exsolution lamellae cf. augite, olivine
Tet.	minute prisms	poor, absent		Z	metamict	I, S, M	cf. apatite
Tet.	small, prismatic					I, M, S	inclusions, pleochroic haloes around grains cf. zircon, sphene, monazite
Tet.	subhedral, veinlets, diamond-shaped x-sect.	prismatic	geniculate, cyclic, common	Z		O, I, S	cf. sphalerite, rutile, sphene
Tet.	prismatic, acicular, reticulate network, inclusions	one good	geniculate			M, I, S	often opaque cf. brookite, anatase, sphene baddeleyite, cassiterite,

BROWN MINERALS	Pleochroism	δ	Relief	2V Dispersion	Extinction	Orientation	Mineral	Composition
Uniaxial –								
	W $\omega>\varepsilon$	1 anom. (iso-tropic)	+M		parallel	length slow	MELILITE	$(Ca,Na)_2\{(Mg,Fe^{+2},Al,Si)_3O_7\}$
		1 anom.	+H		parallel	length fast	VESUVIANITE (IDOCRASE)	$Ca_{10}(Mg,Fe)_2Al_4\{Si_2O_7\}_2\{SiO_4\}_5(OH,F)_4$
	S $\omega>\varepsilon$	3	+M		parallel	length fast	DRAVITE (TOURMALINE)	$NaMg_3Al_6B_3Si_6O_{27}(OH,F)_4$
	S $\gamma=\beta>\alpha$	3 – 5	+L – H	$\simeq 0^{o}$	$\simeq$ parallel	length slow	STILPNOMELANE	$(K,Na,Ca)_{0-1\cdot4}$ $(Fe^{+3},Fe^{+2},Mg,Al,Mn)_{5\cdot9-8\cdot2}$ $\{Si_8O_{20}\}(OH)_4(O,OH,H_2O)_{3\cdot6-8\cdot5}$
	S $\gamma\geqslant\beta>\alpha$ $\beta>\gamma>\alpha$	4 – 5	+M	$0^{o} - 25^{o}$ Mg r$\lessgtr$v Fe r$>$v	$0^{o} - 9^{o}$ wavy	cleavage length slow	BIOTITE	$K_2(Mg,Fe^{+2})_{6-4}(Fe^{+3},Al,Ti)_{0-2}$ $\{Si_{6-5}Al_{2-3}O_{20}\}(OH,F)_4$
	W $\omega>\varepsilon$ rare	5	–L – H		symmetrical to cleavage	subhedral, microcrystalline	MAGNESITE	$MgCO_3$
		5	+L – VH		symmetrical to cleavage		SIDERITE	$FeCO_3$
	W $\omega>\varepsilon,\ \varepsilon>\omega$	5 anom.	E	0° – small			ANATASE	TiO_2

78

System	Form	Cleavage	Twinning	Zoning	Alteration	Occurrence	Remarks
Tet.	tabular, peg structure	moderate, single crack		Z	zeolites, carbonate	I	cf. zoisite, vesuvianite, apatite, nepheline
Tet.	variable, prismatic, fibrous, granular, radial	imperfect	sector	Z		M, I, S	cf. zoisite, clinozoisite, apatite, grossularite, melilite, andalusite
Trig.	large, prismatic	fractures at 90°	rare	Z		M	cf. chondrodite
Mon.	plates, pseudo-hex., micaceous masses	two at 90°, perfect basal	polysynthetic	Z		I, M, O	cf. biotite, chlorite, chloritoid, clintonite
Mon. (Pseudo-Hex.)	tabular, flakes, plates, pseudo-hex.	perfect basal	simple	Z	chlorite, vermiculite, prehnite	M, I, S	pleochroic haloes, inclusions common, birds-eye maple structure cf. stilpnomelane, phlogopite
Trig.	rhombohedral				huntite	M, S	cf. rhombohedral carbonates
Trig.	anhedral, aggregates, oolitic, spherulitic, colloform	rhombohedral	polysynthetic, // long diagonal, uncommon		brown spots	S, O, I, M	twinkling, brown stain around borders and along cleavage cracks cf. rhombohedral carbonates
Tet.	small, prismatic, acicular	two perfect		Z	leucoxene	S, I, M	usually yellow to blue cf. rutile, brookite

BROWN MINERALS	Pleochroism	δ	Relief	2V Dispersion	Extinction	Orientation	Mineral	Composition
Biaxial +								
	W $\gamma>\alpha$	1	E	90° r>v	$\approx 45^{\circ}$		PEROVSKITE (incl. KNOPITE-LOPARITE-DYSANALYTE)	$(Ca,Na,Fe^{+2},Ce)(Ti,Nb)O_3$
	W $\beta>\alpha=\gamma$	2	+M – H	$26^{\circ} - 85^{\circ}$ r<v	$4^{\circ} - 32^{\circ}$	length slow	PUMPELLYITE	$Ca_4(Mg,Fe^{+2})(Al,Fe^{+3})_5O(OH)_3$ $\{Si_2O_7\}_2\{SiO_4\}_2 \cdot 2H_2O$
	W $\gamma=\beta>\alpha$ $\gamma>\beta=\alpha$	2 – 3	+M	$68^{\circ} - 90^{\circ}$ r$\lessgtr$v	parallel, symmetrical	length slow	ANTHOPHYLLITE (Fe-rich)	$(Mg,Fe^{+2})_7\{Si_8O_{22}\}(OH,F)_2$
	M – S (anom.)	2 – 3	+M – H	$25^{\circ} - 50^{\circ}$ r>v	$35^{\circ} - 48^{\circ}$	extinction nearest cleavage fast	TITANAUGITE	$(Ca,Na,Mg,Fe^{+2},Mn,Fe^{+3},Al,Ti)_2$ $\{(SiAl)_2O_6\}$
	N – W	2 – 3	+M – H	$25^{\circ} - 83^{\circ}$ r>v	$35^{\circ} - 48^{\circ}$	extinction nearest cleavage fast	AUGITE-FERROAUGITE (incl. SALITE)	$(Ca,Na,Mg,Fe^{+2},Mn,Fe^{+3},Al,Ti)_2$ $\{(Si,Al)_2O_6\}$
	W $\gamma=\beta>\alpha$ $\gamma>\beta=\alpha$	2 – 3	+M – H	$68^{\circ} - 90^{\circ}$ r$\lessgtr$v	parallel, symmetrical	length slow	GEDRITE	$(Mg,Fe^{+2})_5Al_2\{Si_6Al_2O_{22}\}(OH,F)_2$
	W $\gamma>\alpha$	2 – 3	H – VH	$40^{\circ} - 90^{\circ}$ r$\lessgtr$v	$1^{\circ} - 42^{\circ}$ straight, // elongation	length fast/slow	ALLANITE (ORTHITE)	$(Ca,Ce)_2(Fe^{+2},Fe^{+3})Al_2O\cdot OH$ $\{Si_2O_7\}\{SiO_4\}$
		3	+L	$0^{\circ} - 40^{\circ}$ r>v	21°	elongate twinned crystals, length slow	GIBBSITE	$Al(OH)_3$
	W $\gamma>\alpha$	3	+M	$21^{\circ} - 30^{\circ}$ r>v	parallel	length slow	SILLIMANITE	Al_2SiO_5
		3	+M		parallel	length slow	FIBROLITE (SILLIMANITE)	Al_2SiO_5
	M – S $\gamma\geqslant\beta>\alpha$	3	+M	$67^{\circ} - 90^{\circ}$ r>v	26°	length slow	PARGASITE	$NaCa_2Mg_4Al_3Si_6O_{22}(OH)_2$

System	Form	Cleavage	Twinning	Zoning	Alteration	Occurrence	Remarks
Mon? Pseudo-Cub.	small cubes, skeletal	poor to distinct	polysynthetic, complex, interpenetrant	Z	leucoxene	I, M	cf. melanite, picotite, ilmenite
Mon.	fibrous, needles, bladed, radial aggregates	two moderate	common	Z		M	cf. clinozoisite, zoisite, epidote, lawsonite
Orth.	bladed, prismatic, fibrous, asbestiform	two at $54\frac{1}{2}°$			talc	M	cf. gedrite, cummingtonite
Mon.	prismatic	two at $93°$	polysynthetic, twin seams	Z hour-glass		I	cf. hypersthene
Mon.	prismatic	two at $87°$, parting	simple, polysynthetic, twin seams	Z	amphibole	I, M	herringbone structure, exsolution lamellae cf. pigeonite, diopside, epidote
Orth.	bladed, prismatic, fibrous, asbestiform	two at $54\frac{1}{2}°$			talc	M	cf. Fe-anthophyllite, tremolite, cummingtonite, grunerite, zoisite
Mon.	distinct crystals, columnar, six-sided x-sect., irregular	two, imperfect	uncommon	Z	metamict	I, M	anastomosing cracks, isotropic in metamict state cf. epidote, melanite
Mon.	aggregates, tabular, very small, stalactitic	one perfect	polysynthetic			S, M, I	cf. chalcedony, dahllite, muscovite, kaolinite, boehmite, diaspore
Orth.	needles, slender prisms, fibrous, rhombic x-sect., faserkiesel	one, // long diagonal of rhombic x-sect.				M	cf. andalusite, mullite
Orth.	fibrous, felted, faserkiesel					M	stained brown in fibrous mats
Mon.	prismatic	two at $56°$, parting	simple, polysynthetic			M, I	cf. hornblende, hastingsite, cummingtonite

BROWN MINERALS	Pleochroism	δ	Relief	2V Dispersion	Extinction	Orientation	Mineral	Composition
Biaxial +								
	N – M $\gamma=\alpha>\beta$ $\beta>\alpha=\gamma$	3	+M – H	0° – 30° $r \lessgtr v$	37° – 44°	extinction nearest cleavage slow	PIGEONITE	$(Mg,Fe^{+2},Ca)(Mg,Fe^{+2})\{Si_2O_6\}$
	W $\gamma>\beta>\alpha$	3	+M – H	50° – 62° $r>v$	38° – 48°	cleavage fast	HEDENBERGITE (incl. FERROSALITE)	$Ca(Fe,Mg)\{Si_2O_6\}$
	S $\alpha>\gamma$	3 – 4	H	70° – 90° $r>v$	$\gamma:z$ 70° – 90°	extinction nearest cleavage fast	AEGIRINE-AUGITE	$(Na,Ca)(Fe^{+3},Fe^{+2},Mg)\{Si_2O_6\}$
	W	4	+M – VH	48° – 90° $r \lessgtr v$	parallel	cleavage length slow	OLIVINE	$(Mg,Fe)_2\{SiO_4\}$
	W – M $\gamma>\beta>\alpha$	4	H	25° – 75° $r<v$	parallel		IDDINGSITE	$Mg,Fe_2^{+3}\{Si_3O_{10}\}4H_2O$
	M $\alpha>\beta>\gamma$	5	H	70° – 88° $r>v$	13°	length slow	ASTROPHYLLITE (KUPLETSKITE)	$(K,Na)_3(Fe,Mn)_7Ti_2\{Si_4O_{12}\}_2(O,OH,F)_7$
	M – S $\gamma>\alpha$	5	VH	32° $r<v$	4° – 45°		COSSYRITE (AENIGMATITE)	$Na_2Fe_5^{+2}TiSi_6O_{20}$
	W – S	5	VH – E	0° – 38° (uniaxial)	parallel, oblique to twin plane	length slow	CASSITERITE	SnO_2
	W – M $\alpha>\beta>\gamma$	5 anom.	VH – E	17° – 40° $r>v$	40° symmetrical		SPHENE	$CaTi\{SiO_4\}(O,OH,F)$
	W – M $\varepsilon>\omega$	5	E	anom. (uniaxial)	parallel		RUTILE	TiO_2
	W $\gamma>\alpha$	5 anom.	E	0° – 30° strong	parallel	length slow	BROOKITE	TiO_2

System	Form	Cleavage	Twinning	Zoning	Alteration	Occurrence	Remarks
Mon.	prismatic, anhedral, overgrowths	two at 87°, parting	simple, polysynthetic	Z		I	exsolution lamellae cf. augite, olivine
Mon.	prismatic	two at 87°	simple, polysynthetic	Z		M, I	cf. diopside, aegirine-augite, augite
Mon.	short prismatic, needles, felted aggregates	two at 87°, parting	simple, polysynthetic, twin seams	Z hourglass		I	cf. aegirine, acmite, Na-amphibole
Orth.	anhedral - euhedral, rounded	uncommon, one moderate, irregular fractures	uncommon, vicinal	Z	chlorite, antigorite, serpentine, iddingsite, bowlingite	I, M	deformation lamellae cf. diopside, augite, humite group, epidote
Orth.	tabular, pseudomorphs	reported			limonite	I, M	alteration of olivine
Tr.	tabular plates	perfect				I	cf. biotite, staurolite, ottrelite
Tr.	small prismatic, aggregates	two at 66°	simple, repeated			I	cf. katophorite, kaersutite, basaltic hornblende
Tet.	subhedral veinlets, diamond-shaped x-sect.	prismatic	geniculate, cyclic, common	Z		I, O, S	cf. sphalerite, rutile, sphene
Mon.	rhombs, irregular grains	parting	simple, polysynthetic		leucoxene	I, M, S	cf. monazite, rutile, cassiterite
Tet.	prismatic, acicular, reticulate network, inclusions	one good	geniculate			M, I, S	often opaque cf. brookite, anatase, sphene, baddeleyite, cassiterite,
Orth.	prismatic	poor				I, M, S	cf. rutile, pseudobrookite

BROWN MINERALS	Pleochroism	δ	Relief	2V Dispersion	Extinction	Orientation	Mineral	Composition
Biaxial −								
		1	−L	$0° - 85°$			ANALCITE	$Na\{AlSi_2O_6\}\cdot H_2O$
	N − W	1	+M	small	small	cleavage length slow	CHAMOSITE	$(Mg,Al,Fe)_{12}\{(Si,Al)_8O_{20}\}(OH)_1$
	N − S $\gamma>\beta>\alpha$	1 − 2	+M − H	$50° - 90°$ $r\lessgtr v$	parallel	length slow	HYPERSTHENE (incl. BRONZITE-EULITE)	$(Mg,Fe^{+2})\{SiO_3\}$
	M − S $\gamma>\beta>\alpha$ $\gamma<\beta>\alpha$	1 − 3	+M	$0° - 50°$ $r<v$	$36° - 70°$	length fast	KATOPHORITE-MAGNESIOKATOPHORITE	$Na_2Ca(Mg,Fe^{+2})_4Fe^{+3}\{Si_7AlO_{22}\}(OH,F)_2$
	M $\gamma=\beta>\alpha$	2	+M	$0° - 23°$ $r<v$	parallel	cleavage length slow	XANTHOPHYLLITE	$Ca_2(Mg,Fe)_{4.6}Al_{1.4}\{Si_{2.5}Al_{5.5}O_{20}\}(OH)_4$
	M $\gamma=\beta>\alpha$	2	+M	$32°$ $r<v$	parallel	cleavage length slow	CLINTONITE	$Ca_2(Mg,Fe)_{4.6}Al_{1.4}\{Si_{2.5}Al_{5.5}O_{20}\}(OH)_4$
	S $\gamma>\beta>\alpha$	2	+M	$40° - 50°$ $r>v$	$11° - 18°$	length slow	BARKEVIKITE	$Ca_2(Na,K)(Fe^{+2},Mg,Fe^{+3},Mn)_5\{Si_{6.5}Al_{1.5}O_{22}\}(OH)_2$
	W	2	+M	$63° - 80°$ $r<v$	inclined		AXINITE	$(Ca,Mn,Fe^{+2})_3Al_2BO_3\{Si_4O_{12}\}OH$
	W $\gamma>\beta>\alpha$	2 − 3	+M	$68° - 90°$ $r\lessgtr v$	parallel, symmetrical	length slow	ANTHOPHYLLITE (Mg-rich)	$(Mg,Fe^{+2})_7\{Si_8O_{22}\}(OH,F)_2$
	S $\gamma>\beta>\alpha$ $\beta>\gamma>\alpha$	2 − 3	+M − H	$66° - 85°$ $r>v$	$13° - 34°$	length slow	HORNBLENDE	$(Na,K)_{0-1}Ca_2(Mg,Fe^{+2},Fe^{+3},Al)_5\{Si_{6-7}Al_{2-1}O_{22}\}(OH,F)_2$
	W $\gamma=\beta>\alpha$ $\gamma>\beta=\alpha$	2 − 3	+M − H	$68° - 90°$ $r\lessgtr v$	parallel, symmetrical	length slow	FERROGEDRITE	$Fe_5Al_4Si_6O_{22}(OH)_2$

System	Form	Cleavage	Twinning	Zoning	Alteration	Occurrence	Remarks
Cub.	trapezohedral, rounded radiating, irregular	poor	polysynthetic, complex, interpenetrant			I, S, M	cf. leucite, sodalite, wairakite
Mon.	pseudospherulitic, concentric, tabular, massive	one good, concentric parting				S, O	cf. glauconite, collophane, greenalite, thuringite
Orth.	prismatic, anhedral – euhedral	two at 88°	polysynthetic, twin seams	Z		I, M	exsolution lamellae, schiller inclusions cf. andalusite, titanaugite
Mon.	prismatic	two at 56°, parting	simple	Z		I	cf. barkevikite, kaersutite, basaltic hornblende, arfvedsonite, cossyrite
Mon.	tabular, short prismatic	perfect basal	simple			M	cf. clintonite, chlorite, chloritoid
Mon.	tabular, short prismatic	perfect basal	simple			M	cf. xanthophyllite, chlorite, chloritoid
Mon.	prismatic	two at $\simeq 56^{\circ}$	simple	Z		I	cf. hastingsite, kaersutite, basaltic hornblende
Tr.	bladed, wedge-shaped, clusters	imperfect in several directions		Z		I, M	inclusions common cf. quartz
Orth.	bladed, prismatic, fibrous, asbestiform	two at $54\frac{1}{2}^{\circ}$			talc	M	cf. gedrite, tremolite, zoisite, cummingtonite, grunerite, holmquistite
Mon.	prismatic	two at 56°, parting	simple, polysynthetic	Z	mica, chlorite	M, I	cf. edenite, biotite, aegirine-augite, actinolite, pargasite, ferrohastingsite
Orth.	prismatic, asbestiform	two at $54\frac{1}{2}^{\circ}$				M	cf. anthophyllite, zoisite, cummingtonite, grunerite

BROWN MINERALS	Pleochroism	δ	Relief	2V Dispersion	Extinction	Orientation	Mineral	Composition
Biaxial -								
	W $\gamma>\alpha$	2 - 3	H - VH	40° - 90° $r\lessgtr v$	1° - 42° straight, // elongation	length fast/slow	ALLANITE (ORTHITE)	$(Ca,Ce)_2(Fe^{+2},Fe^{+3})Al_2O\cdot OH$ $\{Si_2O_7\}\{SiO_4\}$
	S $\gamma>\beta>\alpha$	2 - 5	+M - H	60° - 82° $r<v$	0° - 18°	length slow	BASALTIC HORNBLENDE (LAMPROBOLITE, OXYHORNBLENDE)	$(Ca,Na)_{2-3}(Mg,Fe^{+2})_{3-2}$ $(Fe^{+3},Al)_{2-3}O_2\{Si_6Al_2O_{22}\}$
	W $\gamma=\beta>\alpha$	3	+L	0° - 8° $r\lessgtr v$	1° - 2°	cleavage length slow	VERMICULITE	$(Mg,Ca)_{0.7}(Mg,Fe^{+3},Al)_{6.0}$ $\{(Al,Si)_8O_{20}\}(OH)_4\cdot 8H_2O$
	W $\gamma>\beta>\alpha$	3	+L	0° - 40° $r>v$	0° - 2°	cleavage length slow	ZINNWALDITE	$K_2(Fe^{+2}_{2-1},Li_{2-3},Al_2)$ $\{Si_{6-7}Al_{2-1}O_{20}\}(F,OH)_4$
	S $\beta\geqslant\gamma>\alpha$ $\gamma\geqslant\beta>\alpha$	3	+M - H	10° - 90° $r\lessgtr v$	9° - 40°	length slow	HASTINGSITE-FERROHASTINGSITE	$NaCa_2(Mg,Fe^{+2})_4$ $(Al,Fe^{+3})Al_2Si_6O_{22}(OH,F)_2$
	M $\alpha>\beta>\gamma$	3	VH	$\simeq 0^{o}$			CRONSTEDTITE	$(Fe_4^{+2}Fe_2^{+3})(Si_2Fe_2^{+3})O_{10}(OH)_8$
	W	3	VH				DEERITE	$(Fe,Mn)_{13}(Fe,Al)_7Si_{13}O_{44}(OH)_{11}$
	M $\gamma\geqslant\beta>\alpha$ $\alpha>\beta=\gamma$	3 - 4	+L - +M	0° - 15° $r<v$	0° - 5°	cleavage length slow	PHLOGOPITE	$K_2(Mg,Fe^{+2})_6\{Si_6Al_2O_{20}\}(OH,F)_4$
	W $\gamma>\beta=\alpha$	3 - 4	+M - H	84° - 90° $r>v$	10° - 15°	length slow	GRUNERITE	$(Fe^{+2},Mg)_7\{Si_8O_{22}\}(OH)_2$
	S $\alpha>\gamma$	3 - 4	H	70° - 90° $r>v$	$\gamma:z$ 70° - 90°	extinction nearest cleavage fast	AEGIRINE-AUGITE	$(Na,Ca)(Fe^{+3},Fe^{+2},Mg)\{Si_2O_6\}$
	S $\gamma=\beta>\alpha$	3 - 5	+L - H	$\simeq 0^{o}$	$\simeq$ parallel	length slow	STILPNOMELANE	$(K,Na,Ca)_{0-1.4}$ $(Fe^{+3},Fe^{+2},Mg,Al,Mn)_{5.9-8.2}$ $\{Si_8O_{20}\}(OH)_4(O,OH,H_2O)_{3.6-8.5}$
	S $\gamma>\beta>\alpha$	3 - 5	+M - H	66° - 82° $r>v$	0° - 19°	length slow	KAERSUTITE	$Ca_2(Na,K)(Mg,Fe^{+2},Fe^{+3})_4Ti$ $\{Si_6Al_2O_{22}\}(O,OH,F)_2$

System	Form	Cleavage	Twinning	Zoning	Alteration	Occurrence	Remarks
Mon.	distinct crystals, columnar, six-sided x-sect., irregular	two imperfect	uncommon	Z	metamict	I, M	anastomosing cracks, isotropic in metamict state cf. epidote, melanite
Mon.	short prismatic	two at 56°	simple, polysynthetic	Z		I	cf. kaersutite, barkevikite, katophorite, cossyrite
Mon.	minute particles, plates	perfect basal				S, I, M	cf. biotite, smectites
Mon.	tabular, short prismatic, flakes, pseudo-hex.	perfect basal	simple			I	cf. lepidolite, biotite
Mon.	prismatic	two at 56°, parting	simple, polysynthetic	Z		I, M	cf. arfvedsonite, hornblende
Mon. (Pseudo-Hex.)						O	cf. septochlorites
Mon.	acicular, amphibole-like	one good	simple			M	
Mon.	tabular, flakes, plates, pseudo-hex.	perfect basal	inconspicuous	Z colour zoning		I, M	inclusions common, birds-eye maple structure cf. lepidolite, rutile, tourmaline
Mon.	prismatic, subradiating, fibrous, asbestiform	two at 55°, cross fractures	simple, polysynthetic		limonite	M	cf. tremolite, actinolite, cummingtonite, anthophyllite
Mon.	short prismatic, needles, felted aggregates	two at 87°, parting	simple, polysynthetic, twin seams	Z hour-glass		I	cf. aegirine, acmite, Na-amphibole
Mon.	plates, pseudo-hex., micaceous masses	two at 90° perfect basal	polysynthetic	Z		I, M, O	cf. biotite, chlorite, chloritoid, clintonite
Mon.	prismatic	two at 56°, parting	simple, polysynthetic	Z		I	cf. barkevikite, katophorite, basaltic hornblende, cossyrite, titanaugite

BROWN MINERALS	Pleochroism	δ	Relief	2V Dispersion	Extinction	Orientation	Mineral	Composition
Biaxial −								
	W	4	+M − VH	48° − 90° $r \lessgtr v$	parallel	cleavage length slow	OLIVINE	$(Mg,Fe)_2\{SiO_4\}$
	W − M $\gamma \geq \beta > \alpha$	4	+H	25° − 75° $r<v$	parallel		IDDINGSITE (BOWLINGITE)	$Mg,Fe_2^{+3}\{Si_3O_{10}\}4H_2O$
	S $\gamma>\beta>\alpha$ $\beta>\gamma>\alpha$	4 − 5	+M	0° − 25° Mg $r \lessgtr v$ Fe $r>v$	0° − 9° wavy	cleavage length slow	BIOTITE	$K_2(Mg,Fe^{+2})_{6-4}(Fe^{+3},Al,Ti)_{0-2}\{Si_{6-5}Al_{2-3}O_{20}\}(OH,F)_4$
	W $\alpha>\beta\geq\gamma$	4 − 5	H − VH	60° − 70° $r>v$	0° − 10°	extinction nearest cleavage fast	ACMITE	$NaFe^{+3}\{Si_2O_6\}$
	W $\omega>\varepsilon$, $\varepsilon>\omega$	5 anom.	E	(uniaxial)			ANATASE	TiO_2
	W − M $\alpha>\beta>\gamma$	5	E	30° $r>v$ strong	12°		BADDELEYITE	ZrO_2

System	Form	Cleavage	Twinning	Zoning	Alteration	Occurrence	Remarks
Orth.	anhedral – euhedral, rounded	uncommon, one moderate, irregular fractures	uncommon, vicinal	Z	chlorite, antigorite, serpentine, iddingsite, bowlingite	I, M	deformation lamellae cf. diopside, augite, humite group, epidote
Orth.	tabular, pseudomorphs	reported			limonite	I, M	alteration of olivine
Mon.	tabular, flakes, plates, pseudo-hex.	perfect basal	simple	Z	chlorite, vermiculite, prehnite	M, I, S	pleochroic haloes, inclusions common, birds-eye maple structure cf. stilpnomelane, phlogopite
Mon.	short prismatic, needles, pointed terminations	two at 87°, parting	simple, polysynthetic, twin seams	Z		I	cf. aegirine, aegirine-augite, Na-amphibole
Tet.	small, prismatic, acicular	two perfect		Z	leucoxene	S, I, M	usually yellow to blue cf. rutile, brookite
Mon.	tabular	one	simple, polysynthetic			I, S	cf. zircon, rutile, brookite, anatase

ORANGE MINERALS	Pleochroism	δ	Relief	2V Dispersion	Extinction	Orientation	Mineral	Composition
Isometric								
			$-L - +M$				CHLOROPHAEITE	Chloritic mineral
	(anom.)		$H - VH$				GARNET	$(Mg,Fe^{+2},Mn,Ca)_3$ $(Fe^{+3},Ti,Cr,Al)_2\{Si_3O_{12}\}$
Uniaxial +								
	$M - S$ $\alpha=\beta>\gamma$	1	$+L - +M$	$0^{o} - 60^{o}$ $r<v$	0° - small	cleavage length fast	Mn and Cr-CHLORITES	$(Mg,Mn,Al,Cr,Fe)_{12}$ $\{(Si,Al)_8O_{20}\}(OH)_{16}$
	$W - M$ $\varepsilon>\omega$	5	E			parallel	RUTILE	TiO_2
Uniaxial −								
	$M - S$ $\gamma=\beta>\alpha$	1	$+L - +M$	$0^{o} - 20^{o}$ $r>v$	0° - small	cleavage length slow	Mn and Cr-CHLORITES	$(Mg,Mn,Al,Cr,Fe)_{12}$ $\{(Si,Al)_8O_{20}\}(OH)_{16}$
	S $\gamma\geqslant\beta>\alpha$ $\beta>\gamma>\alpha$	4 - 5	$+M$	$0^{o} - 25^{o}$ Mg $r\lesssim v$ Fe $r<v$	$0^{o} - 9^{o}$	cleavage length slow	BIOTITE	$K_2(Mg,Fe^{+2})_{6-4}(Fe^{+3},Al,Ti)_{0-2}$ $\{Si_{6-5}Al_{2-3}O_{20}\}(OH,F)_4$
	W $\omega>\varepsilon,\ \varepsilon>\omega$	5 anom.	E	0° - small			ANATASE	TiO_2

System	Form	Cleavage	Twinning	Zoning	Alteration	Occurrence	Remarks
	massive, cryptocrystalline					I	pseudomorphs after olivine cf. bowlingite, iddingsite
Cub.	four, six, eight-sided, polygonal x-sect., aggregates	parting, irregular fractures	complex, sector	Z		M, I, S	inclusions cf. spinel
Mon.	tabular, radiating	perfect basal				O, I, M	
Tet.	prismatic, acicular, reticulate network, inclusions	one good	geniculate			M, I, S	often opaque, rarely biaxial cf. brookite, anatase, sphene, baddeleyite, cassiterite
Mon.	tabular, radiating	perfect basal				O, I, M	
Mon.	tabular, flakes, plates, pseudo-hex.	perfect basal	simple	Z	chlorite, vermiculite	M, I, S	birds-eye maple structure, pleochroic haloes
Tet.	small prismatic, acicular	two perfect		Z	leucoxene	S, I, M	usually yellow to blue cf. rutile, brookite

ORANGE MINERALS	Pleochroism	δ	Relief	2V Dispersion	Extinction	Orientation	Mineral	Composition
Biaxial +								
	M - S $\alpha=\beta>\gamma$	1	+L - +M	$0^o - 60^o$ r<v	0° - small	cleavage length fast	Mn and Cr-CHLORITES	$(Mg,Mn,Al,Cr,Fe)_{12}\{(Si,Al)_8O_{20}\}(OH)_{16}$
	M $\gamma>\beta>\alpha$	2	H	$82^o - 90^o$ r>v	parallel, symmetrical	length slow	STAUROLITE	$(Fe^{+2},Mg)_2(Al,Fe^{+3})_9O_6\{SiO_4\}_4\{O,OH\}_2$
	W $\alpha>\gamma$	3 - 4 (anom.)	+M	$73^o - 76^o$ r>v	$9^o - 15^o$		CLINOHUMITE (TITANOCLINOHUMITE)	$Mg(OH,F)_2 \cdot 4Mg_2SiO_4$
	W - M $\alpha>\beta>\gamma$	5 anom.	VH - E	$17^o - 40^o$ r>v	40° symmetrical		SPHENE	$CaTi\{SiO_4\}(O,OH,F)$
	W $\gamma>\alpha$	5 anom.	E	$0^o - 30^o$ strong	parallel	length slow	BROOKITE	TiO_2
Biaxial -								
	W $\gamma>\alpha$	1	+L	small			KOCHUBEITE	Cr-chlorite
	M - S $\gamma=\beta>\alpha$	1	+L - +M	$0^o - 20^o$ r>v		cleavage length slow	Mn and Cr-CHLORITES	$(Mg,Mn,Al,Cr,Fe)_{12}\{(Si,Al)_8O_{20}\}(OH)_{16}$
	W $\beta>\gamma>\alpha$	1	H	$50^o - 66^o$ r<v	$6^o - 9^o$	length slow	SAPPHIRINE	$(Mg,Fe)_2Al_4O_6\{SiO_4\}$
	M $\gamma=\beta>\alpha$	2	+M	$0^o - 23^o$ r<v	parallel	cleavage length slow	XANTHOPHYLLITE	$Ca_2(Mg,Fe)_{4.6}Al_{1.4}\{Si_{2.5}Al_{5.5}O_{20}\}(OH)_4$
	$\gamma=\beta>\alpha$	2	+M	32° r<v	parallel	cleavage length slow	CLINTONITE	$Ca_2(Mg,Fe)_{4.6}Al_{1.4}\{Si_{2.5}Al_{5.5}O_{20}\}(OH)_4$
	S $\gamma>\beta>\alpha$	2	+M	$40^o - 50^o$ r>v	$11^o - 18^o$	length slow	BARKEVIKITE	$Ca_2(Na,K)(Fe^{+2},Mg,Fe^{+3},Mn)_5\{Si_{6.5}Al_{1.5}O_{22}\}(OH)_2$
	W	2	+M - H	$30^o - 44^o$ r<v	$15^o - 35^o$		BUSTAMITE	$(Mn,Ca,Fe)\{SiO_3\}$
	M $\beta>\gamma>\alpha$	2 - 3 anom.	+M	$66^o - 87^o$ r<v	$15^o - 40^o$	length slow	RICHTERITE-FERRORICHTERITE	$Na_2Ca(Mg,Fe^{+3},Fe^{+2},Mn)_5\{Si_8O_{22}\}(OH,F)_2$

System	Form	Cleavage	Twinning	Zoning	Alteration	Occurrence	Remarks
Mon.	tabular, radiating	perfect basal				O, I, M	
Mon. (Pseudo-Orth.)	short prismatic, six-sided x-sect., sieve texture	inconspicuous	interpenetrant	Z		M, S	inclusions, cf. chondrodite, vesuvianite, melanite, Fe-olivine
Mon.	rounded	poor	simple, polysynthetic			M, O	cf. humite group, olivine, staurolite
Mon.	rhombs, irregular grains	parting	simple, polysynthetic		leucoxene	I, M, S	cf. monazite, rutile
Orth.	prismatic	poor				I, M, S	cf. rutile, pseudobrookite
Mon.	tabular	perfect				O	cf. chlorite
Mon.	tabular, radiating	perfect basal				O, I, M	
Mon.	tabular	poor	polysynthetic, uncommon			M	cf. corundum, cordierite, kyanite, zoisite, Na-amphibole
Mon.	tabular, short prismatic	perfect basal	simple			M	cf. clintonite, chlorite, chloritoid
Mon.	tabular, short prismatic	perfect basal	simple			M	cf. xanthophyllite, chlorite, chloritoid
Mon.	prismatic	two at $\simeq 56°$	simple	Z		I	cf. hastingsite, kaersutite, basaltic hornblende
Tr.	prismatic, fibrous	three, two at $95°$	simple			M, O	cf. rhodonite, pyroxmangite
Mon.	long prismatic, fibrous	two at $56°$, parting	simple, polysynthetic	Z		M, I	

ORANGE MINERALS	Pleochroism	δ	Relief	2V Dispersion	Extinction	Orientation	Mineral	Composition
Biaxial –								
	M $\gamma \geqslant \beta > \alpha$ $\alpha > \beta = \gamma$	3 – 4	+L – +M	0° – 15° r<v	0° – 5°	cleavage length slow	PHLOGOPITE	$K_2(Mg,Fe^{+2})_6\{Si_6Al_2O_{20}\}(OH,F)$
	S $\gamma > \beta > \alpha$	3 – 5	+M – H	66° – 82° r>v	0° – 19°	length slow	KAERSUTITE	$Ca_2(Na,K)(Mg,Fe^{+2},Fe^{+3})_4Ti$ $\{Si_6Al_2O_{22}\}(O,OH,F)_2$
	S $\gamma \geqslant \beta > \alpha$ $\beta > \gamma > \alpha$	4 – 5	+M	0° – 25° Mg r$\gtrless$v Fe r<v	0° – 9° wavy	cleavage length slow	BIOTITE	$K_2(Mg,Fe^{+2})_{6-4}(Fe^{+3},Al,Ti)_{0-2}$ $\{Si_{6-5}Al_{2-3}O_{20}\}(OH,F)_4$
	W $\alpha > \beta > \gamma$	5	E	0° – 27° extreme	parallel	length fast/slow	GOETHITE	$\alpha\text{-}FeO\cdot OH$
	S $\gamma > \beta > \alpha$	5	E	83° slight	parallel		LEPIDOCROCITE	$\gamma\text{-}FeO\cdot OH$

System	Form	Cleavage	Twinning	Zoning	Alteration	Occurrence	Remarks
Mon.	tabular, flakes, plates, pseudo-hex.	perfect basal	inconspicuous	Z colour zoning		I, M	inclusions common, birds-eye maple structure cf. biotite, muscovite, lepidolite, rutile, tourmaline
Mon.	prismatic	two at 56°, parting	simple, polysynthetic	Z		I	cf. barkevikite, katophorite, basaltic hornblende, cossyrite, titanaugite
Mon. (Pseudo-Hex.)	tabular, flakes, plates, pseudo-hex.	perfect basal	simple	Z	chlorite, vermiculite, prehnite	M, I, S	pleochroic haloes, inclusions common, birds-eye maple structure cf. stilpnomelane, phlogopite
Orth.	pseudomorphs, fibrous, oolitic, concretions	one perfect				O, S, I, M	weathering product cf. bowlingite, limonite
Orth.	tabular	two perfect, moderate				O, S	

YELLOW MINERALS	Pleochroism	δ	Relief	2V Dispersion	Extinction	Orientation	Mineral	Composition
Isometric								
			+M				GREENALITE	$Fe_6^{+2}Si_4O_{10}(OH)_8$
			H				HELVITE	$Mn_4\{Be_3Si_3O_{12}\}S$
			H				PYROPE	$Mg_3Al_2Si_3O_{12}$
			H - VH				ALLANITE (ORTHITE)	$(Ca,Ce)_2(Fe^{+2},Fe^{+3})Al_2O\cdot OH$ $\{Si_2O_7\}\{SiO_4\}$
		(weak)	-L				SODALITE (incl. HACKMANITE)	$Na_8\{Al_6Si_6O_{24}\}Cl_2$
	W $\varepsilon>\omega$	1 (aniso-tropic)	-L - +M				EUDIALYTE-EUCOLITE	$(Na,Ca,Fe)_6Zr\{(Si_3O_9)_2\}$ (OH,F,CL)
		(aniso-tropic)	+M			length slow	MELILITE	$(Ca,Na)_2\{(Mg,Fe^{+2},Al,Si)_3O_7\}$
		(anom.)	H				GROSSULARITE (HESSONITE)	$Ca_3Al_2Si_3O_{12}$
		(anom.)	H - VH				GARNET	$(Mg,Fe^{+2},Mn,Ca)_3$ $(Fe^{+3},Ti,Cr,Al)_2\{Si_3O_{12}\}$
		(anom.)	VH	90° rare			ANDRADITE	$Ca_3(Fe^{+3},Ti)_2Si_3O_{12}$
	W $\gamma>\alpha$	(weak)	E				PEROVSKITE (incl. KNOPITE-LOPARITE-DYSANALYTE)	$(Ca,Na,Fe^{+2},Ce)(Ti,Nb)O_3$

System	Form	Cleavage	Twinning	Zoning	Alteration	Occurrence	Remarks
M'loid.	massive					O	restricted to Iron Formation, Mesabi Range, Minnesota
Cub.	tetrahedral, triangular x-sect., granular	one poor	simple		ochre, manganese oxide	I, M	cf. garnet, danalite
Cub.	four, six, eight-sided, polygonal x-sect., aggregates	parting, irregular fractures		Z		M, I	inclusions cf. garnet group
Mon.	distinct crystals, columnar, six-sided x-sect., irregular	two imperfect	uncommon	Z	metamict	I, M	anastomosing cracks, isotropic in metamict state cf. epidote, melanite
Cub.	hexagonal x-sect., anhedral aggregates	poor	simple		zeolites, diaspore, gibbsite	I	cf. fluorite, leucite
Trig.	rhombohedral, aggregates	one good, one poor		Z		I	cf. catapleite, låvenite, rosenbuschite, garnet
Tet.	tabular, peg structure	moderate, single crack			zeolites, carbonate	I	usually colourless cf. zoisite, vesuvianite, apatite, nepheline
Cub.	four, six, eight-sided, polygonal x-sect., aggregates	parting, irregular fractures	sector	Z		M	cf. garnet group, periclase, vesuvianite
Cub.	four, six, eight-sided, polygonal x-sect., aggregates	parting, irregular fractures	complex, sector	Z	chlorite	M, I, S	inclusions cf. spinel
Cub.	four, six, eight-sided, polygonal x-sect., aggregates	parting, irregular fractures	complex, sector	Z		M	cf. garnet group
Mon? Pseudo-Cub.	small cubes, skeletal	poor to distinct	polysynthetic, complex, interpenetrant	Z	leucoxene	I, M	cf. melanite, picotite, ilmenite

97

YELLOW MINERALS	Pleochroism	δ	Relief	2V Dispersion	Extinction	Orientation	Mineral	Composition
Isometric								
	(weak)		E				SPHALERITE	ZnS
	(weak)		+E				LIMONITE	$FeO \cdot OH \cdot nH_2O$
Uniaxial +								
	W $\varepsilon > \omega$	1 (iso-tropic)	-L - +M				EUDIALYTE	$(Na,Ca,Fe)_6 Zr\{(Si_3O_9)_2\}$ (OH,F,Cl)
	W $\alpha = \beta > \gamma$	1 anom.	+L	$0^o - 20^o$ r<v	$\approx$ parallel	cleavage length fast	PENNINITE	$(Mg,Al,Fe)_{12}\{(Si,Al)_8O_{20}\}(OH)_1$
	W $\varepsilon > \omega$	1 anom. (iso-tropic)	+M		parallel	length slow	MELILITE	$(Ca,Na)_2\{(Mg,Fe^{+2},Al,Si)_3O_7\}$
		1 anom.	H		parallel	length fast	VESUVIANITE (IDOCRASE) (incl. WILUITE)	$Ca_{10}(Mg,Fe)_2Al_4\{Si_2O_7\}_2$ $\{SiO_4\}_5(OH,F)_4$
	W $\alpha = \beta > \gamma$	1 - 2	+L	$0^o - 40^o$ r<v	$0^o - 9^o$	cleavage length fast	CLINOCHLORE	$(Mg,Al,Fe)_{12}\{(Si,Al)_8O_{20}\}(OH)_1$
	W - S $\alpha = \beta > \gamma$	1 - 2 (anom.)	+L - +M	$0^o - 60^o$ r<v	$0^o - 9^o$	cleavage length fast	CHLORITE (Unoxidised)	$(Mg,Al,Fe)_{12}\{(Si,Al)_8O_{20}\}(OH)_1$
	N - M $\gamma = \alpha > \beta$ $\beta > \alpha = \gamma$	3	+M - H	$0^o - 30^o$ $r \lessgtr v$	$37^o - 44^o$	extinction nearest cleavage slow	PIGEONITE	$(Mg,Fe^{+2},Ca)(Mg,Fe^{+2})\{Si_2O_6\}$
	W $\varepsilon > \omega$	4 - 5	VH		parallel	length slow	ZIRCON	$Zr\{SiO_4\}$

System	Form	Cleavage	Twinning	Zoning	Alteration	Occurrence	Remarks
Cub.	irregular, anhedral, curved surfaces	six perfect	polysynthetic, lamellar intergrowths	Z		O	colour variable, (uniaxial) cf. cassiterite
M'loid.	stain or border to other minerals, pseudomorphs					I, M, S, O	opaque to translucent cf. goethite
Trig.	rhombohedral, aggregates	one good, one poor		Z		I	some zones isotropic cf. catapleite, lâvenite, rosenbuschite, garnet, eucolite
Mon.	tabular, vermicular, radiating, pseudomorphs	perfect basal	simple, pennine law	Z		I, M	pleochroic haloes cf. clinochlore, prochlorite
Tet.	tabular, peg structure	moderate, single crack		Z	zeolites, carbonate	I	cf. zoisite, vesuvianite, apatite, nepheline
Tet.	variable, prismatic, fibrous, granular, radial	imperfect	sector	Z		M, I, S	usually colourless cf. zoisite, clinozoisite, apatite, grossularite, melilite, andalusite
Mon.	tabular, fibrous, pseudo-hex.	perfect basal	polysynthetic			M	pleochroic haloes cf. penninite, prochlorite, leuchtenbergite, katschubeite
Mon. (Pseudo-Hex.)	tabular, scaly, radiating, pseudomorphs	perfect basal	simple, polysynthetic	Z		I, M, S, O	pleochroic haloes
Mon.	prismatic, anhedral, overgrowths	two at 87°, parting	simple, polysynthetic	Z		I	exsolution lamellae cf. augite, olivine
Tet.	minute prisms	poor, absent		Z	metamict	I, S, M	cf. apatite

YELLOW MINERALS	Pleochroism	δ	Relief	2V Dispersion	Extinction	Orientation	Mineral	Composition
Uniaxial +								
	W	5	H		straight		XENOTIME	YPO_4
	W - S $\varepsilon>\omega$	5	VH - E	$0°$ - $38°$ anom., strong	parallel, oblique to twin plane	length slow	CASSITERITE	SnO_2
	W - M $\varepsilon>\omega$	5	E		parallel		RUTILE	TiO_2
Uniaxial -								
	W $\varepsilon>\omega$	1 (iso-tropic)	+L - +M				EUCOLITE	$(Na,Ca,Fe)_6Zr\{(Si_3O_9)_2\}(OH,F,Cl)$
	W $\omega>\varepsilon$	1 anom. (iso-tropic)	+M			length slow	MELILITE	$(Ca,Na)_2\{(Mg,Fe^{+2},Al,Si)_3O_7\}$
		1 anom.	H		parallel	length fast	VESUVIANITE (IDOCRASE)	$Ca_{10}(Mg,Fe)_2Al_4\{Si_2O_7\}_2\{SiO_4\}_5(OH,F)_4$
	W $\omega>\varepsilon$	1	H	$0°$ - $30°$	parallel, symmetrical	tabular length slow, prismatic length fast	CORUNDUM	$\alpha-Al_2O_3$
	W - S $\gamma=\beta>\alpha$	1 - 2 (anom.)	+L - +M	$0°$ - $20°$ r>v	$0°$ - small	cleavage length slow	CHLORITE (Unoxidised)	$(Mg,Al,Fe)_{12}\{(Si,Al)_8O_{20}\}(OH)_{16}$
		3	-L	anom.	parallel	length fast	CANCRINITE	$(Na,Ca,K)_{6-8}\{Al_6Si_6O_{24}\}(CO_3,SO_4,Cl)_{1-2}\cdot1-5H_2O$

System	Form	Cleavage	Twinning	Zoning	Alteration	Occurrence	Remarks
Tet.	small, prismatic					I, M, S	inclusions, pleochroic haloes around grains cf. zircon, sphene, monazite
Tet.	subhedral, veinlets, diamond-shaped x-sect.	prismatic	geniculate, cyclic, common	Z		O, I, S	cf. sphalerite, rutile, sphene
Tet.	prismatic, acicular, reticulate network, inclusions	one good	geniculate			M, I, S	often opaque cf. brookite, anatase, sphene, baddeleyite, cassiterite
Trig.	aggregates	one good, one poor		Z		I	some zones isotropic cf. catapleite, lavenite, rosenbuschite, garnet, eudialyte
Tet.	tabular, peg structure	moderate, single crack		Z		I	usually colourless cf. zoisite, vesuvianite, apatite, nepheline
Tet.	variable, prismatic, fibrous, granular, radial	imperfect		Z		M, I	cf. zoisite, clinozoisite, grossularite
Trig.	tabular, prismatic, six-sided x-sect.	parting	simple, polysynthetic	Z		M, I	inclusions
Mon. (Pseudo-Hex.)	tabular, scaly, radiating, pseudomorphs	perfect basal	simple, polysynthetic	Z		I, M, S, O	pleochroic haloes
Hex.	anhedral	good	polysynthetic, rare			I	cf. vishnevite

YELLOW MINERALS	Pleochroism	δ	Relief	2V Dispersion	Extinction	Orientation	Mineral	Composition
Uniaxial −								
	S $\omega>\varepsilon$	3	+M		parallel	length fast	DRAVITE (TOURMALINE)	$NaMg_3Al_6B_3Si_6O_{27}(OH,F)_4$
	S $\omega>\varepsilon$	3	+M		parallel	length fast	SCHORL (TOURMALINE)	$Na(Fe,Mn)_3Al_6B_3Si_6O_{27}(OH,F)_4$
	S $\gamma=\beta>\alpha$	3 – 5	+L – H	$\simeq 0°$	$\simeq$ parallel	length slow	STILPNOMELANE	$(K,Na,Ca)_{0-1.4}$ $(Fe^{+3},Fe^{+2},Mg,Al,Mn)_{5.9-8.2}$ $\{Si_8O_{20}\}(OH)_4(O,OH,H_2O)_{3.6-8.5}$
	S $\gamma\geq\beta>\alpha$ $\beta>\gamma>\alpha$	4 – 5	+M	$0° - 25°$ Mg r$\lessgtr$v Fe r$>$v	$0° - 9°$ wavy	cleavage length slow	BIOTITE	$K_2(Mg,Fe^{+2})_{6-4}(Fe^{+3},Al,Ti)_{0-2}$ $\{Si_{6-5}Al_{2-3}O_{20}\}(OH,F)_4$
	S $\gamma>\alpha$	5	+M – H		parallel or symmetrical		JAROSITE	$KFe_3^{+3}(OH)_6(SO_4)_2$
	W $\omega>\varepsilon$, $\varepsilon>\omega$	5 anom.	E	$0°$ – small			ANATASE	TiO_2
Biaxial +								
	W $\alpha=\beta>\gamma$	1 anom.	+L	$0° - 20°$ r$<$v	$\simeq$ parallel	cleavage length fast	PENNINITE	$(Mg,Al,Fe)_{12}\{(Si,Al)_8O_{20}\}(OH)_1$
	M	1 – 2	−L – +L	$65° - 90°$ r$<$v	parallel		CORDIERITE	$Al_3(Mg,Fe^{+2})_2\{Si_5AlO_{18}\}$
	W $\alpha=\beta>\gamma$	1 – 2	+L	$0° - 40°$ r$<$v	$0° - 9°$	cleavage length fast	CLINOCHLORE	$(Mg,Al,Fe)_{12}\{(Si,Al)_8O_{20}\}(OH)_1$
	W – S $\alpha=\beta>\gamma$	1 – 2 (anom.)	+L – +M	$0° - 60°$ r$<$v	$0° - 9°$	cleavage length fast	CHLORITE (Unoxidised)	$(Mg,Al,Fe)_{12}\{(Si,Al)_8O_{20}\}(OH)_1$

System	Form	Cleavage	Twinning	Zoning	Alteration	Occurrence	Remarks
Trig.	prismatic	fractures at 90°		Z		M	cf. tourmaline group, chondrodite
Trig.	hexagonal, rounded, suns, triangular x-sect.	fractures		Z		I, M, S	cf. tourmaline group, biotite
Mon.	plates, pseudo-hex., micaceous masses	two at 90°, perfect basal	polysynthetic	Z		I, M, O	cf. biotite, chlorite, chloritoid, clintonite
Mon. (Pseudo-Hex.)	tabular, flakes, bent plates, pseudo-hex.	perfect basal	simple	Z	chlorite, vermiculite, prehnite	M, I, S	pleochroic haloes, inclusions common, birds-eye maple structure cf. stilpnomelane, phlogopite
Hex.	tabular, aggregates				limonite	I, O	cf. alunite, natrojarosite
Tet.	small, prismatic, acicular	two perfect		Z	leucoxene	S, I, M	cf. rutile, brookite
Mon.	tabular, vermicular, radiating, pseudomorphs	perfect basal	simple, pennine law	Z		I, M	pleochroic haloes cf. clinochlore, prochlorite
Orth. (Pseudo-Hex.)	pseudo-hex., anhedral	moderate to absent	simple, polysynthetic, cyclic, sector, interpenetrant		pinite, talc	M, I	pleochroic haloes, inclusions (including opaque dust) common cf. quartz, plagioclase
Mon.	tabular, fibrous, pseudo-hex.	perfect basal	polsynthetic			M	pleochroic haloes cf. penninite, prochlorite, leuchtenbergite, katschubeite
Mon. (Pseudo-Hex.)	tabular, scaly, radiating, pseudomorphs	perfect basal	simple, polysynthetic	Z		I, M, S, O	pleochroic haloes

YELLOW MINERALS	Pleochroism	δ	Relief	2V Dispersion	Extinction	Orientation	Mineral	Composition
Biaxial +								
	W	1 - 2	+M	48° - 68° r>v	parallel, symmetrical	cleavage fast	TOPAZ	$Al_2\{SiO_4\}(OH,F)_2$
	W - M β>α>γ	1 - 2 anom.	H	45° - 68° r>v	2° - 30°	length fast	CHLORITOID (OTTRELITE)	$(Fe^{+2},Mg,Mn)_2(Al,Fe^{+3})Al_3O_2\{SiO_4\}_2(OH)_4$
	M	1 - 3	+M - H	0° - 60° r>v	parallel	length fast/slow	THULITE	$Ca_2Mn^{+3}Al_2O\cdot OH\{Si_2O_7\}\{SiO_4\}$
	W γ=β>α γ>β=α	2	+M	68° - 90° r⪋v	parallel, symmetrical	length slow	ANTHOPHYLLITE (Fe-rich)	$(Mg,Fe^{+2})_7\{Si_8O_{22}\}(OH,F)_2$
	W β>α=γ	2	+M - H	26° - 85° r<v	4° - 32°	length slow	PUMPELLYITE	$Ca_4(Mg,Fe^{+2})(Al,Fe^{+3})_5O(OH)_3\{Si_2O_7\}_2\{SiO_4\}_2\cdot 2H_2O$
	M γ>β>α	2	H	82° - 90° r>v	parallel, symmetrical	length slow	STAUROLITE	$(Fe^{+2},Mg)_2(Al,Fe^{+3})_9O_6\{SiO_4\}_4(O,OH)_2$
	W γ=β>α γ>β=α	2 - 3	+M - H	68° - 90° r⪋v	parallel	length slow	GEDRITE	$(Mg,Fe^{+2})_5Al_2\{Si_6Al_2O_{22}\}(OH,F)_2$
	W γ>α	2 - 3	H - VH	40° - 90° r⪋v	1° - 42° straight, // elongation	length fast/slow	ALLANITE (ORTHITE)	$(Ca,Ce)_2(Fe^{+2},Fe^{+3})Al_2O\cdot OH\{Si_2O_7\}\{SiO_4\}$
	W α=β>γ	3	+L	44° - 50° r>v	parallel		NORBERGITE	$Mg(OH,F)_2\cdot Mg_2SiO_4$
	W γ>α	3	+M	21° - 30° r>v	parallel	length slow	SILLIMANITE	Al_2SiO_5
	W α>β=γ	3	+M	65° - 84° r>v	parallel		HUMITE	$Mg(OH,F)\cdot 3Mg_2SiO_4$
	M - S γ⪖β>α	3	+M	67° - 90° r>v	26°	length slow	PARGASITE	$NaCa_2Mg_4Al_3Si_6O_{22}(OH)_2$

System	Form	Cleavage	Twinning	Zoning	Alteration	Occurrence	Remarks
Orth.	prismatic aggregates	one perfect				I, M, S	usually colourless cf. quartz, andalusite
Mon., Tr.	tabular	one perfect, one imperfect, parting	polysynthetic	Z hour-glass		M, I	inclusions common cf. chlorite, clintonite, biotite, stilpnomelane
Orth.	columnar aggregates	one perfect, one imperfect	polysynthetic, rare			M	cf. zoisite, clinozoisite, vesuvianite, sillimanite
Orth.	bladed, prismatic, fibrous, asbestiform	two at $54\frac{1}{2}°$			talc	M	cf. gedrite, tremolite, cummingtonite, holmquistite
Mon.	fibrous, needles, bladed, radial aggregates	two moderate	common	Z		M	cf. clinozoisite, zoisite, epidote, lawsonite
Mon. (Pseudo-Orth.)	short prismatic, six-sided x-sect., sieve texture	inconspicuous	interpenetrant	Z		M, S	inclusions cf. chondrodite, vesuvianite, melanite, Fe-olivine
Orth.	bladed, prismatic, fibrous, asbestiform	two at $54\frac{1}{2}°$			talc	M	cf. Fe-anthophyllite, tremolite, holmquistite, cummingtonite, grunerite, zoisite
Mon.	distinct crystals, columnar, aggregates, six-sided x-sect.	two imperfect	uncommon	Z	metamict	I, M	anastomosing cracks, isotropic in metamict state cf. epidote, melanite
Orth.	massive					M, O	cf. humite group
Orth.	needles, slender prisms, fibrous, rhombic x-sect., faserkiesel	one, // long diagonal of x-sect.				M	cf. andalusite, mullite
Orth.	tabular, rounded	poor				M, O	cf. humite group, olivine, staurolite
Mon.	prismatic	two at $56°$, parting	simple, polysynthetic			M, I	cf. hornblende, hastingsite, cummingtonite

YELLOW MINERALS	Pleochroism	δ	Relief	2V Dispersion	Extinction	Orientation	Mineral	Composition
Biaxial +								
	W $\gamma>\beta=\alpha$	3	+M	$68^\circ - 78^\circ$ r>v	28°	length slow	ROSENBUSCHITE	$(Ca,Na,Mn)_3(Zr,Ti,Fe^{+3})$ $\{SiO_4\}_2(F,OH)$
	W $\alpha>\beta=\gamma$	3	+M	$71^\circ - 85^\circ$ r>v	$22^\circ - 31^\circ$	extinction nearest twin plane fast	CHONDRODITE	$Mg(OH,F)_2 \cdot 2Mg_2SiO_4$
	W $\alpha>\gamma$	3	+M	$76^\circ - 87^\circ$ r>v	parallel, symmetrical	length fast, long diagonal of rhombic sect. slow	LAWSONITE	$CaAl_2(OH)_2\{Si_2O_7\}H_2O$
	N – M $\gamma=\alpha>\beta$ $\beta>\alpha=\gamma$	3	+M – H	$0^\circ - 30^\circ$ $r \lessgtr v$	$37^\circ - 44^\circ$	extinction nearest cleavage slow	PIGEONITE	$(Mg,Fe^{+2},Ca)(Mg,Fe^{+2})\{Si_2O_6\}$
	W $\alpha>\gamma$	3 – 4 (anom.)	+M	$73^\circ - 76^\circ$ r>v	$9^\circ - 15^\circ$		CLINOHUMITE	$Mg(OH,F)_2 \cdot 4Mg_2SiO_4$
	S $\gamma>\alpha>\beta$ $\gamma>\beta>\alpha$	3 – 5	H – VH	$64^\circ - 85^\circ$ $r \lessgtr v$	$2^\circ - 9^\circ$	length fast/slow	PIEMONTITE	$Ca_2(Mn,Fe^{+3},Al)_2AlO\cdot OH$ $\{Si_2O_7\}\{SiO_4\}$
	W	4	+M – VH	$48^\circ - 90^\circ$ $r \lessgtr v$	parallel	cleavage length slow	OLIVINE	$(Mg,Fe)_2\{SiO_4\}$
	W $\beta>\alpha=\gamma$	4 – 5	H – VH	$6^\circ - 19^\circ$ r<v	$2^\circ - 7^\circ$	length fast/slow	MONAZITE	$(Ce,La,Th)PO_4$
	M $\alpha>\beta>\gamma$	5	H	$70^\circ - 88^\circ$ r>v	13°	length slow	ASTROPHYLLITE	$(K,Na)_3(Fe,Mn)_7Ti_2$ $\{Si_4O_{12}\}_2(O,OH,F)_7$
	W – S	5	VH – E	$0^\circ - 38^\circ$ anom. strong	parallel, oblique to twin plane	length slow	CASSITERITE	SnO_2
	W – M $\alpha>\beta>\gamma$	5 anom.	VH – E	$17^\circ - 40^\circ$ r>v	40° symmetrical		SPHENE	$CaTi\{SiO_4\}(O,OH,F)$
	W $\gamma>\alpha$	5 anom.	E	$0^\circ - 30^\circ$ strong	parallel	length slow	BROOKITE	TiO_2

System	Form	Cleavage	Twinning	Zoning	Alteration	Occurrence	Remarks
Tr.	prismatic, fibres, needles	one perfect, two poor				I	cf. lävenite
Mon.	rounded	poor	simple, polysynthetic, twin seams			M, O	cf. humite group, olivine, staurolite
Orth.	rectangular, rhombic	two perfect	simple, polysynthetic			M	cf. zoisite, prehnite, pumpellyite, andalusite, scapolite
Mon.	prismatic, anhedral, overgrowths	two at 87°, parting	simple, polysynthetic	Z		I	exsolution lamellae cf. augite, olivine
Mon.	rounded	poor	simple, polysynthetic			M, O	cf. humite group, olivine, staurolite
Mon.	columnar	one perfect	polysynthetic, uncommon			M, I	cf. thulite, dumortierite
Orth.	anhedral – euhedral, rounded	uncommon, one moderate, irregular fractures	uncommon, vicinal	Z	chlorite, antigorite, serpentine, iddingsite, bowlingite	I, M	deformation lamellae cf. diopside, augite, humite group, epidote
Mon.	small, euhedral	parting	rare		metamict	I, S	cf. sphene, zircon
Tr.	tabular, plates	perfect				I	cf. biotite, staurolite, ottrelite
Tet.	subhedral, veinlets, diamond-shaped x-sect.	prismatic	geniculate, cyclic, common	Z		O, I, S	cf. sphalerite, rutile
Mon.	rhombs, irregular grains	parting	simple, polysynthetic		leucoxene	I, M, S	cf. monazite, rutile, cassiterite
Orth.	prismatic	poor				I, M, S	cf. rutile, pseudobrookite

YELLOW MINERALS	Pleochroism	δ	Relief	2V Dispersion	Extinction	Orientation	Mineral	Composition
Biaxial –								
	W $\gamma=\beta>\alpha$	1 anom.	+L	$0°$ – $40°$ r>v	≈ parallel	cleavage length slow	PENNINITE	$(Mg,Al,Fe)_{12}\{(Si,Al)_8O_{20}\}(OH)_1$
	M	1 – 2	–L – +L	$65°$ – $90°$ r<v	parallel		CORDIERITE	$Al_3(Mg,Fe^{+2})_2\{Si_5AlO_{18}\}$
	W – S $\gamma=\beta>\alpha$	1 – 2 (anom.)	+L – +M	$0°$ – $20°$ r>v	$0°$ – small	cleavage length slow	CHLORITE (Oxidised)	$(Mg,Al,Fe)_{12}\{(Si,Al)_8O_{20}\}(OH)_1$
	W – S $\gamma=\beta>\alpha$	1 – 2 (anom.)	+L – +M	$0°$ – $20°$ r>v	$0°$ – $9°$	cleavage length slow	CHLORITE (Unoxidised)	$(Mg,Al,Fe)_{12}\{(Si,Al)_8O_{20}\}(OH)_1$
	S $\gamma>\beta>\alpha$ $\beta>\gamma>\alpha$	1 – 2 anom.	+M	$12°$ – $65°$ r<v	$2°$ – $30°$	length slow	CROSSITE	$Na_2(Mg_3,Fe_3{}^{+2},Fe_2{}^{+3},Al_2)\{Si_8O_{22}\}(OH)_2$
	S $\alpha>\beta>\gamma$	1 – 2	+M – H	$0°$ – $50°$ r<v	$0°$ – $30°$ anom.	length fast	ARFVEDSONITE	$Na_3(Mg,Fe^{+2})_4Al\{Si_8O_{22}\}(OH,F)_2$
	W – M $\beta>\alpha>\gamma$	1 – 2 anom.	H	$45°$ – $68°$ r>v	$20°$ – $30°$	length fast	CHLORITOID (OTTRELITE)	$(Fe^{+2},Mg,Mn)_2(Al,Fe^{+3})Al_3O_2\{SiO_4\}_2(OH)_4$
	M – S $\gamma>\beta>\alpha$ $\gamma<\beta>\alpha$	1 – 3	+M	$0°$ – $50°$ r<v	$36°$ – $70°$	length fast	KATOPHORITE-MAGNESIOKATOPHORITE	$Na_2Ca(Mg,Fe^{+2})_4Fe^{+3}\{Si_7AlO_{22}\}(OH,F)_2$
	M $\gamma=\beta>\alpha$	2	+M	$0°$ – $23°$ r<v	parallel	cleavage length slow	XANTHOPHYLLITE	$Ca_2(Mg,Fe)_{4.6}Al_{1.4}\{Si_{2.5}Al_{5.5}O_{20}\}(OH)_4$
	M $\gamma=\beta>\alpha$	2	+M	$32°$ r<v	parallel	cleavage length slow	CLINTONITE	$Ca_2(Mg,Fe)_{4.6}Al_{1.4}\{Si_{2.5}Al_{5.5}O_{20}\}(OH)_4$
	S $\gamma>\beta>\alpha$	2	+M	$40°$ – $50°$ r>v	$11°$ – $18°$	length slow	BARKEVIKITE	$Ca_2(Na,K)(Fe^{+2},Mg,Fe^{+3},Mn)_5\{Si_{6.5}Al_{1.5}O_{22}\}(OH)_2$
	W	2	+M	$63°$ – $80°$ r<v	inclined		AXINITE	$(Ca,Mn,Fe^{+2})_3Al_2BO_3\{Si_4O_{12}\}OH$

System	Form	Cleavage	Twinning	Zoning	Alteration	Occurrence	Remarks
Mon.	tabular, vermicular, radiating, pseudomorphs	perfect basal	simple, pennine law	Z		I, M	pleochroic haloes cf. clinochlore, prochlorite
Orth. (Pseudo-Hex.)	pseudo-hex., anhedral	moderate to absent	simple, polysynthetic, cyclic, sector, interpenetrant		pinite, talc	M, I	pleochroic haloes, inclusions (including opaque dust) common cf. quartz, plagioclase
Mon. (Pseudo-Hex.)	tabular, scaly, radiating, pseudomorphs	perfect basal	simple, polysynthetic	Z		I, M, S, O	pleochroic haloes
Mon. (Pseudo-Hex.)	tabular, scaly, radiating, pseudomorphs	perfect basal	simple, polysynthetic	Z		I, M, S, O	pleochroic haloes
Mon.	prismatic, columnar aggregates	two at 58°	simple, polysynthetic	Z		M	cf. glaucophane, riebeckite
Mon.	prismatic	two at 56°, parting	simple, polysynthetic	Z		I	cf. riebeckite, katophorite, ferrohastingsite, glaucophane
Mon., Tr.	tabular	one perfect, one imperfect, parting	polysynthetic	Z hour-glass		M, I	usually positive cf. chlorite, clintonite, biotite, stilpnomelane
Mon.	prismatic	two at 56°, parting	simple	Z		I	cf. barkevikite, kaersutite, basaltic hornblende, arfvedsonite, cossyrite
Mon.	tabular, short prismatic	perfect basal	simple			M	cf. clintonite, chlorite, chloritoid
Mon.	tabular, short prismatic	perfect basal	simple			M	cf. xanthophyllite, chlorite, chloritoid
Mon.	prismatic	two at $\simeq 56^{\circ}$	simple	Z		I	cf. hastingsite, kaersutite, basaltic hornblende
Tr.	bladed, wedge-shaped clusters	imperfect in several directions		Z		I, M	inclusions common cf. quartz

YELLOW MINERALS	Pleochroism	δ	Relief	2V Dispersion	Extinction	Orientation	Mineral	Composition
Biaxial −								
	W patchy $\alpha > \gamma$	2	+M	$73^{\circ} - 86^{\circ}$ r<v	parallel	length fast	ANDALUSITE	Al_2SiO_5
		2 − 3	−L	$40^{\circ} - 60^{\circ}$	$\simeq$ parallel	length slow	SEPIOLITE	$H_4Mg_2\{Si_3O_{10}\}$
	M $\gamma = \beta > \alpha$	2 − 3	+M	$0^{\circ} - 20^{\circ}$ r<v	$0^{\circ} - 3^{\circ}$	cleavage length slow	GLAUCONITE (incl. CELADONITE)	$(K,Na,Ca)_{1.2-2.0}$ $(Fe^{+3},Al,Fe^{+2},Mg)_{4.0}$ $\{Si_{7-7.6}Al_{1-0.4}O_{20}\}(OH)_4 \cdot nH_2O$
	M $\beta > \gamma > \alpha$	2 − 3 anom.	+M	$66^{\circ} - 87^{\circ}$ r<v	$15^{\circ} - 40^{\circ}$	length slow	RICHTERITE- FERRORICHTERITE	$Na_2Ca(Mg,Fe^{+3},Fe^{+2},Mn)_5$ $\{Si_8O_{22}\}(OH,F)_2$
	W $\gamma = \beta > \alpha$ $\gamma > \beta = \alpha$	2 − 3	+M	$68^{\circ} - 90^{\circ}$ r$\lesssim$v	parallel, symmetrical	length slow	ANTHOPHYLLITE (Fe-rich)	$(Mg,Fe^{+2})_7\{Si_8O_{22}\}(OH,F)_2$
	W $\gamma > \beta > \alpha$	2 − 3	+M	$73^{\circ} - 86^{\circ}$ r<v	$10^{\circ} - 17^{\circ}$	length slow	ACTINOLITE- FERROACTINOLITE	$Ca_2(Mg,Fe^{+2})_5\{Si_8O_{22}\}(OH,F)_2$
	W $\gamma = \beta > \alpha$ $\gamma > \beta = \alpha$	2 − 3	+M − H	$68^{\circ} - 90^{\circ}$ r$\lesssim$v	parallel, symmetrical	length slow	FERROGEDRITE	$Fe_5Al_4Si_6O_{22}(OH)_2$
	W $\gamma > \alpha$	2 − 3	H − VH	$40^{\circ} - 90^{\circ}$ r$\lesssim$v	$1^{\circ} - 42^{\circ}$ straight, // elongation	length fast/slow	ALLANITE (ORTHITE)	$(Ca,Ce)_2(Fe^{+2},Fe^{+3})Al_2O \cdot OH$ $\{Si_2O_7\}\{SiO_4\}$
	S $\gamma > \beta > \alpha$	2 − 5	+M − H	$60^{\circ} - 82^{\circ}$ r<v	$0^{\circ} - 18^{\circ}$	length slow	BASALTIC HORNBLENDE (LAMPROBOLITE, OXYHORNBLENDE)	$(Ca,Na)_{2-3}(Mg,Fe^{+2})_{3-2}$ $(Fe^{+3},Al)_{2-3}O_2\{Si_6Al_2O_{22}\}$
	W $\gamma > \beta > \alpha$	3	−L − +M	$0^{\circ} - 30^{\circ}$	small		MONTMORILLONITE (SMECTITE)	$(\frac{1}{2}Ca,Na)_{0.7}(Al,Mg,Fe)_4$ $\{(Si,Al)_8O_{20}\}(OH)_4 \cdot nH_2O$
	S	3	+M	51° r>v	parallel	length slow	HOLMQUISTITE	$Li_2(Mg,Fe^{+2})_3(Al,Fe^{+3})_2$ $\{Si_8O_{22}\}(OH)_2$
	S $\beta \gtrless \gamma > \alpha$ $\gamma \gtrless \beta > \alpha$	3	+M − H	$10^{\circ} - 90^{\circ}$ r$\gtrless$v	$9^{\circ} - 40^{\circ}$	length slow	HASTINGSITE- FERROHASTINGSITE	$NaCa_2(Mg,Fe^{+2})_4$ $(Al,Fe^{+3})Al_2Si_6O_{22}(OH,F)_2$

System	Form	Cleavage	Twinning	Zoning	Alteration	Occurrence	Remarks
Orth.	columnar, square x-sect., inclusions in shape of a cross (chiastolite)	two at $89°$			sericite, kyanite, sillimanite	M, S	cf. sillimanite
Mon. (Orth.)	fibrous aggregates, curved, matted					I, S	
Mon.	grains, pellets, plates, pseudomorphs	perfect basal			limonite, goethite	S	cf. chamosite, biotite, chlorite
Mon.	long prismatic, fibrous	two at $56°$, parting	simple, polysynthetic	Z		M, I	
Orth.	bladed, prismatic, fibrous, asbestiform	two at $54\frac{1}{2}°$			talc	M	cf. gedrite, tremolite, cummingtonite, holmquistite
Mon.	long prismatic, fibrous	two at $56°$, parting	simple, polysynthetic	Z		M, I	cf. tremolite, orthoamphibole, hornblende
Orth.	prismatic, asbestiform	two at $54\frac{1}{2}°$				M	cf. anthophyllite, zoisite, cummingtonite, grunerite
Mon.	distinct crystals, columnar, six-sided x-sect., irregular	two imperfect	uncommon	Z	metamict	I, M	anastomosing cracks, isotropic in metamict state cf. epidote, melanite
Mon.	short prismatic	two at $56°$	simple, polysynthetic	Z		I	cf. kaersutite, barkevikite, katophorite, cossyrite
Mon.	shards, scales, massive, microcrystalline	perfect basal				I, S	cf. nontronite
Orth.	fibrous, bladed, prismatic,	two at $54\frac{1}{2}°$				M, I	cf. anthophyllite, glaucophane
Mon.	prismatic	two at $56°$, parting	simple, polysynthetic	Z		I, M	cf. arfvedsonite, hornblende

YELLOW MINERALS	Pleochroism	δ	Relief	2V Dispersion	Extinction	Orientation	Mineral	Composition
Biaxial −								
	S	3	H	65° r<v			HOWIEITE	$Na(Fe,Mn)_{10}$ $(Fe,Al)_2Si_{12}O_{31}(OH)_{13}$
	W $\gamma=\beta>\alpha$	3 − 4	+L	small	$\simeq$ parallel	cleavage length slow	MINNESOTAITE	$(Fe^{+2},Mg)_3\{(Si,Al)_4O_{10}\}OH_2$
	M $\gamma>\beta>\alpha$ $\alpha>\beta=\gamma$	3 − 4	+L − +M	0° − 15° r<v	0° − 5°	cleavage length slow	PHLOGOPITE	$K_2(Mg,Fe^{+2})_6\{Si_6Al_2O_{20}\}(OH,F)_4$
	W $\gamma=\beta>\alpha$	3 − 4	+L − +M	25° − 70°	indistinct	length slow	NONTRONITE (SMECTITE)	$(\tfrac{1}{2}Ca,Na)_{0.7}(Al,Mg,Fe)_4$ $\{(Si,Al)_8O_{20}\}(OH)_4 \cdot nH_2O$
	W $\gamma>\beta=\alpha$	3 − 4	+M − H	84° − 90° r>v	10° − 15°	length slow	GRUNERITE	$(Fe^{+2},Mg)_7\{Si_8O_{22}\}(OH)_2$
	W $\beta>\gamma>\alpha$	3 − 4 anom.	H	74° − 90° r>v	0° − 15° parallel in elong. sect.	length fast/slow	EPIDOTE	$Ca_2Fe^{+3}Al_2O \cdot OH\{Si_2O_7\}\{SiO_4\}$
	S $\gamma=\beta>\alpha$	3 − 5	+L − H	$\simeq 0^\circ$	$\simeq$ parallel	length slow	STILPNOMELANE	$(K,Na,Ca)_{0-1.4}$ $(Fe^{+3},Fe^{+2},Mg,Al,Mn)_{5.9-8.2}$ $\{Si_8O_{20}\}(OH)_4(O,OH,H_2O)_{3.6-8.5}$
	S $\gamma>\beta>\alpha$	3 − 5	+M − H	66° − 82° r>v	0° − 19°	length slow	KAERSUTITE	$Ca_2(Na,K)(Mg,Fe^{+2},Fe^{+3})_4Ti$ $\{Si_6Al_2O_{22}\}(O,OH,F)_2$
	W	4	+M − VH	48° − 90° r$\lessgtr$v	parallel	cleavage length slow	OLIVINE	$(Mg,Fe)_2\{SiO_4\}$
	W $\gamma>\beta>\alpha$	4	H	73° − 85° r<v	40° − 41°	length fast	LÅVENITE	$(Na,Ca,Mn,Fe^{+2})_3(Zr,Nb,Ti)$ $\{Si_2O_7\}(OH,F)$
	W $\gamma>\beta>\alpha$	4	VH	44° − 70° r>v	parallel	length fast/slow	TEPHROITE	$Mn_2\{SiO_4\}$

System	Form	Cleavage	Twinning	Zoning	Alteration	Occurrence	Remarks
Tr.	bladed	three, one good				M	
Mon.	minute plates, needles, radiating, aggregates	perfect basal				M	cf. talc
Mon.	tabular, flakes, plates, pseudo-hex.	perfect basal	inconspicuous	Z colour zoning		I, M	inclusions common, birds-eye maple structure cf. biotite, muscovite, lepidolite, rutile, tourmaline
Mon.	shards, scales, fibrous, massive	perfect basal				I, M	cf. montmorillonite, kaolinite
Mon.	prismatic, subradiating, fibrous, asbestiform	two at 55°, cross fractures	simple, polysynthetic		limonite	M	cf. tremolite, actinolite, cummingtonite, anthophyllite
Mon.	distinct crystals, columnar, aggregates, six-sided x-sect.	one perfect	uncommon	Z		M, I, S	cf. zoisite, clinozoisite, diopside, augite, sillimanite
Mon.	plates, pseudo-hex., micaceous masses	two at 90°, perfect basal	polysynthetic	Z		I, M, O	cf. biotite, chlorite, chloritoid, clintonite
Mon.	prismatic	two at 56°, parting	simple, polysynthetic	Z		I	cf. barkevikite, katophorite, basaltic hornblende, cossyrite, titanaugite
Orth.	anhedral – euhedral, rounded	uncommon, moderate, irregular fractures	uncommon, vicinal	Z	chlorite, antigorite, serpentine, iddingsite, bowlingite	I, M	deformation lamellae cf. diopside, augite, pigeonite, humite group, epidote
Mon.	prismatic	one good	polysynthetic			I	
Orth.	anhedral – euhedral, rounded	two moderate, imperfect	uncommon			O, M	cf. knebelite, olivine

YELLOW MINERALS	Pleochroism	δ	Relief	2V Dispersion	Extinction	Orientation	Mineral	Composition
Biaxial -								
	W	4	VH	$44^{\circ} - 70^{\circ}$ r>v	parallel	length fast/slow	KNEBELITE	$(Mn,Fe)_2\{SiO_4\}$
	W	4	VH	$48^{\circ} - 52^{\circ}$ r>v	parallel	cleavage length slow	FAYALITE	Fe_2SiO_4
	S $\gamma \geqslant \beta > \alpha$ $\beta > \gamma > \alpha$	4 - 5	+M	$0^{\circ} - 25^{\circ}$ Mg r$\lesssim$v Fe r>v	$0^{\circ} - 9^{\circ}$ wavy	cleavage length slow	BIOTITE	$K_2(Mg,Fe^{+2})_{6-4}(Fe^{+3},Al,Ti)_{0-2}$ $\{Si_{6-5}Al_{2-3}O_{20}\}(OH,F)_4$
	W $\alpha > \beta \geqslant \gamma$	4 - 5	H - VH	$60^{\circ} - 70^{\circ}$ r>v	$0^{\circ} - 10^{\circ}$	extinction nearest cleavage fast	ACMITE	$NaFe^{+3}\{Si_2O_6\}$
	W $\omega > \epsilon, \epsilon > \omega$	5 anom.	E	(uniaxial)			ANATASE	TiO_2
	$\alpha > \gamma > \beta$	5	E	$0^{\circ} - 27^{\circ}$ extreme	parallel	length fast/slow	GOETHITE	$\alpha - FeO \cdot OH$
	W - M $\alpha > \beta > \gamma$	5	E	30° r>v strong	12°		BADDELEYITE	ZrO_2
	S $\gamma > \beta > \alpha$	5	E	83° slight	parallel		LEPIDOCROCITE	$\gamma - FeO \cdot OH$

System	Form	Cleavage	Twinning	Zoning	Alteration	Occurrence	Remarks
Orth.	anhedral – euhedral, rounded	two moderate, imperfect	uncommon			O, M	cf. tephroite, olivine
Orth.	anhedral – euhedral, rounded	one moderate, irregular fractures	uncommon, vicinal		grunerite, serpentine	I, M	cf. olivine, knebelite, pyroxene
Mon. (Pseudo-Hex.)	tabular, flakes, bent plates, pseudo-hex.	perfect basal	simple	Z	chlorite, vermiculite, prehnite	M, I, S	pleochroic haloes, inclusions common, birds-eye maple structure cf. stilpnomelane, phlogopite
Mon.	short prismatic, needles, pointed terminations	two at 87°, parting	simple, polysynthetic, twin seams	Z		I	cf. aegirine, aegirine-augite
Tet.	small, prismatic, acicular	two perfect		Z	leucoxene	S, I, M	cf. rutile, brookite
Orth.	pseudomorphs, fibrous, oolitic concretions	one perfect				O, S, I, M	weathering product cf. bowlingite, limonite
Mon.	tabular	one	simple, polysynthetic			I, S	cf. zircon, rutile, brookite, anatase
Orth.	tabular	two perfect, moderate				O, S	

GREEN MINERALS	Pleochroism	δ	Relief	2V Dispersion	Extinction	Orientation	Mineral	Composition
Isometric								
			$-L - +M$				CHLOROPHAEITE	Chloritic mineral
			$-M$				FLUORITE	CaF_2
			$+M$				GREENALITE	$Fe_6^{+2}Si_4O_{10}(OH)_8$
			$H - E$				SPINEL (HERCYNITE, GAHNITE, PLEONASTE, CEYLONITE)	$(Mg,Fe^{+2},Zn)Al_2O_4$
	(weak)		$-L - +M$				PALAGONITE	Altered glass
	(anom.)		H				GROSSULARITE	$Ca_3Al_2Si_3O_{12}$
	(anom.)		$H - VH$				GARNET (GROSSULARITE, UVAROVITE)	$Ca_3(Al,Cr)_2Si_3O_{12}$
	anom.		VH				UVAROVITE	$Ca_3Cr_2Si_3O_{12}$
	W $\gamma>\alpha$	(weak)	$+E$				PEROVSKITE (DYSANALYTE)	$(Ca,Na,Fe^{+2},Ce)(Ti,Nb)O_3$
Uniaxial +								
	1 anom.		H		parallel	length fast	VESUVIANITE (IDOCRASE)	$Ca_{10}(Mg,Fe)_2Al_4\{Si_2O_7\}_2\{SiO_4\}_5(OH,F)_4$
	W - S $\alpha=\beta>\gamma$	1 - 2 (anom.)	$+L - +M$	$0^o - 60^o$ $r<v$	$0^o - 9^o$	cleavage length fast	CHLORITE (Unoxidised)	$(Mg,Al,Fe)_{12}\{(Si,Al)_8O_{20}\}(OH)_1$

System	Form	Cleavage	Twinning	Zoning	Alteration	Occurrence	Remarks
	massive, cryptocrystalline					I	pseudomorphs after olivine cf. bowlingite, iddingsite
Cub.	anhedral, hexagonal x-sect.	two or three perfect	interpenetrant	Z		I, S, O, M	colour spots cf. cryolite, halite
M'loid.	massive					O	restricted to Iron Formation, Mesabi Range, Minnesota
Cub.	small grains, cubes, octahedra, rhombic x-sect.	parting		Z		M, I	colour variable according to composition cf. garnet
M'loid.	amorphous, oolitic				chlorite	I	cf. volcanic glass, opal, collophane
Cub.	four, six, eight-sided, polygonal x-sect., aggregates	parting, irregular fractures	sector	Z		M	cf. uvarovite
Cub.	four, six, eight-sided, x-sect., grains, aggregates	parting, irregular fractures		Z		M, I	inclusions
Cub.	four, six, eight-sided, polygonal x-sect., aggregates	parting, irregular fractures	complex, sector	Z		M	cf. grossularite
Mon? Pseudo-Cub.	small cubes, skeletal	one poor to distinct	polysynthetic, complex, interpenetrant	Z	leucoxene	I, M	cf. melanite, picotite, ilmenite
Tet.	variable, prismatic, fibrous, granular, radial	imperfect	sector	Z		M, I, S	cf. zoisite, clinozoisite, apatite, grossularite, melilite, andalusite
Mon. (Pseudo-Hex.)	tabular, scaly, radiating, pseudomorphs	perfect basal	simple, polysynthetic	Z		I, M, S, O	pleochroic haloes cf. bowlingite

GREEN MINERALS	Pleochroism	δ	Relief	2V Dispersion	Extinction	Orientation	Mineral	Composition
Uniaxial +								
	W – M $\beta>\alpha>\gamma$	1 – 2 anom.	H	45° – 68°	2° – 30°	length fast	CHLORITOID (OTTRELITE)	$(Fe^{+2},Mg,Mn)_2(Al,Fe^{+3})Al_3O_2\{SiO_4\}_2(OH)_4$
	N – M $\gamma=\alpha>\beta$ $\beta>\alpha=\gamma$	3	+M – H	0° – 30° $r\lessgtr v$	37° – 44°	extinction nearest cleavage slow	PIGEONITE	$(Mg,Fe^{+2},Ca)(Mg,Fe^{+2})\{Si_2O_6\}$
	W	5	H		straight		XENOTIME	YPO_4
	W – M $\varepsilon>\omega$	5	E		parallel		RUTILE	TiO_2
Uniaxial –								
	W	1	+L			long. sect. length fast, x-sect. length slow	BERYL (incl. EMERALD)	$Be_3Al_2\{Si_6O_{18}\}$
		1 anom.	H		parallel	length fast	VESUVIANITE (IDOCRASE)	$Ca_{10}(Mg,Fe)_2Al_4\{Si_2O_7\}_2\{SiO_4\}_5(OH,F)_4$
	W $\omega>\varepsilon$	1	H	0° – 30°	parallel	tabular length slow, prismatic length fast	CORUNDUM	$\alpha-Al_2O_3$
	W – S $\gamma=\beta>\alpha$	1 – 2 (anom.)	+L – +M	0° – 20° $r>v$	0° – small	cleavage length slow	CHLORITE (Oxidised)	$(Mg,Al,Fe)_{12}\{(Si,Al)_8O_{20}\}(OH)_{16}$
	W – S $\gamma=\beta>\alpha$	1 – 2 (anom.)	+L – +M	0° – 20° $r>v$	0° – small	cleavage length slow	CHLORITE (Unoxidised)	$(Mg,Al,Fe)_{12}\{(Si,Al)_8O_{20}\}(OH)_{16}$
	S $\omega>\varepsilon$	2 – 3	+M		parallel	length fast	ELBAITE (TOURMALINE)	$Na(Li,Al)_3Al_6B_3Si_6O_{27}(OH,F)_4$

System	Form	Cleavage	Twinning	Zoning	Alteration	Occurrence	Remarks
Mon., Tr.	tabular	one perfect, one imperfect, parting	polysynthetic	Z hour-glass		M, I	inclusions common cf. chlorite, clintonite, biotite, stilpnomelane
Mon.	prismatic, anhedral, overgrowths	two at 87°, parting	simple, polysynthetic	Z		I	exsolution lamellae cf. augite, olivine
Tet.	small, prismatic					I, M, S	inclusions, pleochroic haloes around grains cf. zircon, sphene, monazite
Tet.	prismatic, acicular, reticulate network, inclusions	good	geniculate			M, I, S	often opaque cf. brookite, anatase, sphene, baddeleyite, cassiterite
Hex.	prismatic, inclusions zoned	imperfect		Z	kaolin	I, M	liquid inclusions cf. apatite, quartz
Tet.	variable, prismatic, fibrous, granular, radial	imperfect		Z		M, I, S	cf. zoisite, clinozoisite, apatite, grossularite, melilite, andalusite
Trig.	tabular, prismatic, six-sided x-sect.	parting	simple, lamellar seams	Z colour banding		M, I	inclusions cf. sapphirine
Mon. (Pseudo-Hex.)	tabular, scaly, radiating, pseudomorphs	perfect basal	simple, polysynthetic	Z		I, S, M, O	pleochroic haloes cf. bowlingite
Mon. (Pseudo-Hex.)	tabular, scaly, radiating, pseudomorphs	perfect basal	simple, polysynthetic	Z		I, S, M, O	pleochroic haloes cf. bowlingite
Trig.	prismatic, radiating	fractures at 90°	rare	Z		M	cf. tourmaline group, biotite, hornblende

GREEN MINERALS	Pleochroism	δ	Relief	2V Dispersion	Extinction	Orientation	Mineral	Composition
Uniaxial −								
	S $\omega>\varepsilon$	3	+M		parallel	length fast	SCHORL (TOURMALINE)	$Na(Fe,Mn)_3Al_6B_3Si_6O_{27}(OH,F)_4$
	W	3	+M		parallel		ZUSSMANITE	$K(Fe,Mg,Mn)_{13}(Si,Al)_{18}O_{42}(OH)$
	S $\gamma=\beta>\alpha$	3 − 5	+L − H	$\simeq 0^\circ$	$\simeq$ parallel	length slow	STILPNOMELANE	$(K,Na,Ca)_{0-1.4}$ $(Fe^{+3},Fe^{+2},Mg,Al,Mn)_{5.9-8.2}$ $\{Si_8O_{20}\}(OH)_4(O,OH,H_2O)_{3.6-8.4}$
	S $\gamma\geqq\beta>\alpha$ $\beta>\gamma>\alpha$	4 − 5	+M	$0^\circ - 25^\circ$ Mg r$\lessgtr$v Fe r$>$v	$0^\circ - 9^\circ$ wavy	cleavage length slow	BIOTITE	$K_2(Mg,Fe^{+2})_{6-4}(Fe^{+3},Al,Ti)_{0-2}$ $\{Si_{6-5}Al_{2-3}O_{20}\}(OH,F)_4$
Biaxial +								
	W $\alpha=\beta>\gamma$	1 anom.	+L	$0^\circ - 20^\circ$ r$<$v	$\simeq$ parallel	cleavage length fast	PENNINITE	$(Mg,Al,Fe)_{12}\{(Si,Al)_8O_{20}\}(OH)$
	W $\alpha=\beta>\gamma$	1	+L	$\simeq 20^\circ$ r$<$v	$\simeq$ parallel	cleavage length fast	SHERIDANITE	$(Mg,Al,Fe)_{12}\{(Si,Al)_8O_{20}\}(OH)$
	W $\alpha=\beta>\gamma$	1 anom.	+L − +M	$0^\circ - 30^\circ$ r$<$v	$\simeq$ parallel	cleavage length fast	RIPIDOLITE- KLEMENTITE (PROCHLORITE)	$(Mg,Al,Fe)_{12}\{(Si,Al)_8O_{20}\}(OH)$
	M − S $\alpha=\beta>\gamma$	1	+L − +M	$0^\circ - 60^\circ$ r$<$v	0° − small	cleavage length fast	Mn and Cr- CHLORITES	$(Mg,Mn,Al,Cr,Fe)_{12}$ $\{(Si,Al)_8O_{20}\}(OH)_{16}$
	W $\gamma>\alpha$	1	+E	90° r$>$v	$\simeq 45^\circ$		PEROVSKITE (DYSANALYTE)	$(Ca,Na,Fe^{+2},Ce)(Ti,Nb)O_3$
	M	1 − 2	−L − +L	$65^\circ - 90^\circ$ r$<$v	parallel		CORDIERITE	$Al_3(Mg,Fe^{+2})_2\{Si_5AlO_{18}\}$
	W $\alpha=\beta>\gamma$	1 − 2	+L	$0^\circ - 40^\circ$ r$<$v	$0^\circ - 9^\circ$	cleavage length fast	CLINOCHLORE	$(Mg,Al,Fe)_{12}\{(Si,Al)_8O_{20}\}(OH)$

System	Form	Cleavage	Twinning	Zoning	Alteration	Occurrence	Remarks
Trig.	hexagonal, rounded, suns, triangular x-sect.	fractures		Z		I, M	cf. tourmaline group, biotite, hornblende
Hex.	tabular	one perfect				M	
Mon.	plates, pseudo-hex., micaceous masses	two at 90°, perfect basal	polysynthetic	Z		I, M, O	cf. biotite, chlorite, chloritoid, clintonite
Mon. (Pseudo-Hex.)	tabular, flakes, bent plates, pseudo-hex.	perfect basal	simple	Z	chlorite, vermiculite, prehnite	M, I, S	pleochroic haloes, inclusions common, birds-eye maple structure cf. stilpnomelane, phlogopite
Mon.	tabular, vermicular, radiating, pseudomorphs	perfect basal	simple, pennine law	Z		I, M	cf. clinochlore, prochlorite
Mon.	tabular, spherulites	perfect basal	simple			I	
Mon.	tabular, scaly, vermicular, fan-shaped aggregates	perfect basal				M, I, O	cf. clinochlore, penninite
Mon.	tabular	perfect basal				O	
Mon? Pseudo-Cub.	small cubes, skeletal	poor to distinct	polysynthetic, complex, interpenetrant	Z	leucoxene	I, M	
Orth. (Pseudo-Hex.)	pseudo-hex., anhedral	moderate to absent	simple, polysynthetic, cyclic, sector, interpenetrant		pinite, talc	M, I	pleochroic haloes, inclusions (including opaque dust) common cf. quartz, plagioclase
Mon.	tabular, fibrous, pseudo-hex.	perfect basal	polysynthetic			M	pleochroic haloes cf. penninite, prochlorite, leuchtenbergite, katschubeite

GREEN MINERALS	Pleochroism	δ	Relief	2V Dispersion	Extinction	Orientation	Mineral	Composition

Biaxial +

	Pleochroism	δ	Relief	2V Dispersion	Extinction	Orientation	Mineral	Composition
	W - S $\alpha=\beta>\gamma$	1 - 2 (anom.)	+L - +M	$0^o - 60^o$ r<v	$0^o - 9^o$	cleavage length fast	CHLORITE (Unoxidised)	$(Mg,Al,Fe)_{12}\{(Si,Al)_8O_{20}\}(OH)_1$
	S $\alpha>\beta>\gamma$ $\alpha>\gamma\geqslant\beta$	1 - 2	+M - H	$40^o - 50^o$	$\beta : z$ $15^o - 30^o$	length fast	MAGNESIORIEBECKITE	$Na_2Mg_3Fe_2^{+3}\{Si_8O_{22}\}(OH)_2$
	S $\alpha>\beta>\gamma$ $\alpha>\gamma\geqslant\beta$	1 - 2	+M - H	$40^o - 90^o$ r$\lesssim$v	$3^o - 21^o$	length fast	RIEBECKITE	$Na_2Fe_3^{+2}Fe_2^{+3}\{Si_8O_{22}\}(OH)_2$
	W - M $\beta>\alpha>\gamma$	1 - 2 anom.	H	$45^o - 68^o$ r>v	$2^o - 30^o$	length fast	CHLORITOID (OTTRELITE)	$(Fe^{+2},Mg,Mn)_2(Al,Fe^{+3})Al_3O_2\{SiO_4\}_2(OH)_4$
	W $\alpha=\beta>\gamma$	2	+L - +M	31° r<v	$8^o - 10^o$	cleavage length fast	CORUNDOPHILITE	$(Mg,Al,Fe)_{12}\{(Si,Al)_8O_{20}\}(OH)_1$
	S $\gamma\geqslant\beta>\alpha$	2	+M	$52^o - 83^o$ r>v	$17^o - 27^o$	length slow	EDENITE	$NaCa_2(Mg,Fe^{+2})_5AlSi_7O_{22}(OH)_2$
	W	2	+M	$67^o - 70^o$ r>v	$33^o - 40^o$	extinction nearest cleavage slow	JADEITE	$NaAl\{Si_2O_6\}$
	W $\alpha>\beta=\gamma$	2	+M	$73^o - 86^o$ r<v	parallel	length slow	VIRIDINE (ANDALUSITE)	$(Al,Fe^{+3})_2SiO_5$
	W $\beta>\alpha=\gamma$	2	+M - H	$26^o - 85^o$ r<v	$4^o - 32^o$	length slow	PUMPELLYITE	$Ca_4(Mg,Fe^{+2})(Al,Fe^{+3})_5O(OH)_3\{Si_2O_7\}_2\{SiO_4\}_2\cdot2H_2O$
	W	2 - 3	+M	$58^o - 68^o$ r<v	$22^o - 26^o$	extinction nearest cleavage slow	HIDDENITE (SPODUMENE)	$LiAl\{Si_2O_6\}$
	W $\gamma=\beta>\alpha$ $\gamma>\beta=\alpha$	2 - 3	+M	$68^o - 90^o$ r$\lesssim$v	parallel, symmetrical	length slow	ANTHOPHYLLITE	$(Mg,Fe^{+2})_7\{Si_8O_{22}\}(OH,F)_2$
	M - S	2 - 3 (anom.)	+M - H	$25^o - 50^o$ r>v	$35^o - 48^o$	extinction nearest cleavage fast	TITANAUGITE	$(Ca,Na,Mg,Fe^{+2},Mn,Fe^{+3},Al,Ti)_2\{(Si,Al)_2O_6\}$

System	Form	Cleavage	Twinning	Zoning	Alteration	Occurrence	Remarks
Mon. (Pseudo-Hex.)	tabular, scaly, radiating, pseudomorphs	perfect basal	simple, polysynthetic	Z		I, M, S, O	pleochroic haloes cf. bowlingite
Mon.	prismatic, fibrous, columnar aggregates	two at 56°	simple, polysynthetic	Z		I, M	cf. glaucophane, crossite, arfvedsonite, hastingsite
Mon.	prismatic, fibrous, columnar aggregates	two at 56°	simple, polysynthetic	Z		I, M	cf. glaucophane, crossite, arfvedsonite, hastingsite
Mon., Tr.	tabular	one perfect, one imperfect, parting	polysynthetic	Z hour-glass		M	inclusions common cf. chlorite, clintonite, biotite, stilpnomelane
Mon.	tabular	perfect basal				M	
Mon.	prismatic	two at 56°		Z		M, I	cf. hornblende
Mon.	prismatic	two at 87°	simple, polysynthetic	Z	tremolite-actinolite	M	cf. nephrite, diopside, omphacite, fassaite
Orth.	columnar, square x-sect.	two at 89°				M	cf. andalusite
Mon.	fibrous, needles, bladed, radial aggregates	two moderate	common	Z		M	cf. clinozoisite, zoisite, epidote, lawsonite
Mon.	tabular	two at 87°, parting	simple		muscovite, cymatolite, kaolinite	I	cf. aegirine-augite, diopside, eucryptite
Orth.	bladed, prismatic, fibrous, asbestiform	two at $54\frac{1}{2}^{\circ}$			talc	M	cf. gedrite, tremolite, cummingtonite
Mon.	prismatic	two at 93°	polysynthetic, twin seams	Z hour-glass		I	cf. hypersthene

GREEN MINERALS	Pleochroism	δ	Relief	2V Dispersion	Extinction	Orientation	Mineral	Composition
Biaxial +								
	N – W	2 – 3	+M – H	$25° - 83°$ r>v	$35° - 48°$	extinction nearest cleavage fast	AUGITE-FERROAUGITE (incl. SALITE)	$(Ca,Na,Mg,Fe^{+2},Mn,Fe^{+3},Al,Ti)_2\{(Si,Al)_2O_6\}$
	W	2 – 3	+M – H	$51° - 62°$	$35° - 48°$	extinction nearest cleavage fast	FASSAITE	$(Ca,Na,Mg,Fe^{+2},Mn,Fe^{+3},Al,Ti)_2\{(Si,Al)_2O_6\}$
	W	2 – 3	+M – H	$60° - 67°$	$39° - 41°$	extinction nearest cleavage fast	OMPHACITE	$(Ca,Na,Mg,Fe^{+2},Mn,Fe^{+3},Al,Ti)_2\{(Si,Al)_2O_6\}$
	W $\gamma=\beta>\alpha$ $\gamma>\beta=\alpha$	2 – 3	+M – H	$68° - 90°$ r$\lessgtr$v	parallel, symmetrical	length slow	FERROGEDRITE	$Fe_5Al_4Si_6O_{22}(OH)_2$
	W $\gamma>\alpha$	3	+M	$21° - 30°$ r>v	parallel	length slow	SILLIMANITE	Al_2SiO_5
	W	3	+M	$50° - 62°$ r>v	$38° - 48°$	extinction nearest cleavage slow	DIOPSIDE (SALITE)	$Ca(Mg,Fe)\{Si_2O_6\}$
	M – S $\gamma\gtrsim\beta>\alpha$	3	+M	$67° - 90°$ r>v	$26°$	length slow	PARGASITE	$NaCa_2Mg_4Al_3Si_6O_{22}(OH)_2$
	W $\alpha>\gamma$	3	+M	$76° - 87°$ r>v	parallel, symmetrical	length slow, long diagonal of rhombic sect. slow	LAWSONITE	$CaAl_2(OH)_2\{Si_2O_7\}H_2O$
	N – M $\gamma=\alpha>\beta$ $\beta>\alpha=\gamma$	3	+M – H	$0° - 30°$ r$\lessgtr$v	$37° - 44°$	extinction nearest cleavage slow	PIGEONITE	$(Mg,Fe^{+2},Ca)(Mg,Fe^{+2})\{Si_2O_6\}$
	W $\gamma>\beta>\alpha$	3	+M – H	$50° - 62°$ r>v	$38° - 48°$	cleavage fast	HEDENBERGITE (incl. FERROSALITE)	$Ca(Fe,Mg)\{Si_2O_6\}$
	N – W $\gamma>\beta=\alpha$	3 – 4	+M	$65° - 90°$ Mg r<v Fe r>v	$15° - 21°$	length slow	CUMMINGTONITE	$(Mg,Fe^{+2})_7\{Si_8O_{22}\}(OH)_2$

System	Form	Cleavage	Twinning	Zoning	Alteration	Occurrence	Remarks
Mon.	prismatic	two at 87°, parting	simple, polysynthetic, twin seams	Z	amphibole	I, M	herringbone structure, exsolution lamellae cf. pigeonite, diopside, epidote
Mon.	prismatic	two at 87°	simple, polysynthetic	Z	amphibole	I, M	cf. augite, diopside, omphacite, jadeite
Mon.	prismatic	two at 87°	simple, polysynthetic			M	inclusions, cf. fassaite, diopside, jadeite
Orth.	bladed, prismatic, fibrous, asbestiform	two at $54\frac{1}{2}^{\circ}$			talc	M	cf. anthophyllite, zoisite, cummingtonite, grunerite
Orth.	needles, slender prisms fibrous, rhombic x-sect., faserkiesel	one, // long diagonal of x-sect.				M	cf. andalusite, mullite
Mon.	short prismatic	two at 87°	simple, polysynthetic	Z	tremolite-actinolite	M, I	cf. hedenbergite, tremolite, omphacite, wollastonite, epidote
Mon.	prismatic	two at 56°, parting	simple, polysynthetic			M, I	cf. hornblende, hastingsite, cummingtonite
Orth.	rectangular, rhombic	two perfect, one imperfect	simple, polysynthetic			M	cf. zoisite, prehnite, pumpellyite, andalusite, scapolite
Mon.	prismatic, anhedral, overgrowths	two at 87°, parting	simple, polysynthetic	Z		I	exsolution lamellae cf. augite, olivine
Mon.	prismatic	two at 87°	simple, polysynthetic	Z		M, I	cf. diopside, aegirine-augite, augite
Mon.	prismatic, subradiating, fibrous, asbestiform	two at 55°	simple, polysynthetic			M, I	cf. tremolite, actinolite, grunerite, anthophyllite

GREEN MINERALS	Pleochroism	δ	Relief	2V Dispersion	Extinction	Orientation	Mineral	Composition
Biaxial +								
	S $\alpha>\gamma$	3 – 4	H	$70^o - 90^o$ $r>v$	$\gamma : z$ $70^o - 90^o$	extinction nearest cleavage fast	AEGIRINE-AUGITE	$(Na,Ca)(Fe^{+3},Fe^{+2},Mg)\{Si_2O_6\}$
	W – M $\alpha>\beta>\gamma$	5 anom.	VH – E	$17^o - 40^o$ $r>v$	40° symmetrical		SPHENE	$CaTi\{SiO_4\}(O,OH,F)$
Biaxial –								
		1	+L			cleavage slow	LIZARDITE (SERPENTINE)	$Mg_3\{Si_2O_5\}(OH)_4$
	W $\gamma=\beta>\alpha$	1 anom.	+L	$0^o - 40^o$ $r>v$	$\simeq$ parallel	cleavage length slow	PENNINITE	$(Mg,Al,Fe)_{12}\{(Si,Al)_8O_{20}\}(OH)$
	W $\gamma>\alpha$	1 anom.	+L	$37^o - 61^o$ $r>v$	parallel	length slow	ANTIGORITE (SERPENTINE)	$Mg_3\{Si_2O_5\}(OH)_4$
	W $\gamma>\beta>\alpha$	1	+L – +M	$0^o - 20^o$	$0^o - 7^o$	cleavage length slow	DELESSITE	$(Mg,Al,Fe)_{12}\{(Si,Al)_8O_{20}\}(OH)$
	W – M $\gamma=\beta>\alpha$	1 anom.	+L – +M	$0^o - 20^o$	$\simeq$ parallel $r>v$	cleavage length slow	DIABANTITE- BRUNSVIGITE	$(Mg,Al,Fe)_{12}\{(Si,Al)_8O_{20}\}(OH)$
	M – S $\gamma=\beta>\alpha$	1	+L – +M	$0^o - 20^o$	0° – small $r>v$	cleavage length slow	Mn and Cr- CHLORITES	$(Mg,Mn,Al,Cr,Fe)_{12}$ $\{(Si,Al)_8O_{20}\}(OH)_{16}$
	W $\gamma=\beta>\alpha$	1 anom.	+M	0° – small	$\simeq$ parallel $r>v$	cleavage length slow	RIPIDOLITE PROCHLORITE	$(Mg,Al,Fe)_{12}\{(Si,Al)_8O_{20}\}(OH)$
	W $\gamma=\beta>\alpha$	1 anom.	+M	$0^o - 20^o$	small	cleavage length slow	DAPHNITE	$(Mg,Al,Fe)_{12}\{(Si,Al)_8O_{20}\}(OH)$
	N – W	1 anom.	+M	small	small	cleavage length slow	CHAMOSITE	$(Mg,Al,Fe)_{12}\{(Si,Al)_8O_{20}\}(OH)$

System	Form	Cleavage	Twinning	Zoning	Alteration	Occurrence	Remarks
Mon.	short prismatic, needles, felted aggregates	two at 87°, parting	simple, polysynthetic, twin seams	Z hour-glass		I	cf. aegirine, acmite
Mon.	rhombic, irregular grains	parting	simple, polysynthetic		leucoxene	I, M, S	cf. monazite, rutile
Mon.	tabular, fine grained aggregates	perfect basal	occasional			I, M	mesh, hour-glass structure cf. chlorite group
Mon.	tabular, vermicular, radiating, pseudomorphs	perfect basal	simple	Z		I, M	pleochroic haloes cf. clinochlore, prochlorite
Mon.	anhedral, fibrolamellar, flakes, laths, plates	perfect basal	occasional			I, M	mesh, hour-glass structure cf. chrysotile, serpophite, chlorite, amphibole
Mon.	spherulitic	perfect basal	simple			I	cf. thuringite, greenalite
Mon.	tabular, fibrous	perfect basal				M, I, O	cf. chlorite group
Mon.	tabular, radiating	perfect basal				O, I, M	
Mon.	tabular, scaly, vermicular, fan-shaped aggregates	perfect basal				M, I, O	cf. clinochlore, penninite
Mon.	concentric aggregates, fibrous plates	perfect basal				O	
Mon.	pseudospherulitic, concentric, tabular, massive	one good, concentric parting				S, O	cf. glauconite, collophane, greenalite, thuringite

GREEN MINERALS	Pleochroism	δ	Relief	2V Dispersion	Extinction	Orientation	Mineral	Composition
Biaxial −								
	W	1	H	0° − 30° (uniaxial) moderate	parallel	tabular length slow, prismatic length fast	CORUNDUM	$\alpha\text{-}Al_2O_3$
	M	1 − 2	−L − +L	65° − 90° r<v	parallel		CORDIERITE	$Al_3(Mg,Fe^{+2})_2\{Si_5AlO_{18}\}$
	W − S $\gamma=\beta>\alpha$	1 − 2 (anom.)	+L − +M	0° − 20° r>v	0° − small	cleavage length slow	CHLORITE (Oxidised)	$(Mg,Al,Fe)_{12}\{(Si,Al)_8O_{20}\}(OH)$
	W − S $\gamma=\beta>\alpha$	1 − 2 (anom.)	+L − +M	0° − 20° r>v	0° − small	cleavage length slow	CHLORITE (Unoxidised)	$(Mg,Al,Fe)_{12}\{(Si,Al)_8O_{20}\}(OH)$
	S $\alpha>\beta>\gamma$	1 − 2	+M − H	0° − 50° r<v	0° − 30° anom.	length fast	ARFVEDSONITE	$Na_3(Mg,Fe^{+2})_4Al\{Si_8O_{22}\}(OH,F)$
	S $\alpha>\beta>\gamma$ $\alpha>\gamma\gtrsim\beta$	1 − 2	+M − H	40° − 50°	β : z 15° − 30°	length fast	MAGNESIORIEBECKITE	$Na_2Mg_3Fe_2^{+3}\{Si_8O_{22}\}(OH)_2$
	S $\alpha>\beta>\gamma$ $\alpha>\gamma\gtrsim\beta$	1 − 2	+M − H	40° − 90° r$\lesssim$v	3° − 21°	length fast	RIEBECKITE	$Na_2Fe_3^{+2}Fe_2^{+3}\{Si_8O_{22}\}(OH)_2$
	N − S $\gamma>\beta>\alpha$	1 − 2	+M − H	50° − 90° r$\lesssim$v	parallel	length slow	HYPERSTHENE (incl. BRONZITE-EULITE)	$(Mg,Fe^{+2})\{SiO_3\}$
	W − M $\beta>\alpha>\gamma$	1 − 2 anom.	H	45° − 68° r>v	2° − 30°	length fast	CHLORITOID (OTTRELITE)	$(Fe^{+2},Mg,Mn)_2(Al,Fe^{+3})Al_3O_2\{SiO_4\}_2(OH)_4$
	M − S $\gamma>\beta>\alpha$ $\gamma<\beta>\alpha$	1 − 3	+M	0° − 50° r<v	36° − 70°	length fast	KATOPHORITE-MAGNESIOKATOPHORITE	$Na_2Ca(Mg,Fe^{+2})_4Fe^{+3}\{Si_7AlO_{22}\}(OH,F)_2$
		2	+L	30° − 32°	parallel	length slow	CHRYSOTILE (SERPENTINE)	$Mg_3\{Si_2O_5\}(OH)_4$

System	Form	Cleavage	Twinning	Zoning	Alteration	Occurrence	Remarks
Trig.	tabular, prismatic, six-sided x-sect.	parting	simple, lamellar seams	Z colour banding		M, I	inclusions
Orth. (Pseudo-Hex.)	pseudo-hex., anhedral	moderate to absent	simple, polysynthetic, cyclic, sector, interpenetrant		pinite, talc	M, I	pleochroic haloes, inclusions (including opaque dust) common cf. quartz, plagioclase
Mon. (Pseudo-Hex.)	tabular, scaly, radiating, pseudomorphs	perfect basal	simple, polysynthetic	Z		I, S, M, O	pleochroic haloes cf. bowlingite
Mon. (Pseudo-Hex.)	tabular, scaly, radiating, pseudomorphs	perfect basal	simple, polysynthetic	Z		I, M, S, O	pleochroic haloes cf. bowlingite
Mon.	prismatic	two at 56°, parting	simple, polysynthetic	Z		I	cf. riebeckite, katophorite, glaucophane, tourmaline
Mon.	prismatic, fibrous, columnar aggregates	two at 56°	simple, polysynthetic	Z		I, M	cf. glaucophane, crossite, arfvedsonite, hastingsite
Mon.	prismatic, fibrous, columnar aggregates	two at 56°	simple, polysynthetic	Z		I, M	cf. glaucophane, crossite, arfvedsonite, hastingsite
Orth.	prismatic, anhedral - euhedral	two at 88°	polysynthetic, twin seams	Z		I, M	exsolution lamellae, schiller inclusions cf. andalusite
Mon., Tr.	tabular	one perfect, one imperfect, parting	polysynthetic	Z hour-glass		M, I	inclusions common cf. chlorite, clintonite, biotite, stilpnomelane
Mon.	prismatic	two at 56°, parting	simple	Z		I	cf. barkevikite, kaersutite, basaltic hornblende, arfvedsonite, cossyrite
Mon.	fibrous, cross-fibre, veinlets	fibrous				I, M	mesh, hour-glass structure cf. antigorite, serpophite, chlorite, amphibole

GREEN MINERALS	Pleochroism	δ	Relief	2V Dispersion	Extinction	Orientation	Mineral	Composition
Biaxial −								
	M $\gamma=\beta>\alpha$	2	+M	0° − 23° r<v	parallel	cleavage length slow	XANTHOPHYLLITE	$Ca_2(Mg,Fe)_{4.6}Al_{1.4}$ $\{Si_{2.5}Al_{5.5}O_{20}\}(OH)_4$
	W	2	+M	0° − 20°	small	cleavage length slow	THURINGITE	$(Mg,Al,Fe)_{12}\{(Si,Al)_8O_{20}\}(OH)_1$
	S $\alpha>\beta>\gamma$	2	+M	15° − 80° r>v	18° − 53° (40°) 'flamy'	length slow	ECKERMANNITE	$Na_3(Mg,Fe^{+2})_4Al\{Si_8O_{22}\}(OH,F)_2$
	M $\gamma=\beta>\alpha$	2	+M	32° r<v	parallel	cleavage length slow	CLINTONITE	$Ca_2(Mg,Fe)_{4.6}Al_{1.4}$ $\{Si_{2.5}Al_{5.5}O_{20}\}(OH)_4$
	S $\gamma \geqslant \beta>\alpha$	2	+M	52° − 83° r>v	17° − 27°	length slow	EDENITE	$NaCa_2(Mg,Fe^{+2})_5AlSi_7O_{22}(OH)_2$
	W patchy $\alpha>\gamma$	2	+M	73° − 86° r<v	parallel	length fast	ANDALUSITE	Al_2SiO_5
	M $\gamma=\beta>\alpha$	2 − 3	+M	0° − 20° r<v	0° − 3°	cleavage length slow	GLAUCONITE (incl. CELADONITE)	$(K,Na,Ca)_{1.2-2.0}$ $(Fe^{+3},Al,Fe^{+2},Mg)_{4.0}$ $\{Si_{7-7.6}Al_{1-0.4}O_{20}\}(OH)_4 \cdot nH_2O$
	W $\gamma=\beta>\alpha$ $\gamma>\beta=\alpha$	2 − 3	+M	68° − 90° r$\lessgtr$v	parallel, symmetrical	length slow	ANTHOPHYLLITE (Mg-rich) − FERROGEDRITE	$(Mg,Fe^{+2})_7\{Si_8O_{22}\}(OH,F)_2$
	W $\gamma>\beta>\alpha$	2 − 3	+M	73° − 86° r<v	10° − 17°	length slow	ACTINOLITE-FERROACTINOLITE	$Ca_2(Mg,Fe^{+2})_5\{Si_8O_{22}\}(OH,F)_2$
	S $\gamma \geqslant \beta>\alpha$ $\beta>\gamma>\alpha$	2 − 3	+M − H	66° − 85° r>v	13° − 34°	length slow	HORNBLENDE	$(Na,K)_{0-1}Ca_2(Mg,Fe^{+2},Fe^{+3},Al)_5$ $\{Si_{6-7}Al_{2-1}O_{22}\}(OH,F)_2$
	W $\gamma>\beta>\alpha$	3	−L − +M	0° − 30°	small		MONTMORILLONITE-BEIDELLITE (SMECTITE)	$(\tfrac{1}{2}Ca,Na)_{0.7}(Al,Mg,Fe)_4$ $\{(Si,Al)_8O_{20}\}(OH)_4 \cdot nH_2O$
	W $\gamma=\beta>\alpha$	3	+L	0° − 8° r$\leqslant$v	1° − 22°	cleavage length slow	VERMICULITE	$(Mg,Ca)_{0.7}(Mg,Fe^{+3},Al)_{6.0}$ $\{(Al,Si)_8O_{20}\}(OH)_4 \cdot 8H_2O$

System	Form	Cleavage	Twinning	Zoning	Alteration	Occurrence	Remarks
Mon.	tabular, short prismatic	perfect basal	simple			M	cf. clintonite, chlorite, chloritoid
Mon. (Pseudo-Hex.)	tabular, radiating	perfect basal				O	
Mon.	prismatic	two at 56°, parting	simple, polysynthetic			I	cf. glaucophane, arfvedsonite, cummingtonite, tremolite, hornblende
Mon.	tabular, short prismatic	perfect basal	simple			M	cf. xanthophyllite, chlorite, chloritoid
Mon.	prismatic	two at 56°		Z		M, I	cf. hornblende
Orth.	columnar, square x-sect.	two at 89°	rare		sericite, kyanite, sillimanite	M	variety chiastolite, inclusions in shape of cross
Mon.	grains, pellets, plates, pseudomorphs	perfect basal			limonite, goethite	S	cf. chamosite, biotite, chlorite
Orth.	bladed, prismatic, fibrous, asbestiform	two at 54½°			talc	M	cf. gedrite, actinolite, zoisite, cummingtonite, grunerite, holmquistite
Mon.	long, prismatic, fibrous	two at 56°, parting	simple, polysynthetic	Z		M, I	cf. tremolite, orthoamphibole, hornblende
Mon.	prismatic	two at 56°, parting	simple, polysynthetic	Z	mica, chlorite	M, I	cf. edenite, biotite, aegirine-augite, actinolite, pargasite, ferrohastingsite
Mon.	shards, scales, massive, microcrystalline	perfect basal				I, S	cf. nontronite
Mon.	minute particles, plates	perfect basal				S, I, M	cf. biotite, smectites

GREEN MINERALS	Pleochroism	δ	Relief	2V Dispersion	Extinction	Orientation	Mineral	Composition
Biaxial –								
	S $\beta \geq \gamma > \alpha$ $\gamma \geq \beta > \alpha$	3	+M – H	$10° - 90°$ $r \lesssim v$	$9° - 40°$	length slow	HASTINGSITE-FERROHASTINGSITE	$NaCa_2(Mg,Fe^{+2})_4$ $(Al,Fe^{+3})Al_2Si_6O_{22}(OH,F)_2$
	S	3	H	$65°$ $r < v$			HOWIEITE	$Na(Fe,Mn)_{10}$ $(Fe,Al)_2Si_{12}O_{31}(OH)_{13}$
	M $\alpha > \beta = \gamma$	3	VH	$\simeq 0°$			CRONSTEDTITE	$(Fe_4^{+2}Fe_2^{+3})(Si_2Fe_2^{+3})O_{10}(OH)_8$
	W $\gamma = \beta > \alpha$	3 – 4	–L – +M	$25° - 70°$	indistinct	length slow	NONTRONITE (SMECTITE)	$(\frac{1}{2}Ca,Na)_{0.7}(Al,Mg,Fe)_4$ $\{(Si,Al)_8O_{20}\}(OH)_4 \cdot nH_2O$
	W $\gamma = \beta > \alpha$	3 – 4	+L	small	$\simeq$ parallel	cleavage length slow	MINNESOTAITE	$(Fe^{+2},Mg)_3\{(Si,Al)_4O_{10}\}OH_2$
	M $\gamma \geq \beta > \alpha$ $\alpha > \beta = \gamma$	3 – 4	+L – +M	$0° - 15°$ $r < v$	$0° - 5°$	cleavage length slow	PHLOGOPITE	$K_2(Mg,Fe^{+2})_6\{Si_6Al_2O_{20}\}(OH,F)_4$
	W $\gamma > \beta = \alpha$	3 – 4	+M – H	$84° - 90°$ $r > v$	$10° - 15°$	length slow	GRUNERITE	$(Fe^{+2},Mg)_7\{Si_8O_{22}\}(OH)_2$
	S $\alpha > \gamma$	3 – 4	H	$70° - 90°$ $r > v$	$\gamma : z$ $70° - 90°$	extinction nearest cleavage fast	AEGIRINE-AUGITE	$(Na,Ca)(Fe^{+3},Fe^{+2},Mg)\{Si_2O_6\}$
	W $\beta > \gamma > \alpha$ anom.	3 – 4	H	$74° - 90°$ $r > v$	$0° - 15°$ parallel in elong. sect.	length fast/slow	EPIDOTE	$Ca_2Fe^{+3}Al_2O \cdot OH\{Si_2O_7\}\{SiO_4\}$
	M $\gamma \geq \beta > \alpha$	4	+L	$32° - 46°$ $r > v$	$1° - 3°$	cleavage length slow	FUCHSITE	Cr- muscovite
	M $\gamma > \beta > \alpha$	4	VH	$44° - 70°$ $r > v$	parallel	length fast/slow	TEPHROITE	$Mn_2\{SiO_4\}$
	S $\gamma \geq \beta > \alpha$ $\beta > \gamma > \alpha$	4 – 5	+M	$0° - 25°$ Mg $r \lesssim v$ Fe $r > v$	$0° - 9°$ wavy	cleavage length slow	BIOTITE	$K_2(Mg,Fe^{+2})_{6-4}(Fe^{+3},Al,Ti)_{0-2}$ $\{Si_{6-5}Al_{2-3}O_{20}\}(OH,F)_4$

System	Form	Cleavage	Twinning	Zoning	Alteration	Occurrence	Remarks
Mon.	prismatic	two at 56°, parting	simple, polysynthetic	Z		I, M	cf. arfvedsonite, hornblende
Tr.	bladed	three, one good				M	
Mon. (Pseudo-Hex.)	tabular	perfect				O	green to opaque cf. septochlorites
Mon.	shards, scales, fibrous, massive	perfect basal				I, M	cf. montmorillonite, kaolinite
Mon.	minute plates, needles, radiating aggregates	perfect basal				M	cf. talc
Mon.	tabular, flakes, plates, pseudo-hex.	perfect basal	inconspicuous	Z colour banding		I, M	inclusions common, birds-eye maple structure cf. biotite, muscovite, lepidolite, rutile, tourmaline
Mon.	prismatic, subradiating, fibrous, asbestiform	two at 55°, cross fractures	simple, polysynthetic		limonite	M	cf. tremolite, actinolite, cummingtonite, anthophyllite
Mon.	short, prismatic, needles, felted aggregates	two at 87°, parting	simple, polysynthetic, twin seams	Z hour-glass		I	cf. aegirine, acmite
Mon.	distinct crystals, columnar, aggregates, six-sided x-sect.	one perfect	uncommon	Z		M, I, S	cf. zoisite, clinozoisite, diopside, augite, sillimanite
Mon.	thin tablets, shreds	perfect basal	simple			M	birds-eye maple structure cf. muscovite
Orth.	anhedral – euhedral, rounded	moderate				O, M	pleochroic in reds and black cf. monticellite, olivine
Mon. (Pseudo-Hex.)	tabular, flakes, plates, pseudo-hex.	perfect basal	simple	Z	chlorite, vermiculite, prehnite	M, I, S	pleochroic haloes, inclusions common, birds-eye maple structure cf. stilpnomelane, phlogopite

GREEN MINERALS	Pleochroism	δ	Relief	2V Dispersion	Extinction	Orientation	Mineral	Composition
Biaxial –								
	S $\alpha>\beta\geqslant\gamma$	4 – 5	H – VH	$60^{o} - 70^{o}$ r>v	$0^{o} - 10^{o}$	extinction nearest cleavage fast	**AEGIRINE**	$NaFe^{+3}\{Si_2O_6\}$

System	Form	Cleavage	Twinning	Zoning	Alteration	Occurrence	Remarks
Mon.	prismatic, needles, felted aggregates, blunt terminations	two at 87°, parting	simple, polysynthetic, twin seams	Z hour-glass		I	borders may be black cf. aegirine-augite, acmite

BLUE MINERALS	Pleochroism	δ	Relief	2V Dispersion	Extinction	Orientation	Mineral	Composition
Isometric								
	(weak)		−L				SODALITE (incl. HACKMANITE)	$Na_8\{Al_6Si_6O_{24}\}Cl_2$
	1 (weak)		−L				NOSEAN	$Na_8\{Al_6Si_6O_{24}\}SO_4$
	1 (weak)		−L				HAÜYNE	$(Na,Ca)_{4-8}\{Al_6Si_6O_{24}\}(SO_4,S)_{1-2}$
Uniaxial +								
	1 anom.		H		parallel	length fast	VESUVIANITE (IDOCRASE) (incl. WILUITE)	$Ca_{10}(Mg,Fe)_2Al_4\{Si_2O_7\}_2\{SiO_4\}_5(OH,F)_4$
Uniaxial −								
	W	1	+L		parallel	long. sect. length fast, x-sect. length slow	BERYL	$Be_3Al_2\{Si_6O_{18}\}$
	W $\epsilon>\omega$	1	+M		parallel	length fast, tabular length slow	APATITE (incl. DAHLLITE, FRANCOLITE)	$Ca_5(PO_4)_3(OH,F,Cl)$
	1 anom.		H		parallel	length fast	VESUVIANITE (IDOCRASE)	$Ca_{10}(Mg,Fe)_2Al_4\{Si_2O_7\}_2\{SiO_4\}_5(OH,F)_4$
	W $\omega>\epsilon$	1	H	$0° - 30°$	parallel	tabular length slow, prismatic length fast	CORUNDUM	$\alpha-Al_2O_3$

System	Form	Cleavage	Twinning	Zoning	Alteration	Occurrence	Remarks
Cub.	hexagonal x-sect., anhedral aggregates	poor	simple		zeolites, diaspore, gibbsite	I	cf. fluorite, leucite
Cub.	hexagonal x-sect., anhedral aggregates	poor	simple	Z	zeolites, diaspore, gibbsite, limonite	I	clouded with inclusions
Cub.	hexagonal x-sect., anhedral aggregates	imperfect	simple	Z	zeolites, diaspore, gibbsite	I	
Tet.	variable, prismatic, fibrous, granular, radial	imperfect		Z		M, I, S	cf. zoisite, clinozoisite, apatite, grossularite, melilite, andalusite
Hex.	prismatic, inclusions zoned	imperfect		Z	kaolin	I, M	liquid inclusions cf. apatite, quartz
Hex.	small, prismatic, hexagonal	poor basal		Z		I, S, M, O	cf. beryl, topaz, dahllite
Tet.	variable	poor		Z		M, I, S	cf. zoisite, clinozoisite, apatite, grossularite
Trig.	tabular, prismatic, six-sided x-sect.	parting	simple, lamellar seams	Z colour banding		M, I	inclusions cf. sapphirine

BLUE MINERALS	Pleochroism	δ	Relief	2V Dispersion	Extinction	Orientation	Mineral	Composition
Uniaxial −								
	S $\omega > \varepsilon$	3	+M		parallel	length fast	SCHORL (TOURMALINE)	$Na(Fe,Mn)_3Al_6B_3Si_6O_{27}(OH,F)_4$
	W $\omega > \varepsilon,\ \varepsilon > \omega$	5 anom.	E	0° − small			ANATASE	TiO_2
Biaxial +								
	W $\gamma > \beta > \alpha$	1	+M	50° r<v	parallel	length slow	CELESTINE	$SrSO_4$
	1 anom.		H	5° − 65° strong	parallel	length fast	VESUVIANITE (IDOCRASE) (incl. WILUITE)	$Ca_{10}(Mg,Fe)_2Al_4\{Si_2O_7\}_2$ $\{SiO_4\}_5(OH,F)_4$
	W $\beta > \gamma > \alpha$	1	H	50° − 66° r<v	6° − 9°	length slow	SAPPHIRINE	$(Mg,Fe)_2Al_4O_6\{SiO_4\}$
	M	1 − 2	−L − +L	65° − 90° r<v	parallel		CORDIERITE	$Al_3(Mg,Fe^{+2})_2\{Si_5AlO_{18}\}$
	S $\alpha > \beta > \gamma$ $\alpha > \gamma \geq \beta$	1 − 2	+M − H	40° − 50°	$\beta : z$ 15° − 30°	length fast	MAGNESIORIEBECKITE	$Na_2Mg_3Fe_2^{+3}\{Si_8O_{22}\}(OH)_2$
	S $\alpha > \beta > \gamma$ $\alpha > \gamma \geq \beta$	1 − 2	+M − H	40° − 90° r$\lessgtr$v	3° − 21°	length fast	RIEBECKITE (CROCIDOLITE)	$Na_2Fe_3^{+2}Fe_2^{+3}\{Si_8O_{22}\}(OH)_2$
	W − M $\beta > \alpha > \gamma$	1 − 2	H	45° − 68° r>v	2° − 30°	length fast	CHLORITOID (OTTRELITE)	$(Fe^{+2},Mg,Mn)_2(Al,Fe^{+3})Al_3O_2$ $\{SiO_4\}_2(OH)_4$
	W $\gamma > \alpha$	3	+M	21° − 30° r>v	parallel	length slow	SILLIMANITE	Al_2SiO_5

System	Form	Cleavage	Twinning	Zoning	Alteration	Occurrence	Remarks
Trig.	hexagonal, rounded, suns, triangular x-sect.	fractures		Z		I, M, S	cf. tourmaline group, Na-amphibole
Tet.	small, prismatic, acicular	two perfect		Z	leucoxene	S, I, M	often yellow cf. rutile, brookite
Orth.	tabular, fibrous	three				S	cf. barytes
Tet.	variable	poor		Z		M, I, S	cf. zoisite, clinozoisite, apatite, grossularite
Mon.	tabular	poor	polysynthetic, uncommon			M	cf. corundum, cordierite, kyanite, zoisite, Na-amphibole
Orth. (Pseudo-Hex.)	pseudo-hex., anhedral	moderate to absent	simple, polysynthetic, cyclic, sector, interpenetrant		pinite, talc	M, I	pleochroic haloes, inclusions (including opaque dust) common cf. quartz, plagioclase
Mon.	prismatic, fibrous, columnar aggregates	two at 56°	simple, polysynthetic	Z		I, M	cf. glaucophane, crossite, arfvedsonite, hastingsite
Mon.	prismatic, fibrous, columnar aggregates	two at 56°	simple, polysynthetic	Z		I, M	cf. glaucophane, crossite, arfvedsonite, hastingsite
Mon., Tr.	tabular	one perfect, one imperfect, parting	polysynthetic	Z hour-glass		M, I	often green, inclusions common
Orth.	needles, slender prisms, fibrous, rhombic x-sect., faserkiesel	one, // long diagonal x-sect.				M	cf. andalusite, mullite

BLUE MINERALS	Pleochroism	δ	Relief	2V Dispersion	Extinction	Orientation	Mineral	Composition
Biaxial +								
	W $\alpha>\gamma$	3	+M	$76^o - 87^o$ r>v	parallel, symmetrical	length slow, long diagonal of rhombic sect. slow	LAWSONITE	$CaAl_2(OH)_2\{Si_2O_7\}H_2O$
	W $\gamma>\beta>\alpha$	4	H	$84^o - 86^o$ r<v	parallel	length fast	DIASPORE	$\alpha-AlO(OH)$
Biaxial -								
	W $\omega>\epsilon$	1	H	$0^o - 30^o$ (uniaxial) moderate	parallel	tabular length slow, prismatic length fast	CORUNDUM	$\alpha-Al_2O_3$
	W $\beta>\gamma>\alpha$	1	H	$50^o - 66^o$ r<v	$6^o - 9^o$	length slow	SAPPHIRINE	$(Mg,Fe)_2Al_4O_6\{SiO_4\}$
	M	1 - 2	-L - +L	$65^o - 90^o$ r<v	parallel		CORDIERITE	$Al_3(Mg,Fe^{+2})_2\{Si_5AlO_{18}\}$
	S $\gamma>\beta>\alpha$ $\beta>\gamma>\alpha$	1 - 2 anom.	+M	$12^o - 65^o$ r<v	$2^o - 30^o$	length slow	CROSSITE	$Na_2(Mg_3,Fe_3^{+2},Fe_2^{+3},Al_2)\{Si_8O_{22}\}(OH)_2$
	S $\alpha>\beta>\gamma$	1 - 2	+M - H	$0^o - 50^o$ r<v	$0^o - 30^o$ anom.	length fast	ARFVEDSONITE	$Na_3(Mg,Fe^{+2})_4Al\{Si_8O_{22}\}(OH,F)_2$
	S $\alpha>\beta>\gamma$ $\alpha>\gamma\gtrless\beta$	1 - 2	+M - H	$40^o - 50^o$	$\beta : z$ $15^o - 30^o$	length fast	MAGNESIORIEBECKITE	$Na_2Mg_3Fe_2^{+3}Si_8O_{22}(OH)_2$
	S $\alpha>\beta>\gamma$ $\alpha>\gamma\gtrless\beta$	1 - 2	+M - H	$40^o - 90^o$ r$\lessgtr$v	$3^o - 21^o$	length fast	RIEBECKITE (CROCIDOLITE)	$Na_2Fe_3^{+2}Fe_2^{+3}Si_8O_{22}$
	W - M $\beta>\alpha>\gamma$	1 - 2 anom.	H	$45^o - 68^o$ r>v	$2^o - 30^o$	length fast	CHLORITOID (OTTRELITE)	$(Fe^{+2},Mg,Mn)_2(Al,Fe^{+3})Al_3O_2\{SiO_4\}_2(OH)_4$

System	Form	Cleavage	Twinning	Zoning	Alteration	Occurrence	Remarks
Orth.	rectangular, rhombic	two perfect, one imperfect	simple, polysynthetic			M	cf. zoisite, prehnite, pumpellyite, andalusite, scapolite
Orth.	large tablets	simple				S, M, I	cf. anhydrite, boehmite, gibbsite, corundum, sillimanite
Trig.	tabular, prismatic, six-sided x-sect.	parting	simple, lamellar seams	Z colour banding		M, I	inclusions cf. sapphirine
Mon.	tabular	poor	polysynthetic, uncommon			M	cf. corundum, cordierite, kyanite, zoisite, Na-amphibole
Orth. (Pseudo-Hex.)	pseudo-hex., anhedral	moderate to absent	simple, polysynthetic, cyclic, sector, interpenetrant		pinite, talc	M, I	pleochroic haloes, inclusions (including opaque dust) common cf. quartz, plagioclase
Mon.	prismatic, columnar, aggregates	two at 58°	simple, polysynthetic	Z		M	cf. glaucophane, riebeckite
Mon.	prismatic	two at 56°, parting	simple, polysynthetic	Z		I	cf. riebeckite, katophorite, glaucophane, tourmaline
Mon.	prismatic, fibrous, columnar aggregates	two at 56°	simple, polysynthetic	Z		I, M	cf. glaucophane, crossite, arfvedsonite, hastingsite
Mon.	prismatic, fibrous, columnar aggregates	two at 56°	simple, polysynthetic	Z		I, M	cf. glaucophane, crossite, arfvedsonite, hastingsite
Mon., Tr.	tabular	one perfect, one imperfect, parting	polysynthetic	Z hour-glass		M, I	inclusions common

BLUE MINERALS	Pleochroism	δ	Relief	2V Dispersion	Extinction	Orientation	Mineral	Composition
Biaxial -								
	S $\gamma>\beta>\alpha$	1 - 3 anom.	+M	$0^{o} - 50^{o}$ r>v	$4^{o} - 14^{o}$	length slow	GLAUCOPHANE	$Na_2Mg_3Al_2\{Si_8O_{22}\}(OH)_2$
	S patchy $\alpha>\beta\gtrsim\gamma$	2	+M	$20^{o} - 40^{o}$ r≲v	parallel	length fast	DUMORTIERITE	$HBAl_8Si_3O_{20}$
	W $\gamma>\beta>\alpha$	2	H	$82^{o} - 83^{o}$ r>v	$0^{o} - 32^{o}$	length slow	KYANITE	Al_2SiO_5
	M $\beta>\gamma>\alpha$	2 - 3 anom.	+M	$66^{o} - 87^{o}$ r<v	$15^{o} - 40^{o}$	length slow	RICHTERITE-FERRORICHTERITE	$Na_2Ca(Mg,Fe^{+3},Fe^{+2},Mn)_5\{Si_8O_{22}\}(OH,F)_2$
	S	3	+M	51° r>v	parallel	length slow	HOLMQUISTITE	$Li_2(Mg,Fe^{+2})_3(Al,Fe^{+3})_2\{Si_8O_{22}\}(OH)_2$
	S $\beta\gtrsim\gamma>\alpha$ $\gamma\gtrsim\beta>\alpha$	3	+M - H	$10^{o} - 90^{o}$ r≲v	$9^{o} - 40^{o}$	length slow	HASTINGSITE-FERROHASTINGSITE	$NaCa_2(Mg,Fe^{+2})_4(Al,Fe^{+3})Al_2Si_6O_{22}(OH,F)_2$
	S $\gamma\gtrsim\beta>\alpha$	4	+M	$61^{o} - 70^{o}$ r<v	$9^{o} - 12^{o}$	long diagonal fast	LAZULITE	$Al_2(Mg,Fe)(OH)_2(PO_4)_2$
	W $\gamma>\beta>\alpha$	4	VH	$44^{o} - 70^{o}$ r>v	parallel	length fast/slow	TEPHROITE (incl. ROEPPERITE)	$Mn_2\{SiO_4\}$
	W $\gamma>\beta>\alpha$	4	VH	$44^{o} - 70^{o}$ r>v	parallel	length fast/slow	KNEBELITE	$(Mn,Fe)_2\{SiO_4\}$
	W $\omega>\varepsilon,\ \varepsilon>\omega$	5 anom.	E	(uniaxial)			ANATASE	TiO_2

System	Form	Cleavage	Twinning	Zoning	Alteration	Occurrence	Remarks
Mon.	prismatic, columnar aggregates	two at 58°	simple, polysynthetic	Z		M	resembles holmquistite in pleochroism cf. arfvedsonite, eckermannite, riebeckite
Orth.	prismatic, acicular	imperfect, cross fractures	penetration, trillings		sericite	M, I	cf. tourmaline, sillimanite
Tr.	bladed, prismatic	two, parting	simple, polysynthetic			M	cf. sillimanite, pyroxene
Mon.	long prismatic, fibrous	two at 56°, parting	simple, polysynthetic	Z		M, I	
Orth.	fibrous, bladed, prismatic	two at 54$\frac{1}{2}$^{o}				M, I	cf. anthophyllite, glaucophane
Mon.	prismatic	two at 56°, parting	simple, polysynthetic	Z		I, M	cf. arfvedsonite, hornblende
Mon.	diamond-shaped, anhedral	two	polysynthetic			M	cf. tourmaline, dumortierite
Orth.	anhedral − euhedral, rounded	two moderate, imperfect	uncommon			O, M	cf. knebelite, olivine
Orth.	anhedral − euhedral, rounded	two moderate, imperfect	uncommon			O, M	cf. tephroite, olivine
Tet.	small, prismatic, acicular	two perfect		Z	leucoxene	S, I, M	often yellow cf. rutile, brookite

PURPLE MINERALS	Pleochroism	δ	Relief	2V Dispersion	Extinction	Orientation	Mineral	Composition
Isometric								
			$-M$				FLUORITE	CaF_2
Uniaxial +								
	M – S $\alpha=\beta>\gamma$	1	+L – +M	0° – 60° r<v	0° – small	cleavage length fast	Mn and Cr- CHLORITES	$(Mg,Mn,Al,Cr,Fe)_{12}$ $\{(Si,Al)_8O_{20}\}(OH)_{16}$
Uniaxial –								
	M – S $\gamma=\beta>\alpha$	1	+L – +M	0° – 20° r>v	0° – small	cleavage length slow	Mn and Cr- CHLORITES	$(Mg,Mn,Al,Cr,Fe)_{12}$ $\{(Si,Al)_8O_{20}\}(OH)_{16}$
Biaxial +								
	M – S $\alpha=\beta>\gamma$	1	+L – +M	0° – 60° r<v	0° – small	length fast	Mn and Cr- CHLORITES	$(Mg,Mn,Al,Cr,Fe)_{12}$ $\{(Si,Al)_8O_{20}\}(OH)_{16}$
	W $\gamma>\beta>\alpha$	1	+M	50°	parallel	length slow	CELESTINE	$SrSO_4$
	M	1 – 2	-L – +L	65° – 90° r<v	parallel		CORDIERITE	$Al_3(Mg,Fe^{+2})_2\{Si_5AlO_{18}\}$
	S $\alpha>\beta>\gamma$ $\alpha>\gamma\geqq\beta$	1 – 2	+M – H	40° – 50°	$\beta:z$ 15° – 30°	length fast	MAGNESIORIEBECKITE	$Na_2Mg_3Fe_2^{+3}\{Si_8O_{22}\}(OH)_2$
	S $\alpha>\beta>\gamma$ $\alpha>\gamma\geqq\beta$	1 – 2	+M – H	40° – 90° r$\lessgtr$v	3° – 21°	length fast	RIEBECKITE	$Na_2Fe_3^{+2}Fe_2^{+3}\{Si_8O_{22}\}(OH)_2$

System	Form	Cleavage	Twinning	Zoning	Alteration	Occurrence	Remarks
Cub.	anhedral, hexagonal x-sect.	two or three perfect	interpenetrant	Z		I, S, M, O	colour spots cf. cryolite, halite
Mon.	tabular, radiating	perfect basal				O, I, M	
Mon.	tabular, radiating	perfect basal				O, I, M	
Mon.	tabular, radiating	perfect basal				O, I, M	
Orth.	tabular, fibrous	three				S	cf. barytes
Orth. (Pseudo-Hex.)	pseudo-hex., anhedral	moderate to absent	simple, polysynthetic, cyclic, sector, interpenetrant			M, I	pleochroic haloes, inclusions (including opaque dust) common cf. quartz, plagioclase
Mon.	prismatic, fibrous, columnar aggregates	two at 56°	simple, polysynthetic	Z		I, M	cf. glaucophane, crossite, arfvedsonite, hastingsite
Mon.	prismatic, columnar aggregates, fibrous	two at 56°	simple, polysynthetic			I, M	cf. glaucophane, crossite, arfvedsonite, hastingsite

PURPLE MINERALS	Pleochroism	δ	Relief	2V Dispersion	Extinction	Orientation	Mineral	Composition
Biaxial +								
		2	H	$35° - 46°$ r>v	$64°$		PYROXMANGITE	$(Mn,Fe)\{SiO_3\}$
	W	2 – 3	+M	$58° - 68°$ r<v	$22° - 26°$	extinction nearest cleavage slow	KUNZITE (SPODUMENE)	$LiAl\{Si_2O_6\}$
	M – S (anom.)	2 – 3	+M – H	$25° - 50°$ r>v	$35° - 48°$	extinction nearest cleavage fast	TITANAUGITE	$(Ca,Na,Mg,Fe^{+2},Mn,Fe^{+3},Al,Ti)_2$ $\{(Si,Al)_2O_6\}$
	S γ>α>β γ>β>α	3 – 5	H – VH	$64° - 85°$ r≶v	$2° - 9°$	length fast/slow	PIEMONTITE	$Ca_2(Mn,Fe^{+3},Al)_2AlO\cdot OH$ $\{Si_2O_7\}\{SiO_4\}$
Biaxial −								
	W γ>α	1	+L	small			KÄMMERERITE (KOCHUBEITE)	$(Mg,Al,Cr,Fe)_{12}$ $\{(Si,Al)_8O_{20}\}(OH)_{16}$
	M – S γ=β>α	1	+L – +M	$0° - 20°$ r>v	$0°$ – small	length slow	Mn and Cr-CHLORITES	$(Mg,Mn,Al,Cr,Fe)_{12}$ $\{(Si,Al)_8O_{20}\}(OH)_{16}$
	M	1 – 2	−L – +L	$65° - 90°$ r<v	parallel		CORDIERITE	$Al_3(Mg,Fe^{+2})_2\{Si_5AlO_{18}\}$
	S γ>β>α β>γ>α	1 – 2 anom.	+M	$2° - 65°$ r<v	$12° - 30°$	length slow	CROSSITE	$Na_2(Mg_3,Fe_3^{+2},Fe_2^{+3},Al_2)$ $\{Si_8O_{22}\}(OH)_2$
	S α>β>γ	1 – 2	+M – H	$0° - 50°$ r<v	$0° - 30°$ anom.	length fast	ARFVEDSONITE	$Na_3(Mg,Fe^{+2})_4Al\{Si_8O_{22}\}(OH,F)_2$
	S α>β>γ α>γ≳β	1 – 2	+M – H	$40° - 50°$	$15° - 30°$	length fast	MAGNESIORIEBECKITE	$Na_2Mg_3Fe_2^{+3}\{Si_8O_{22}\}(OH)_2$

System	Form	Cleavage	Twinning	Zoning	Alteration	Occurrence	Remarks
Tr.	prismatic	four, two at 92°	simple, polysynthetic			O, M	cf. rhodonite, bustamite
Mon.	tabular	two at 87°, parting	simple		muscovite, cymatolite, kaolinite, eucryptite	I	
Mon.	prismatic	two at 93°	polysynthetic, twin seams	Z hour-glass		I	
Mon.	columnar, six-sided x-sect.	one perfect	polysynthetic, uncommon			M, I	cf. thulite, titanaugite, dumortierite
Mon.	tabular	perfect				O	cf. chlorite
Mon.	tabular, radiating	perfect basal				O, I, M	
Orth. (Pseudo-Hex.)	pseudo-hex., anhedral	moderate to absent	simple, polysynthetic, cyclic, sector, interpenetrant		pinite, talc	M, I	pleochroic haloes, inclusions (including opaque dust) common cf. quartz, plagioclase
Mon.	prismatic, columnar aggregates	two at 58°	simple, polysynthetic	Z		M	cf. glaucophane, riebeckite
Mon.	prismatic	two at 56°, parting	simple, polysynthetic	Z		I	cf. riebeckite, katophorite, glaucophane, tourmaline
Mon.	prismatic, fibrous, columnar aggregates	two at 56°	simple, polysynthetic	Z		I, M	cf, glaucophane, crossite, arfvedsonite, hastingsite

PURPLE MINERALS	Pleochroism	δ	Relief	2V Dispersion	Extinction	Orientation	Mineral	Composition
Biaxial –								
	S $\alpha > \beta > \gamma$ $\alpha > \gamma \geqslant \beta$	1 – 2	+M – H	$40° - 90°$ $r \lessgtr v$	$3° - 21°$	length fast	RIEBECKITE	$Na_2Fe_3^{+2}Fe_2^{+3}\{Si_8O_{22}\}(OH)_2$
	N – S $\gamma > \beta > \alpha$	1 – 2	+M – H	$50° - 90°$ $r \lessgtr v$	parallel	length slow	HYPERSTHENE	$(Mg,Fe^{+2})\{SiO_3\}$
	S $\gamma > \beta > \alpha$	1 – 3 anom.	+M	$0° - 50°$ $r > v$	$4° - 14°$	length slow	GLAUCOPHANE	$Na_2Mg_3Al_2\{Si_8O_{22}\}(OH)_2$
	S patchy $\alpha > \beta > \gamma$	2	+M	$20° - 40°$ $r \lessgtr v$	parallel	length fast	DUMORTIERITE	$HBAl_8Si_3O_{20}$
	W	2	+M	$63° - 80°$ $r < v$	inclined		AXINITE	$(Ca,Mn,Fe^{+2})_3Al_2BO_3\{Si_4O_{12}\}OH$
	M $\beta > \gamma > \alpha$	2 – 3 anom.	+M	$66° - 87°$ $r < v$	$15° - 40°$	length slow	RICHTERITE-FERRORICHTERITE	$Na_2Ca(Mg,Fe^{+3},Fe^{+2},Mn)_5$ $\{Si_8O_{22}\}(OH,F)_2$
	S	3	+M	$51°$ $r > v$	parallel	length slow	HOLMQUISTITE	$Li_2(Mg,Fe^{+2})_3(Al,Fe^{+3})_2$ $\{Si_8O_{22}\}(OH)_2$
	S	3	H	$65°$ $r < v$			HOWIEITE	$Na(Fe,Mn)_{10}$ $(Fe,Al)_2Si_{12}O_{31}(OH)_{13}$

System	Form	Cleavage	Twinning	Zoning	Alteration	Occurrence	Remarks
Mon.	prismatic, fibrous, columnar aggregates	two at 56°	simple, polysynthetic	Z		I, M	cf. glaucophane, crossite, arfvedsonite, hastingsite
Orth.	prismatic, anhedral - euhedral	two at 88°	polysynthetic, twin seams	Z		I, M	exsolution lamellae, schiller inclusions cf. andalusite
Mon.	prismatic, columnar aggregates	two at 58°	simple, polysynthetic	Z		M	resembles holmquistite in pleochroism cf. arfvedsonite, eckermannite, riebeckite
Orth.	prismatic, acicular	imperfect, cross fractures	interpenetrant, trillings		sericite	M, I	cf. tourmaline, sillimanite
Tr. Mon.	bladed, wedge-shaped, clusters	imperfect in several directions		Z		I, M	inclusions common cf. quartz
Mon.	long prismatic, fibrous	two at 56°, partings	simple, polysynthetic	Z		M, I	
Orth.	fibrous, bladed, prismatic	two at $54\frac{1}{2}^{\circ}$				M, I	cf. anthophyllite, glaucophane
Tr.	bladed	three, one good				M	

BLACK MINERALS	Pleochroism	δ	Relief	2V Dispersion	Extinction	Orientation	Mineral	Composition
Isometric								
			H – E				SPINEL (incl. PLEONASTE)	$(Mg,Fe^{+2})Al_2O_4$
			H – VH				GARNET	$Ca_3(Fe^{+3},Ti)_2Si_3O_{12}$
Uniaxial								
	W $\omega>\varepsilon,\ \varepsilon>\omega$	5	E	0° – small			ANATASE	TiO_2
Biaxial +								
	W $\gamma>\alpha$	(weak)	E	90° r>v	$\simeq 45^o$		PEROVSKITE (incl. LOPARITE-DYSANALYTE)	$(Ca,Na,Fe^{+2},Ce)(Ti,Nb)O_3$
	M – S $\gamma>\alpha$	5	VH	32° r<v	4° – 45°		COSSYRITE (AENIGMATITE)	$Na_2Fe_5^{+2}TiSi_6O_{20}$
Biaxial –								
	W – M $\beta>\alpha>\gamma$	1 – 2	H	45° – 68° r>v	2° – 30°	length fast	CHLORITOID (OTTRELITE)	$(Fe^{+2},Mg,Mn)_2(Al,Fe^{+3})Al_3O_2$ $\{SiO_4\}_2(OH)_4$
	M – S $\gamma>\beta>\alpha$ $\gamma<\beta>\alpha$	1 – 3	+M	0° – 50° r<v	36° – 70°	length fast	KATOPHORITE-MAGNESIOKATOPHORITE	$Na_2Ca(Mg,Fe^{+2})_4Fe^{+3}$ $\{Si_7AlO_{22}\}(OH,F)_2$
	W	3	VH				DEERITE	$(Fe,Mn)_{13}(Fe,Al)_7Si_{13}O_{44}(OH)_{11}$
	S $\gamma=\beta>\alpha$	3 – 5	+L – H	$\simeq 0^o$	$\simeq$ parallel	length slow	STILPNOMELANE	$(K,Na,Ca)_{0-1\cdot4}$ $(Fe^{+3},Fe^{+2},Mg,Al,Mn)_{5\cdot9-8\cdot2}$ $\{Si_8O_{20}\}(OH)_4(O,OH,H_2O)_{3\cdot6-8\cdot5}$

System	Form	Cleavage	Twinning	Zoning	Alteration	Occurrence	Remarks
Cub.	small grains, cubes, octahedra, rhombic x-sect.	parting				M, I	
Cub.	four, six, eight-sided, polygonal x-sect., aggregates	parting, irregular fractures	complex, sector	Z	chlorite	M, I, S	inclusions cf. spinel
Tet.	small, prismatic, acicular	two perfect		Z	leucoxene	S, I, M	usually yellow to blue cf. rutile, brookite
Mon? Pseudo-Cub.	small cubes, skeletal	poor to distinct	polysynthetic, complex, interpenetrant	Z	leucoxene	I, M	cf. melanite, picotite, ilmenite
Tr.	small, prismatic, aggregates	two at 66°	simple, repeated			I	cf. katophorite, kaersutite, basaltic hornblende
Mon., Tr.	tabular	one perfect, one imperfect, parting	polysynthetic	Z hour-glass		M, I	inclusions common cf. chlorite, clintonite, biotite, stilpnomelane
Mon.	prismatic	two at 56°, parting	simple	Z		I	cf. barkevikite, kaersutite, basaltic hornblende, arfvedsonite, cossyrite
Mon.	acicular, amphibole-like	one good	simple			M	
Mon.	plates, pseudo-hex., micaceous masses	two at 90°, perfect basal	polysynthetic	Z		I, M, O	cf. biotite, chlorite, chloritoid, clintonite

BLACK MINERALS	Pleochroism	δ	Relief	2V Dispersion	Extinction	Orientation	Mineral	Composition
Biaxial −								
	S $\alpha > \beta \geqslant \gamma$	4 − 5	H VH	$60° - 70°$ r>v	$0° - 10°$	extinction nearest cleavage fast	AEGIRINE	$NaFe^{+3}\{Si_2O_6\}$
	W	4	VH	$44° - 70°$ r<v	parallel	length fast/slow	KNEBELITE	$(Mn,Fe)_2\{SiO_4\}$
	W $\omega > \varepsilon, \ \varepsilon > \omega$	5 anom.	E	(uniaxial)			ANATASE	TiO_2

The following common opaque minerals also appear black in thin sections:

			H − E				SPINEL GROUP	$(Mg,Mn,Zn,Ni,Al_2,Fe^{+2})Fe_2{}^{+3}O_4$
							CHALCOPYRITE	$CuFeS_2$
			E				CHROMITE (MAGNESIOCHROMITE)	$(Fe^{+2},Mg)Cr_2O_4$
			E				GALENA	PbS
			E				GRAPHITE	C
			E				MAGNETITE	$Fe^{+2}Fe^{+3}{}_2O_4$
							PYRRHOTITE	Fe_7S_8-FeS
	(weak)						PYRITE	FeS_2
	(weak)		E				LIMONITE	$FeO \cdot OH \cdot nH_2O$
		5	E				HAEMATITE	$\alpha-Fe_2O_3$
		5	E				ILMENITE	$FeTiO_3$

System	Form	Cleavage	Twinning	Zoning	Alteration	Occurrence	Remarks
Mon.	prismatic, needles, felted aggregates, blunt terminations	two at 87°, parting	simple, polysynthetic, twin seams	Z hour-glass		I	usually green cf. aegirine-augite, acmite
Orth.	anhedral - euhedral, rounded	two moderate, imperfect	uncommon	Z		O, M	cf. tephroite
Tet.	small, prismatic, acicular	two perfect		Z	leucoxene	S, I, M	usually yellow to blue cf. rutile, brookite
Cub.	small grains, cubes, octahedra, rhombic x-sect.	parting				M, I, O	opaque
Tet.	aggregates					O, S	opaque
Cub.	subhedral grains, octahedra, aggregates					I, O	
Cub.	cubes, octahedra	perfect, parting	interpenetrant			O	opaque
Hex.	thin ragged flakes, scales	parting	perfect basal			M	
Cub.	small grains, octahedra	parting				I, M, S, O	opaque, exsolved ulvöspinel
Hex.	grains, irregular masses	parting				O, I, M	opaque
Cub.	cubes, octahedra, irregular masses		interpenetrant			O, I, M, S	opaque, amorphous = melnikovite
M'loid.	stain or border to other minerals, pseudomorphs					I, M, S, O	opaque to translucent cf. goethite
Trig.	scales, flakes, grains and irregular masses	parting	polysynthetic			M, I, O, S	opaque to translucent red cf. goethite, limonite
Trig.	skeletal, grains and irregular masses				leucoxene	I, M, S	opaque

GREY MINERALS	Pleochroism	δ	Relief	2V Dispersion	Extinction	Orientation	Mineral	Composition
Isometric								
			−M				FLUORITE	CaF_2
		rarely weak	−M − −L				OPAL	SiO_2
		(weak)	−L				SODALITE (incl. HACKMANITE)	$Na_8\{Al_6Si_6O_{24}\}Cl_2$
		(weak)	−L				NOSEAN	$Na_8\{Al_6Si_6O_{24}\}SO_4$
		(weak)	−L				HAÜYNE	$(Na,Ca)_{4-8}\{Al_6Si_6O_{24}\}(SO_4,S)_{1-}$
		(weak)	−L − +M				VOLCANIC GLASS	
		(weak)	E				SPHALERITE	ZnS
	W $\gamma>\alpha$	(weak)	E				PEROVSKITE (KNOPITE- LOPARITE- DYSANALYTE)	$(Ca,Na,Fe^{+2},Ce)(Ti,Nb)O_3$
Uniaxial +								
		1 anom.	H		parallel	length fast	VESUVIANITE (IDOCRASE) (incl. WILUITE)	$Ca_{10}(Mg,Fe)_2Al_4\{Si_2O_7\}_2\{SiO_4\}_5(OH,F)_4$
	W $\varepsilon>\omega$	4 − 5	VH − E		parallel	length slow	ZIRCON	$Zr\{SiO_4\}$

System	Form	Cleavage	Twinning	Zoning	Alteration	Occurrence	Remarks
Cub.	hexagonal x-sect.	two or three perfect	interpenetrant	Z		I, S, O, M	colour spots cf. cryolite, halite
M'loid.	colloform, veinlets, cavity fillings	irregular fractures				I	
Cub.	hexagonal x-sect., anhedral aggregates	poor	simple	Z	zeolites, diaspore, gibbsite	I	cf. fluorite, leucite
Cub.	hexagonal x-sect., anhedral aggregates	imperfect	simple	Z	zeolites, diaspore, gibbsite, ilmonite	I	clouded with inclusions
Cub.	hexagonal x-sect., anhedral aggregates	imperfect	polysynthetic	Z	zeolites, diaspore, gibbsite	I	
M'loid.	amorphous, massive	perlitic, parting			frequent, devit- rification	I	often with crystallites and phenocrysts cf. tachylyte, lechatelierite
Cub.	irregular, anhedral, curved faces	six perfect	polysynthetic, lamellar intergrowths	Z		O	colour variable, (uniaxial) cf. cassiterite
Mon? Pseudo- Cub.	small cubes	poor to distinct	polysynthetic, complex, interpenetrant	Z	leucoxene	I, M	cf. melanite, picotite, ilmenite
Tet.	variable, prismatic, fibrous, granular, radial	imperfect	sector	Z		M, I, S	cf. zoisite, clinozoisite, apatite, grossularite, melilite, andalusite
Tet.	minute prisms	poor, absent		Z	metamict	I, S, M	cf. apatite

GREY MINERALS	Pleochroism	δ	Relief	2V Dispersion	Extinction	Orientation	Mineral	Composition
Uniaxial +								
	W $\varepsilon > \omega$	5	VH – E		parallel, oblique to twin plane	length slow	CASSITERITE	SnO_2
Uniaxial –								
	W $\varepsilon > \omega$	1	+M		parallel	length fast, tabular length slow	APATITE (incl. DAHLLITE, FRANCOLITE)	$Ca_5(PO_4)_3(OH,F,Cl)$
		1 anom.	+H		parallel	length fast	VESUVIANITE (IDOCRASE)	$Ca_{10}(Mg,Fe)_2Al_4\{Si_2O_7\}_2\{SiO_4\}_5(OH,F)_4$
	W $\omega > \varepsilon$	1	H	$0^o - 30^o$	parallel, symmetrical	tabular length slow, prismatic length fast	CORUNDUM	$\alpha\text{-}Al_2O_3$
	S $\omega > \varepsilon$	3	+M		parallel	length fast	SCHORL (TOURMALINE)	$Na(Fe,Mn)_3Al_6B_3Si_6O_{27}(OH,F)_4$
		5	−L – +M		symmetrical to cleavage		CALCITE	$CaCO_3$
		5	−L – +M		symmetrical to cleavage		DOLOMITE	$CaMg(CO_3)_2$
		5	+L – VH		symmetrical to cleavage		SIDERITE	$FeCO_3$
	W $\omega > \varepsilon, \varepsilon > \omega$	5 anom.	E	0° – small			ANATASE	TiO_2

System	Form	Cleavage	Twinning	Zoning	Alteration	Occurrence	Remarks
Tet.	subhedral, veinlets, diamond-shaped x-sect.	prismatic	geniculate, cyclic, common	Z		O, I, S	cf. sphalerite, rutile
Hex.	small, prismatic, hexagonal	poor basal				I, S, M, O	cf. beryl, topaz, dahllite
Tet.	variable, prismatic, fibrous, granular, radial	imperfect		Z		M, I, S	cf. zoisite, clinozoisite, apatite, grossularite, melilite, andalusite
Trig.	tabular, prismatic, six-sided x-sect.	parting	simple, lamellar seams	Z colour banding		M, I	inclusions cf. sapphirine
Trig.	hexagonal, rounded, suns, triangular x-sect.	fractures		Z		I, M, S	cf. tourmaline group
Trig.	anhedral, oolitic, spherulitic	rhombohedral	polysynthetic, // long diagonal			S, M, I, O	twinkling cf. rhombohedral carbonates
Trig.	anhedral, rhombohedral	rhombohedral	polysynthetic, // long and short diagonals	Z	huntite	S, M, I	cf. twinkling rhombohedral carbonates
Trig.	rhombohedral	rhombohedral	polysynthetic, uncommon, // long diagonal			S, O, I, M	twinkling, brown stain around borders and along cleavage cracks cf. rhombohedral carbonates
Tet.	small, prismatic, acicular	simple		Z	leucoxene	S, I, M	usually yellow to blue cf. rutile, brookite

GREY MINERALS	Pleochroism	δ	Relief	2V Dispersion	Extinction	Orientation	Mineral	Composition
Biaxial +								
	W – M $\beta > \alpha > \gamma$	1 – 2 anom.	H	$45^\circ - 68^\circ$ r>v	$2^\circ - 30^\circ$	length fast	CHLORITOID (OTTRELITE)	$(Fe^{+2},Mg,Mn)_2(Al,Fe^{+3})Al_3O_2$ $\{SiO_4\}_2(OH)_4$
	W	3	+M	$50^\circ - 62^\circ$ r>v	$38^\circ - 48^\circ$	extinction nearest cleavage slow	DIOPSIDE (SALITE)	$Ca(Mg,Fe)\{Si_2O_6\}$
		4	+M	$82^\circ - 90^\circ$ r>v	parallel	cleavage length slow	FORSTERITE	Mg_2SiO_4
Biaxial –								
	N – W	1	+M	small	small	length slow	CHAMOSITE	$(Mg,Al,Fe)_{12}\{(Si,Al)_8O_{20}\}(OH)$
	S variable $\alpha > \beta > \gamma$	1 – 2	+M – H	$0^\circ - 50^\circ$ r<v	$0^\circ - 30^\circ$ anom.	length fast	ARFVEDSONITE	$Na_3(Mg,Fe^{+2})_4Al\{Si_8O_{22}\}(OH,F)$
	N – S $\gamma > \beta > \alpha$	1 – 2	+M – H	$50^\circ - 90^\circ$ r$\lessgtr$v	parallel	length slow	HYPERSTHENE	$(Mg,Fe^{+2})\{SiO_3\}$
	W – M $\beta > \alpha > \gamma$	1 – 2	H	$45^\circ - 68^\circ$ r>v	$2^\circ - 30^\circ$	length fast	CHLORITOID (OTTRELITE)	$(Fe^{+2},Mg,Mn)_2(Al,Fe^{+3})Al_3O_2$ $\{SiO_4\}_2(OH)_4$
		2	+M	$38^\circ - 60^\circ$ r>v	$\alpha{:}z\ 30^\circ{-}44^\circ$ $\simeq$ parallel	length fast/slow	WOLLASTONITE	$Ca\{SiO_3\}$
	W $\gamma > \beta > \alpha$	2	H	$82^\circ - 83^\circ$ r>v	$0^\circ - 32^\circ$	length slow	KYANITE	Al_2SiO_5
		2 – 3	–L	$40^\circ - 60^\circ$	$\simeq$ parallel	length slow	SEPIOLITE	$H_4Mg_2\{Si_3O_{10}\}$

System	Form	Cleavage	Twinning	Zoning	Alteration	Occurrence	Remarks
Mon., Tr.	tabular	perfect basal, one imperfect, parting	polysynthetic	Z hour-glass		M, I	inclusions common cf. clintonite, biotite, stilpnomelane
Mon.	short prismatic	two at 87°	simple, polysynthetic	Z	tremolite-actinolite	M, I	cf. hedenbergite, tremolite, omphacite, wollastonite, epidote
Orth.	anhedral - euhedral, rounded	uncommon, irregular fractures	uncommon	Z	chlorite, antigorite, serpentine, iddingsite, bowlingite	I, M	deformation lamellae cf. diopside, augite, pigeonite, humite group, epidote
Mon.	pseudo-spherulitic, concentric, tabular, massive	one good, concentric parting				S, O	cf. glauconite, collophane, greenalite, thuringite
Mon.	prismatic	two at 56°, parting	simple, polysynthetic	Z		I	cf. riebeckite, katophorite, glaucophane, tourmaline
Orth.	prismatic, anhedral - subhedral	two at 88°	polysynthetic, twin seams	Z		I, M	exsolution lamellae, schiller inclusions cf. andalusite
Mon., Tr.	tabular	one perfect, one imperfect, parting	polysynthetic	Z hour-glass		M, I	inclusions common cf. chlorite, clintonite, biotite, stilpnomelane
Tr.	subhedral - euhedral, columnar, fibrous	three	polysynthetic		pectolite, calcite	M, I	cf. tremolite, pectolite
Tr.	bladed, prismatic	two, parting	simple, polysynthetic			M	cf. sillimanite, pyroxene
Mon. (Orth.)	fibrous aggregates, curved, matted					I, S	

GREY MINERALS	Pleochroism	δ	Relief	2V Dispersion	Extinction	Orientation	Mineral	Composition
Biaxial -								
	S	3	H	65° r<v	0° - 15°		HOWIEITE	$Na(Fe,Mn)_{10}$ $(Fe,Al)_2Si_{12}O_{31}(OH)_{13}$
	W $\beta>\gamma>\alpha$	3 - 4 anom.	H	74° - 90° r>v	parallel in elong. sect.	length fast/slow	EPIDOTE	$Ca_2Fe^{+3}Al_2O\cdot OH\{Si_2O_7\}\{SiO_4\}$
	W	4	+M – VH	48° - 90°	parallel	cleavage length slow	OLIVINE	$(Mg,Fe)_2\{SiO_4\}$
		5	–L – +M	4° - 14° (uniaxial)	symmetrical to cleavage		CALCITE	$CaCO_3$
	W – M $\alpha>\beta>\gamma$	5 anom.	VH – E	17° - 40° r>v	40° symmetrical		SPHENE	$CaTi\{SiO_4\}(O,OH,F)$

System	Form	Cleavage	Twinning	Zoning	Alteration	Occurrence	Remarks
Tr.	bladed	three, one good				M	
Mon.	distinct crystals, columnar aggregates	one perfect	uncommon			M, I, S	cf. zoisite, clinozoisite, diopside, augite, sillimanite
Orth.	anhedral - euhedral, rounded	uncommon, one moderate, irregular fractures	uncommon, vicinal	Z	chlorite, antigorite, serpentine, iddingsite, bowlingite	I, M	deformation lamellae cf. diopside, augite, pigeonite, humite group, epidote
Trig.	anhedral	rhombohedral	polysynthetic, // long diagonal			M	twinkling cf. rhombohedral carbonates
Mon.	rhombic, irregular grains	parting	simple, polysynthetic		leucoxene	I, M, S	cf. monazite, calcite

WHITE MINERALS	Pleochroism	δ	Relief	2V Dispersion	Extinction	Orientation	Mineral	Composition
							LEUCOXENE	$TiO_2 \cdot nH_2O$
	(anom.)	H					GROSSULARITE	$Ca_3Al_2Si_3O_{12}$

System	Form	Cleavage	Twinning	Zoning	Alteration	Occurrence	Remarks
Amor-phous	finely crystalline, pseudomorphs, alteration of ilmenite					I, M, O	opaque, finely crystalline rutile or brookite
Cub.	four, six, eight-sided, polygonal x-sect.	parting, irregular fractures		Z		M	cf. garnet group, periclase, vesuvianite

Table 2

Colour, optical group and relief

COLOURLESS MINERALS	Pleochroism	Relief	δ	2V Dispersion	Extinction	Orientation	Mineral	Composition
Isometric								
		$-M$	rarely weak				OPAL	SiO_2
		$-M$					FLUORITE	CaF_2
		$-L$					LECHATELIERITE	SiO_2
		$-L$	1			length fast/slow	CRISTOBALITE	SiO_2
		$-L$	(weak)				FAUJASITE	$(Na_2,Ca)\{Al_2Si_4O_{12}\}\cdot 8H_2O$
		$-L$	(weak)				SODALITE (incl. HACKMANITE)	$Na_8\{Al_6Si_6O_{24}\}Cl_2$
		$-L$	(weak)				NOSEAN	$Na_8\{Al_6Si_6O_{24}\}SO_4$
		$-L$	(weak)				ANALCITE	$Na\{AlSi_2O_6\}\cdot H_2O$
		$-L$	(weak)				HAÜYNE	$(Na,Ca)_{4-8}\{Al_6Si_6O_{24}\}(SO_4,S)_{1-}$
		$-L$	(aniso-tropic)		wavy		LEUCITE	$K\{AlSi_2O_6\}$
		$-L - +M$	(weak)				VOLCANIC GLASS	
		$-L - +M$	(weak)				PALAGONITE	Altered glass

System	Form	Cleavage	Twinning	Zoning	Alteration	Occurrence	Remarks
M'loid.	cryptocrystalline, colloform, veinlets, cavity fillings	irregular fractures				I, S	cf. lechatelierite
Cub.	anhedral, hexagonal x-sect.	two or three perfect	interpenetrant	Z		I, S, O, M	colour spots cf. cryolite, halite
M'loid.	amorphous, vesicular					fulgurites	cf. opal
Tet. (Pseudo-Cub.)	minute square crystals, aggregates, spherulitic	curved fractures	polysynthetic, interpenetrant			I	cf. tridymite
Cub.	octahedral, rounded	one distinct				I	cf. zeolites
Cub.	hexagonal x-sect., anhedral aggregates	poor	simple		zeolites, diaspore, gibbsite	I	cf. fluorite, leucite
Cub.	hexagonal x-sect., anhedral aggregates	imperfect	simple	Z	zeolites, diaspore, gibbsite, limonite	I	clouded with inclusions
Cub.	trapezohedral, rounded, radiating, irregular	poor	polysynthetic, complex, interpenetrant			I, S, M	cf. leucite, sodalite, wairakite
Cub.	hexagonal x-sect., anhedral aggregates	imperfect	polysynthetic	Z	zeolites, diaspore, gibbsite	I	
Tet. (Pseudo-Cub.)	always euhedral, octagonal	poor	polysynthetic, complex	Z		I	inclusions common cf. analcite, microcline
M'loid.	amorphous, massive	perlitic parting			frequent, devit-rification	I	often with crystallites and phenocrysts cf. tachylyte, lechatelierite
M'loid.	amorphous, oolitic				chlorite	I	cf. volcanic glass, opal, collophane

COLOURLESS MINERALS	Pleochroism	Relief	δ	2V Dispersion	Extinction	Orientation	Mineral	Composition
Isometric								
		N					HALITE	$NaCl$
		+L					CLIACHITE	$Al_2O_3(H_2O)x$
		+L	1	small		length slow	HALLOYSITE	$Al_4\{Si_4O_{10}\}(OH)_8 \cdot 8H_2O$
		+L	1			length fast/slow	COLLOPHANE	$Ca_5(PO_4)_3(OH,F,Cl)$
	W $\varepsilon > \omega$	+L – +M	1 (iso-tropic)				EUDIALYTE-EUCOLITE	$(Na,Ca,Fe)_6Zr\{(Si_3O_9)_2\}$ (OH,F,Cl)
		+M					PLAZOLITE	$3CaO \cdot Al_2O_3 \cdot 2SiO_2 \cdot 2H_2O$
		+M	anom.				HIBSCHITE	$3CaO \cdot Al_2O_3 \cdot 2SiO_2 \cdot 2H_2O$
		+M	(aniso-tropic)				MELILITE	$(Ca,Na)_2\{Mg,Fe^{+2},Al,Si)_3O_7\}$
		H					HELVITE	$Mn_4\{Be_3Si_3O_{12}\}S$
		H					DANALITE	$Fe_4\{Be_3Si_3O_{12}\}S$
		H					GENTHELVITE	$Zn_4\{Be_3Si_3O_{12}\}S$
		H					PERICLASE	MgO

System	Form	Cleavage	Twinning	Zoning	Alteration	Occurrence	Remarks
Cub.	anhedral	perfect cubic				S, I	inclusions common cf. sylvite
M'loid.	pisolitic, massive	contraction cracks				S	often with gibbsite and siderite
Tr. Mon(?)	fine aggregates, colloform, clay, fibrous	shatter cracks			sericite	S, I	cf. hydrohalloysite (endellite)
M'loid.	amorphous, pisolitic, massive, cryptocrystalline	irregular fractures				S, I	contraction cracks cf. opal
Trig.	rhombohedral aggregates	one good, one poor		Z		I	normally uniaxial, some zones isotropic cf. catapleite, låvenite, rosenbuschite, garnet
Cub.	dodecahedra	parting, irregular fractures				M	cf. grossularite, hydrogrossular
Cub.	octahedra	parting, irregular fractures				M	cf. grossularite
Tet.	tabular, peg structure	moderate, single crack		Z	zeolites carbonate	I	cf. zoisite, vesuvianite, apatite, nepheline
Cub.	tetrahedral, triangular x-sect., granular	one poor	simple		ochre, manganese oxide	I, M	cf. garnet, danalite
Cub.	tetrahedral, triangular x-sect., granular	one poor				I, M	cf. garnet, helvite
Cub.	tetrahedral, triangular x-sect., granular	one poor				I, M	cf. garnet
Cub.	small crystals, aggregates	perfect cubic			brucite	M	cf. garnet, spinel

COLOURLESS MINERALS	Pleochroism	Relief	δ	2V Dispersion	Extinction	Orientation	Mineral	Composition
Isometric								
		H					PYROPE	$Mg_3Al_2Si_3O_{12}$
		H	(anom.)				GROSSULARITE	$Ca_3Al_2Si_3O_{12}$
		H – VH	(anom.)				GARNET	$(Mg,Fe^{+2},Mn,Ca)_3$ $(Fe^{+3},Ti,Cr,Al)_2\{Si_3O_{12}\}$
		VH					ALMANDINE	$Fe_3{}^{+2}Al_2Si_3O_{12}$
		H – E					SPINEL	(Mg,Fe^{+2},Zn,Mn,Ni) $(Al_2Fe_2{}^{+3},Cr_2,Ti)O_4$
		E	(weak)				SPHALERITE	ZnS
	W $\gamma>\alpha$	E	(weak)				PEROVSKITE (incl. KNOPITE-LOPARITE-DYSANALYTE)	$(Ca,Na,Fe^{+2},Ce)(Ti,Nb)O_3$
Uniaxial +								
		–L	1		wavy		LEUCITE	$K\{AlSi_2O_6\}$
		–L	1		parallel	length slow	ERIONITE	$(Na_2,K_2,Ca,Mg)_{4.5}$ $\{Al_9Si_{27}O_{72}\}\cdot27H_2O$
		–L	1		parallel		DAVYNE-NATRODAVYNE	K-cancrinite

System	Form	Cleavage	Twinning	Zoning	Alteration	Occurrence	Remarks
Cub.	four, six, eight-sided, polygonal x-sect., aggregates	parting, irregular fractures				M, I	cf. garnet group
Cub.	four, six, eight-sided, polygonal x-sect., aggregates	parting, irregular fractures	sector	Z		M	cf. garnet group, periclase, vesuvianite
Cub.	four, six, eight-sided, polygonal x-sect., aggregates	parting, irregular fractures	complex, sector	Z	chlorite	M, I, S	inclusions cf. spinel
Cub.	four, six, eight-sided, polygonal x-sect., aggregates	parting, irregular fractures			chlorite	M, I	inclusions cf. garnet group
Cub.	small grains, cubes, octahedra, rhombic x-sect.	parting		Z		M, I	colour variable according to composition cf. periclase, garnet
Cub.	irregular, anhedral, curved surfaces	six perfect	polysynthetic, lamellar intergrowths	Z		O	colour variable, (uniaxial) cf. cassiterite
Mon? Pseudo-Cub.	small cubes, skeletal	poor to distinct	polysynthetic, complex, interpenetrant	Z	leucoxene	I, M	cf. melanite, picotite, ilmenite
Tet. (Pseudo-Cub.)	always euhedral, octagonal	poor	polysynthetic, complex	Z		I	inclusions common cf. analcite, microcline
Hex.	fibrous, radiating					I	cf. zeolites
Hex.	anhedral	two perfect				I	cf. cancrinite, microsommite

COLOURLESS MINERALS	Pleochroism	Relief	δ	2V Dispersion	Extinction	Orientation	Mineral	Composition
Uniaxial +								
		-L	1	0° - 32°	symmetrical		CHABAZITE	$Ca\{Al_2Si_4O_{12}\}\cdot 6H_2O$
		-L	1 - 3 (iso-tropic)	anom.	parallel	length slow	MICROSOMMITE	$K,NaAlSiO_4\cdot Ca(Cl_2,SO_4)$
		-L - N	1 anom.		parallel	length slow	APOPHYLLITE	$KFCa_4\{Si_8O_{20}\}\cdot 8H_2O$
		-L - N	1		parallel	length slow	ASHCROFTINE	$KNaCa\{Al_4Si_5O_{18}\}\cdot 8H_2O$
	N	1		parallel, symmetrical	length slow	QUARTZ	SiO_2	
	W $\alpha=\beta>\gamma$	+L	1 anom.	0° - 20° r<v	$\simeq$ parallel	cleavage length fast	PENNINITE	$(Mg,Al,Fe)_{12}\{(Si,Al)_8O_{20}\}(OH)_1$
	W $\alpha=\beta>\gamma$	+L	1	$\simeq 20^{\circ}$ r<v	$\simeq$ parallel	cleavage length fast	SHERIDANITE	$(Mg,Al,Fe)_{12}\{(Si,Al)_8O_{20}\}(OH)_1$
	W $\alpha=\beta>\gamma$	+L	1 - 2	0° - 40° r<v	0° - 9°	cleavage length fast	CLINOCHLORE	$(Mg,Al,Fe)_{12}\{(Si,Al)_8O_{20}\}(OH)_1$
		+L	2 anom.		parallel	x-sect. of plates length fast	BRUCITE	$Mg(OH)_2$
		+L	2 - 3	0° - 30° r<v	3°	cleavage fast	CATAPLEITE	$(Na,Ca)_2Zr\{Si_3O_9\}\cdot 2H_2O$
	W $\epsilon>\omega$	+L - +M	1 (iso-tropic)				EUDIALYTE	$(Na,Ca,Fe)_6Zr\{(Si_3O_9)_2\}(OH,F,Cl)$

System	Form	Cleavage	Twinning	Zoning	Alteration	Occurrence	Remarks
Trig.	rhombohedral, approaching cube, anhedral, granular	one poor	interpenetrant	Z		I	basal section in six segments cf. gmelinite, analcite
Hex.	anhedral, prismatic	one perfect, one poor	polysynthetic, rare	Z		I	some zones isotropic cf. nepheline, quartz, cancrinite, vishnevite
Tet.	prismatic, granular, anhedral - euhedral	two, one perfect				I, O	cf. zeolites
Tet.	prismatic, granular, anhedral - euhedral, needles	two at 90°, one perfect				I	cf. zeolites
Trig.	variable, anhedral, flamboyant, intergrowths		rare in thin sections			S, I, M, O	inclusions, Boehm lamellae cf. cordierite, beryl, scapolite, feldspar
Mon. (Pseudo-Hex.)	tabular, vermicular, radiating, pseudomorphs	perfect basal	simple, pennine law	Z		I, M	pleochroic haloes cf. clinochlore, prochlorite
Mon. (Pseudo-Hex.)	tabular, spherulites	perfect basal	simple			I	
Mon.	tabular, fibrous, pseudo-hex.	perfect basal	polysynthetic			M	pleochroic haloes cf. penninite, prochlorite, leuchtenbergite, katschubeite
Trig.	fibrous, plates, whorls, scaly aggregates	one perfect			hydro-magnesite	M	alteration of periclase cf. alunite, muscovite, talc
Mon. (Pseudo-Hex.)	tabular	one perfect, one poor				I	cf. eudialyte
Trig.	rhombohedral, aggregates	one good, one poor		Z		I	cf. catapleite, lavenite, rosenbuschite, garnet, eucolite

COLOURLESS MINERALS	Pleochroism	Relief	δ	2V Dispersion	Extinction	Orientation	Mineral	Composition
Uniaxial +								
	W $\alpha=\beta>\gamma$	+L – +M	1 anom.	$0^o - 30^o$ r<v	$\simeq$ parallel	cleavage length fast	RIPIDOLITE (PROCHLORITE) KLEMENTITE	$(Mg,Al,Fe)_{12}\{(Si,Al)_8O_{20}\}(OH)_{16}$
	W – S $\alpha=\beta>\gamma$	+L – +M	1 – 2 (anom.)	$0^o - 60^o$ r<v	$0^o - 9^o$	cleavage length fast	CHLORITE (Unoxidised)	$(Mg,Al,Fe)_{12}\{(Si,Al)_8O_{20}\}(OH)_{16}$
	W $\alpha=\beta>\gamma$	+L – +M	2	31° r<v	$8^o - 10^o$	cleavage length fast	CORUNDOPHILITE	$(Mg,Al,Fe)_{12}\{(Si,Al)_8O_{20}\}(OH)_{16}$
	W $\omega>\epsilon$	+M	1 anom. (iso-tropic)		parallel	length slow	MELILITE	$(Ca,Na)_2\{(Mg,Fe^{+2},Al,Si)_3O_7\}$
		+M	1			length fast	ÅKERMANITE	$Ca_2\{MgSi_2O_7\}$
		H	1 anom.		parallel	length fast	VESUVIANITE (IDOCRASE) (incl. WILUITE)	$Ca_{10}(Mg,Fe)_2Al_4\{Si_2O_7\}_2\{SiO_4\}_5(OH,F)_4$
		H	5		straight		XENOTIME	YPO_4
		VH	4				STISHOVITE	SiO_2
	W $\epsilon>\omega$	VH	4 – 5		parallel	length slow	ZIRCON	$Zr\{SiO_4\}$
	W – S $\epsilon>\omega$	VH – E	5	$0^o - 38^o$ anom. strong	parallel, oblique to twin plane	length slow	CASSITERITE	SnO_2

System	Form	Cleavage	Twinning	Zoning	Alteration	Occurrence	Remarks
Mon. (Pseudo-Hex.)	tabular, scaly, vermicular, fan-shaped, aggregates	perfect basal				M, I, O	cf. clinochlore, penninite
Mon. (Pseudo-Hex.)	tabular, scaly, radiating pseudomorphs	perfect basal	simple, polysynthetic	Z		I, M, S, O	pleochroic haloes
Mon. (Pseudo-Hex.)	tabular, radiating	perfect basal				M	
Tet.	tabular, peg structure	moderate, single crack		Z	zeolites, carbonate	I	cf. zoisite, vesuvianite, apatite, nepheline
Tet.	tabular, peg structure	moderate, single crack		Z	zeolites, carbonate	I	cf. zoisite, vesuvianite, apatite, nepheline
Tet.	variable, prismatic, fibrous, granular, radial	imperfect	sector	Z		M, I, S	cf. zoisite, clinozoisite, apatite, grossularite, melilite, andalusite
Tet.	small, prismatic					I, M, S	inclusions, pleochroic haloes around grains cf. zircon, sphene, monazite
Tet.	fine-grained					M	impact metamorphism
Tet.	minute prisms	poor, absent		Z	metamict	I, S, M	cf. apatite
Tet.	subhedral, veinlets, diamond-shaped x-sect.	prismatic	geniculate, cyclic, common	Z		O, I, S	cf. sphalerite, rutile, sphene

COLOURLESS MINERALS	Pleochroism	Relief	δ	2V Dispersion	Extinction	Orientation	Mineral	Composition
Uniaxial −								
		−L	1				CRISTOBALITE	SiO_2
		−L	1				LEVYNE	$Ca\{Al_2Si_4O_{12}\}\cdot 6H_2O$
		−L	1	parallel			GARRONITE	$NaCa_{2.5}\{Al_6Si_{10}O_{32}\}\cdot 13\cdot 5H_2O$
		−L	1		parallel		KALSILITE-KALIOPHILITE	$K\{AlSiO_4\}$
		−L	1		parallel	length fast	DAVYNE	K-cancrinite
		−L	1	anom.	parallel	length fast	VISHNEVITE	$(Na,Ca,K)_{6-8}\{Al_6Si_6O_{24}\}(CO_3,SO_4,Cl)_{1-2}\cdot 1\text{-}5H_2O$
		−L	1 (anom.)	$0^o - 25^o$ r>v	$\alpha_o : (001)$ $5^o - 8^o$	optic plane $\perp (010)$	SANIDINE	$(K,Na)\{AlSi_3O_8\}$
		−L	1	$0^o - 32^o$	symmetrical		CHABAZITE	$Ca\{Al_2Si_4O_{12}\}\cdot 6H_2O$
		−L	1	$0^o -$ moderate	symmetrical		GMELINITE	$(Na_2,Ca)\{Al_2Si_4O_{12}\}\cdot 6H_2O$
		−L	1 − 3	anom.	parallel	length fast	CANCRINITE-VISHNEVITE	$(Na,Ca,K)_{6-8}\{Al_6Si_6O_{24}\}(CO_3,SO_4,Cl)_{1-2}\cdot 1\text{-}5H_2O$
		−L	3	anom.	parallel	length fast	CANCRINITE	$(Na,Ca,K)_{6-8}\{Al_6Si_6O_{24}\}(CO_3,SO_4,Cl)_{1-2}\cdot 1\text{-}5H_2O$
		−L − N	1		parallel	rectangular sect. length fast	NEPHELINE	$Na_3(Na,K)\{Al_4Si_4O_{16}\}$
		−L − N	1 anom.		parallel	length fast	APOPHYLLITE	$KFCa_4\{Si_8O_{20}\}\cdot 8H_2O$

System	Form	Cleavage	Twinning	Zoning	Alteration	Occurrence	Remarks
Tet. (Pseudo-Cub.)	minute square crystals, aggregates, spherulitic	curved fractures	interpenetrant			I	cf. tridymite
Trig.	tabular, sheaf-like aggregates, rhombohedral	indistinct, rhombohedral	interpenetrant			I	
Tet.						I	
Hex.	prismatic, hexagonal	two poor	rare	Z		I	cf. nepheline
Hex.	anhedral	two perfect				I	cf. natrodayne, cancrinite, microsommite
Hex.	anhedral	one perfect, one poor	polysynthetic, rare			I	cf. cancrinite, microsommite, nepheline
Mon.	clear, distinct, crystals, tabular, microlites	two, parting	Carlsbad, Baveno, Manebach	Z		I, M	perthitic cf. orthoclase, nepheline
Trig.	rhombohedral, approaching cube, anhedral, granular	one poor	interpenetrant	Z		I	basal section in six segments cf. gmelinite, analcite
Trig.	tabular, prismatic, rhombohedral, approaching cube, radiating	one good, one imperfect, parting	interpenetrant			I	cf. chabazite
Hex.	anhedral	one perfect, one poor	polysynthetic, rare			I	cf. microsommite, muscovite
Hex.	anhedral	one perfect, one poor	polysynthetic, rare			I	cf. vishnevite, microsommite, muscovite
Hex.	prismatic, hexagonal	two poor	rare	Z	zeolites, cancrinite, muscovite	I	inclusions cf. alkali feldspars, analcite, sodalite, leucite, scapolite
Tet.	prismatic, granular, anhedral - euhedral	two, one perfect				I, O	cf. zeolites

COLOURLESS MINERALS	Pleochroism	Relief	δ	2V Dispersion	Extinction	Orientation	Mineral	Composition
Uniaxial -								
		-L - +M	5		symmetrical to cleavage		CALCITE	$CaCO_3$
		-L - +M	5		symmetrical to cleavage		DOLOMITE	$CaMg(CO_3)_2$
	W $\omega>\epsilon$ rare	-L - H	5		symmetrical to cleavage		MAGNESITE	$MgCO_3$
		N	1		parallel		EUCRYPTITE	$LiAlSiO_4$
	W $\omega>\epsilon$	N - VH	5		symmetrical to cleavage		RHODOCHROSITE	$MnCO_3$
	W	+L	1		parallel	long. sect. length fast, x-sect. length slow	BERYL	$Be_3Al_2\{Si_6O_{18}\}$
		+L	1		parallel	cleavage length fast	MARIALITE (SCAPOLITE)	$Na_4\{Al_3Si_9O_{24}\}Cl$
	W $\gamma=\beta>\alpha$	+L anom.	1	$0° - 40°$ r>v	$\simeq$ parallel	cleavage length slow	PENNINITE	$(Mg,Al,Fe)_{12}\{(Si,Al)_8O_{20}\}(OH)$
		+L	1 - 4		parallel	cleavage length fast	SCAPOLITE (MARIALITE, DIPYRE, MIZZONITE, MEIONITE, WERNERITE)	$(Na,Ca,K)_4\{Al_3(Al,Si)_3Si_6O_{24}\}$ (Cl,CO_3,SO_4,OH)
		+L	3 - 4		parallel	cleavage length fast	MEIONITE (SCAPOLITE)	$Ca_4\{Al_6Si_6O_{24}\}CO_3$
	W $\epsilon>\omega$	+L - +M	1 (iso-tropic)				EUCOLITE	$(Na,Ca,Fe)_6Zr$ $\{(Si_3O_9)_2\}(OH,F,Cl)$

System	Form	Cleavage	Twinning	Zoning	Alteration	Occurrence	Remarks
Trig.	anhedral, oolitic, spherulitic	rhombohedral	polysynthetic, // long diagonal			S, M, I, O	twinkling cf. rhombohedral carbonates
Trig.	rhombohedral	rhombohedral	polysynthetic, // long and short diagonals	Z	huntite	S, M, I	twinkling cf. rhombohedral carbonates
Trig.	microcrystalline, subhedral	rhombohedral			huntite	S, I, M	twinkling cf. rhombohedral carbonates
Hex.	anhedral	one distinct				I	cf. spodumene
Trig.	rhombohedral, aggregates, bands	rhombohedral	polysynthetic, rare	Z	manganese oxide	O, I, M	twinkling cf. rhombohedral carbonates
Hex.	prismatic, inclusions zoned	imperfect		Z	kaolin	I, M	liquid inclusions cf. apatite, quartz
Tet.	columnar aggregates	two good			muscovite, calcite, zeolites	M	cf. feldspar, quartz, cancrinite
Mon. (Pseudo-Hex.)	tabular, vermicular, radiating, pseudomorphs	perfect basal	simple, pennine law	Z		I, M	cf. clinochlore, prochlorite
Tet.	columnar aggregates	two good			muscovite, calcite, zeolites	M	cf. feldspar, quartz, cancrinite
Tet.	columnar aggregates	two good			muscovite, ill-defined fibrous aggregates	M	cf. scapolite, feldspar, quartz, cancrinite
Trig.	rhombohedral aggregates	one good, one poor		Z		I	some zones isotropic cf. catapleite, rosenbuschite, lavenite, garnet, eudialyte

COLOURLESS MINERALS	Pleochroism	Relief	δ	2V Dispersion	Extinction	Orientation	Mineral	Composition
Uniaxial −								
	W − S $\gamma=\beta>\alpha$	+L − +M	1 − 2 anom.	0° − 20° r>v	0° − small	cleavage length slow	CHLORITE (Oxidised)	$(Mg,Al,Fe)_{12}\{(Si,Al)_8O_{20}\}(OH)_1$
	W − S $\gamma=\beta>\alpha$	+L − +M	1 − 2 anom.	0° − 20° r>v	0° − small	cleavage length slow	CHLORITE (Unoxidised)	$(Mg,Al,Fe)_{12}\{(Si,Al)_8O_{20}\}(OH)_1$
	M $\gamma\geqslant\beta>\alpha$ $\alpha>\beta=\gamma$	+L − +M	3 − 4	0° − 15° r<v	0° − 5°	cleavage length slow	PHLOGOPITE	$K_2(Mg,Fe^{+2})_6\{Si_6Al_2O_{20}\}(OH,F)_4$
		+L − VH	5			symmetrical to cleavage	SIDERITE	$FeCO_3$
	W $\varepsilon>\omega$	+M	1		parallel	length fast, tabular length slow	APATITE (incl. DAHLLITE, FRANCOLITE)	$Ca_5(PO_4)_3(OH,F,Cl)$
	W $\omega>\varepsilon$	+M	1 anom. (isotropic)		parallel	length slow	MELILITE	$(Ca,Na)_2\{(Mg,Fe^{+2},Al,Si)_3O_7\}$
	W $\gamma=\beta>\alpha$	+M	1 anom.	0° − small r>v	≃ parallel	cleavage length slow	RIPIDOLITE (PROCHLORITE)	$(Mg,Al,Fe)_{12}\{(Si,Al)_8O_{20}\}(OH)_1$
	W $\gamma=\beta>\alpha$	+M	1 anom.	0° − 20°	small	cleavage length slow	DAPHNITE	$(Mg,Al,Fe)_{12}\{(Si,Al)_8O_{20}\}(OH)_1$
		+M	2		parallel	length slow	GEHLENITE (MELILITE)	$Ca_2\{Al_2SiO_7\}$
	W	+M	2	0° − 20°	small	cleavage length slow	THURINGITE	$(Mg,Al,Fe)_{12}\{(Si,Al)_8O_{20}\}(OH)_1$
	S $\omega>\varepsilon$	+M	2 − 3		parallel	length fast	ELBAITE (TOURMALINE)	$Na(Li,Al)_3Al_6B_3Si_6O_{27}(OH,F)_4$

System	Form	Cleavage	Twinning	Zoning	Alteration	Occurrence	Remarks
Mon. (Pseudo- Hex.)	tabular, scaly, radiating, pseudomorphs	perfect basal	simple, polysynthetic	Z		I, M, S, O	pleochroic haloes
Mon. (Pseudo- Hex.)	tabular, scaly, radiating pseudomorphs	perfect basal	simple, polysynthetic	Z		I, M, S, O	pleochroic haloes
Mon.	tabular, flakes, plates, pseudo-hex.	perfect basal	inconspicuous	Z colour zoning		I, M	inclusions common, birds-eye maple structure cf. muscovite, lepidolite
Trig.	anhedral, aggregates, oolitic, spherulitic, colloform	rhombohedral	polysynthetic, // long diagonal, uncommon		brown spots	S, O, I, M	twinkling, brown stain around borders and along cleavage cracks cf. rhombohedral carbonates
Hex.	small, prismatic, hexagonal	poor basal				I, S, M, O	cf. beryl, topaz, dahllite
Tet.	tabular, peg structure	moderate, single crack		Z	zeolites, carbonate	I	cf. zoisite, vesuvianite, apatite, nepheline
Mon. (Pseudo- Hex.)	tabular, scaly, vermicular, fan-shaped aggregates	perfect basal				M, I, O	cf. clinochlore, penninite
Mon. (Pseudo- Hex.)	concentric aggregates, fibrous plates	perfect basal				O	
Tet.	tabular, peg structure	moderate, single crack		Z	zeolites, carbonate	I	cf. zoisite, vesuvianite, apatite, nepheline
Mon. (Pseudo- Hex.)	tabular, radiating	perfect basal				O	
Trig.	prismatic, radiating	fractures at 90°	rare	Z		I, M	cf. tourmaline group, apatite

COLOURLESS MINERALS	Pleochroism	Relief	δ	2V Dispersion	Extinction	Orientation	Mineral	Composition
Uniaxial –								
	S $\omega>\epsilon$	+M	3		parallel	length fast	DRAVITE (TOURMALINE)	$NaMg_3Al_6B_3Si_6O_{27}(OH,F)_4$
	W	+M	3		parallel		ZUSSMANITE	$K(Fe,Mg,Mn)_{13}(Si,Al)_{18}O_{42}(OH)_{14}$
		+M	very strong				HUNTITE	$Mg_3Ca(CO_3)_4$
		+M – H	5		symmetrical to cleavage		ANKERITE	$Ca(Mg,Fe^{+2},Mn)(CO_3)_2$
	H	1 anom.				length fast	VESUVIANITE (IDOCRASE)	$Ca_{10}(Mg,Fe)_2Al_4\{Si_2O_7\}_2\{SiO_4\}_5(OH,F)_4$
	W $\omega>\epsilon$	H	1	$0^{\circ}-30^{\circ}$	parallel	tabular length slow, prismatic length fast	CORUNDUM	$\alpha-Al_2O_3$
Biaxial +								
		–L	1	$0^{\circ}-$ moderate	symmetrical		GMELINITE	$(Na_2,Ca)\{Al_2Si_4O_{12}\}\cdot6H_2O$
		–L	1	$0^{\circ}-32^{\circ}$	symmetrical		CHABAZITE	$Ca\{Al_2Si_4O_{12}\}\cdot6H_2O$
		–L	1	$0^{\circ}-48^{\circ}$ variable r>v	6° variable	cleavage length fast	HEULANDITE	$(Ca,Na_2)\{Al_2Si_7O_{18}\}\cdot6H_2O$
		–L	1	$0^{\circ}-63^{\circ}$			ISOSANIDINE	$(K,Na)\{AlSi_3O_8\}$
		–L	1	small, anom.	wavy		LEUCITE	$K\{AlSi_2O_6\}$
		–L	1	$40^{\circ}-90^{\circ}$			TRIDYMITE	SiO_2

System	Form	Cleavage	Twinning	Zoning	Alteration	Occurrence	Remarks
Trig.	prismatic	fractures at 90°	rare	Z		M	cf. tourmaline group, chondrodite
Hex.	tabular	one perfect				M	
Trig.	compact, porous					S	alteration of dolomite and magnesite
Trig.	rhombohedral	rhombohedral	polysynthetic	Z		S, O, M	twinkling cf. rhombohedral carbonates
Tet.	variable, prismatic, fibrous, granular, radial	imperfect	sector	Z		M, I, S	cf. zoisite, clinozoisite, apatite, grossularite, melilite, andalusite
Trig.	tabular, prismatic, six-sided x-sect.	parting	simple, lamellar seams	Z colour banding		M, I	inclusions, cf. sapphirine
Trig.	tabular, prismatic, rhombohedral, approaching cube, radiating	one good, one imperfect, parting	interpenetrant			I	cf. chabazite
Trig.	rhombohedral, approaching cube, anhedral, granular	one poor	interpenetrant	Z		I	basal section in six segments cf. gmelinite, analcite
Pseudo-Mon.	tabular, aggregates	one perfect				I, M	cf. stilbite, clinoptilite, epistilbite, mordenite brewsterite
Mon.	clear distinct crystals, tabular	two	Carlsbad			I	cf. sanidine
Tet. (Pseudo-Cub.)	always euhedral, octagonal	poor	polysynthetic, complex	Z		I	inclusions common cf. analcite, microcline
Orth.	tabular, radiating, aggregates	poor	wedge-shaped			I, M	cf. cristobalite

COLOURLESS MINERALS	Pleochroism	Relief	δ	2V Dispersion	Extinction	Orientation	Mineral	Composition
Biaxial +								
		-L	1	50°	parallel	length slow	FERRIERITE	$(Na,K)_4Mg_2\{Al_6Si_{30}O_{72}\}$ $(OH)_2 \cdot 18H_2O$
		-L	1	$65^{\circ} - 73^{\circ}$	38°		DACHIARDITE	$(\frac{1}{2}Ca,Na,K)_5\{Al_5Si_{19}O_{48}\} \cdot 12H_2O$
		-L	1	70°			YUGAWARALITE	$Ca\{Al_2Si_5O_{14}\} \cdot 4H_2O$
		-L	1	$76^{\circ} - 90^{\circ}$	parallel	length slow	MORDENITE	$(Na_2,K_2,Ca)\{Al_2Si_{10}O_{24}\} \cdot 7H_2O$
		-L	1 (iso- tropic) r>v	80°	8°	length fast/slow	MESOLITE	$Na_2Ca_2\{Al_2Si_3O_{10}\}_3 \cdot 8H_2O$
		-L	1	80° weak x- dispersion	$63^{\circ} - 67^{\circ}$		HARMOTOME	$Ba\{Al_2Si_6O_{16}\} \cdot 6H_2O$
		-L	1	$\simeq 90^{\circ}$ variable			ISOORTHOCLASE	$(K,Na)\{AlSi_3O_8\}$
		-L	1 - 2	$42^{\circ} - 75^{\circ}$ r>v	parallel	length fast/slow	THOMSONITE	$NaCa_2\{(Al,Si)_5O_{10}\}_2 \cdot 6H_2O$
		-L	1 - 2	$60^{\circ} - 80^{\circ}$ r<v	$46^{\circ} - 85^{\circ}$	length slow	PHILLIPSITE	$(\frac{1}{2}Ca,Na,K)_3\{Al_3Si_5O_{16}\} \cdot 6H_2O$
		-L	1 - 2	$64^{\circ} - 90^{\circ}$	$15^{\circ} - 20^{\circ}$	cleavage length fast	ISOMICROCLINE	$(K,Na)\{AlSi_3O_8\}$
		-L	1 - 2	$77^{\circ} - 82^{\circ}$ r<v	$12^{\circ} - 19^{\circ}$ (in albite twins)		ALBITE	$Na\{AlSi_3O_8\}$
		-L	2	47° r>v	22°	length slow	BREWSTERITE	$(Sr,Ba,Ca)\{Al_2Si_6O_{16}\} \cdot 5H_2O$
		-L	2	58° r>v	52°	cleavage length fast/slow	GYPSUM (SELENITE)	$CaSO_4 \cdot 2H_2O$

System	Form	Cleavage	Twinning	Zoning	Alteration	Occurrence	Remarks
Orth.	tabular, laths, radiating	one perfect				I	cf. zeolites
Mon.	prismatic	two perfect	cyclic, sector			I	
Mon.						I	
Orth.	tabular, acicular, fibrous	one perfect				I	
Mon. (Pseudo-Orth.)	fibrous, aggregates, needles	two perfect at 90°	universal but inconspicuous			I	cf. natrolite, thomsonite, scolecite
Mon. (Orth.)	groups, radiating	two, one good	complex, sector, interpenetrant, cruciform			I, O	
Mon.	anhedral – subhedral, phenocrysts	three	Carlsbad			I	cf. orthoclase
Orth. (Pseudo-Tet.)	fibrous, columnar, radiating	two at 90°				I	cf. natrolite, scolecite, mesolite, cancrinite
Mon. (Orth.)	groups, radiating	two at 90°	complex, sector, interpenetrant			I, S	
Tr.	anhedral – subhedral	two, parting	albite, Carlsbad, pericline, tartan	Z	cloudy, sericite, kaolinite	I, M, S	cf. microcline, orthoclase, albite, plagioclase
Tr.	anhedral – euhedral, laths	three	albite, Carlsbad, pericline, complex	Z	sericite, calcite, kaolinite, zeolites	I, S, M	peristerite, perthitic, antiperthitic cf. cordierite
Mon.	prismatic	one perfect		Z		I	cf. heulandite, epistilbite
Mon.	aggregates, fibrous, acicular, lozenge-shaped	three perfect	polysynthetic			S	cf. celestine, anhydrite

COLOURLESS MINERALS	Pleochroism	Relief	δ	2V Dispersion	Extinction	Orientation	Mineral	Composition
Biaxial +								
		$-L$	2	$58^{\circ} - 64^{\circ}$ r<v	parallel, symmetrical	length slow	NATROLITE	$Na_2\{Al_2Si_3O_{10}\}\cdot 2H_2O$
		$-L$	2	$58^{\circ} - 64^{\circ}$	$7\frac{1}{2}^{\circ}$	length slow	METANATROLITE	$Na_2\{Al_2Si_3O_{10}\}$
		$-L$	2	$82^{\circ} - 84^{\circ}$ r>v	$2^{\circ} - 8^{\circ},$ $24^{\circ} - 30^{\circ}$	length fast	PETALITE	$Li\{AlSi_4O_{10}\}$
		$-L - N$	1	$82^{\circ} - 90^{\circ}$ r>v	$0^{\circ} - 12^{\circ}$ (in albite twins)		OLIGOCLASE	$Na\{AlSi_3O_8\}-Ca\{Al_2Si_2O_8\}$
		$-L - +L$	1	$76^{\circ} - 90^{\circ}$ r$\lessgtr$v	$0^{\circ} - 70^{\circ}$ (in albite twins)		PLAGIOCLASE	$Na\{AlSi_3O_8\}-Ca\{Al_2Si_2O_8\}$
	M	$-L - +L$	1 - 2	$65^{\circ} - 90^{\circ}$ r<v	parallel		CORDIERITE	$Al_3(Mg,Fe^{+2})_2\{Si_5AlO_{18}\}$
	N	1		$0^{\circ} - 10^{\circ}$ (uniaxial)	parallel	length slow	QUARTZ	SiO_2
	W $\alpha=\beta>\gamma$	$+L$	1 anom.	$0^{\circ} - 20^{\circ}$ r<v	$\simeq$ parallel	cleavage length fast	PENNINITE	$(Mg,Al,Fe)_{12}\{(Si,Al)_8O_{20}\}(OH)$
	W $\alpha=\beta>\gamma$	$+L$	1	$\simeq 20^{\circ}$ r<v	$\simeq$ parallel	cleavage length fast	SHERIDANITE	$(Mg,Al,Fe)_{12}\{(Si,Al)_8O_{20}\}(OH)$
		$+L$	1	$52^{\circ} - 80^{\circ}$ r<v	$14^{\circ} - 20^{\circ}$ undulatory	length slow	DICKITE	$Al_4\{Si_4O_{10}\}(OH)_8$

System	Form	Cleavage	Twinning	Zoning	Alteration	Occurrence	Remarks
Orth. (Pseudo-Tet.)	prismatic, fibrous, needles, radiating	// length				I	cf. scolecite, thomsonite
Orth. (Pseudo-Tet.)	prismatic, fibrous, needles, radiating	// length	sector			I	cf. natrolite
Mon.	prismatic	two good	polysynthetic			I	cf. quartz, feldspar
Tr.	anhedral - euhedral, laths, perthite	three	albite, Carlsbad, pericline, complex	Z	sericite, calcite, kaolinite, zeolites	I, M	peristerite, antiperthitic cf. quartz, cordierite
Tr.	anhedral - euhedral, laths	three	albite, Carlsbad, pericline, complex	Z	sericite, calcite, kaolinite, zeolites, saussurite	I, M, S	peristerite, schiller, antiperthitic cf. cordierite
Orth. (Pseudo-Hex.)	pseudo-hex., anhedral	moderate to absent	simple, polysynthetic, cyclic, sector, interpenetrant		pinite, talc	M, I	pleochroic haloes, inclusions (including opaque dust) common cf. quartz, plagioclase
Trig.	variable, anhedral, flamboyant, intergrowths		rare in thin sections			S, I, M, O	Boehm lamellae cf. cordierite, beryl, scapolite, feldspar
Mon.	tabular, vermicular, radiating, pseudomorphs	perfect basal	simple, pennine law	Z		I, M	pleochroic haloes cf. clinochlore, prochlorite
Mon.	tabular, spherulites	perfect basal	simple			I	
Mon.	flakes, pseudo-hex., radial, aggregates	perfect basal				O, I, S	cf. kaolinite, nacrite

COLOURLESS MINERALS	Pleochroism	Relief	δ	2V Dispersion	Extinction	Orientation	Mineral	Composition
Biaxial +								
		+L	1	$76^{\circ} - 86^{\circ}$ r<v	$27\frac{1}{2}^{\circ} - 39^{\circ}$ (in albite twins)		LABRADORITE	$Na\{AlSi_3O_8\}-Ca\{Al_2Si_2O_8\}$
	W $\alpha=\beta>\gamma$	+L	1 - 2	$0^{\circ} - 40^{\circ}$ r<v	$0^{\circ} - 9^{\circ}$	cleavage length fast	CLINOCHLORE	$(Mg,Al,Fe)_{12}\{(Si,Al)_8O_{20}\}(OH)$
		+L	2	small			AMESITE	$(Mg_4Al_2)(Si_2Al_2)O_{10}(OH)_8$
		+L	2	$83^{\circ} - 90^{\circ}$	$\alpha : z$ $3^{\circ} - 5^{\circ}$		CELSIAN	$Ba\{Al_2Si_2O_8\}$
		+L	2 - 3	$0^{\circ} - 30^{\circ}$ r<v	3°	cleavage fast	CATAPLEITE	$(Na,Ca)_2Zr\{Si_3O_9\}\cdot 2H_2O$
		+L	3	$0^{\circ} - 40^{\circ}$ r>v	21°	elongate twinned crystals, length slow	GIBBSITE	$Al(OH)_3$
	W $\alpha=\beta>\gamma$	+L	3	$44^{\circ} - 50^{\circ}$ r>v	parallel		NORBERGITE	$Mg(OH,F)_2\cdot Mg_2SiO_4$
		+L	4	$42^{\circ} - 44^{\circ}$ r<v	parallel		ANHYDRITE	$CaSO_4$
	W $\alpha=\beta>\gamma$	+L - +M	1 anom.	$0^{\circ} - 30^{\circ}$ r<v	$\simeq$ parallel	cleavage length fast	RIPIDOLITE (PROCHLORITE)- KLEMENTITE	$(Mg,Al,Fe)_{12}\{(Si,Al)_8O_{20}\}(OH)$
		+L - +M	1	$54^{\circ} - 64^{\circ}$	unsymmetrical		COESITE	SiO_2
	W - S $\alpha=\beta>\gamma$	+L - +M	1 - 2	$0^{\circ} - 60^{\circ}$ r<v	$0^{\circ} - 9^{\circ}$	cleavage length fast	CHLORITE (Unoxidised)	$(Mg,Al,Fe)_{12}\{(Si,Al)_8O_{20}\}(OH)$

System	Form	Cleavage	Twinning	Zoning	Alteration	Occurrence	Remarks
Tr.	anhedral - euhedral, laths	three	albite, Carlsbad, pericline, complex	Z	sericite, calcite, kaolinite, zeolites, saussurite	I, M	schiller, antiperthitic cf. cordierite
Mon.	tabular, fibrous, pseudo-hex.	perfect basal	polysynthetic			M	pleochroic haloes cf. penninite, prochlorite, leuchtenbergite, katschubeite
Mon. (Pseudo-Hex.)	tabular	perfect basal				M	
Mon.	prismatic	three	Carlsbad, Baveno, Manebach			O, I	cf. orthoclase, hyalophane
Mon. (Pseudo-Hex.)	tabular	one perfect, one poor				I	cf. eudialyte
Mon.	aggregates, tabular, very small, stalactitic	one perfect	polysynthetic			S, M, I	cf. chalcedony, dahllite, muscovite, kaolinite, boehmite, diaspore
Orth.	massive					M, O	cf. humite group
Orth.	aggregates, anhedral - subhedral	rectangular, three	polysynthetic, two at $83\frac{1}{2}°$		gypsum	S	twinkling cf. gypsum, barytes
Mon.	tabular, scaly, vermicular, fan-shaped aggregates	perfect basal				M, I, O	cf. clinochlore, penninite
Mon. (Pseudo-Hex.)	pseudo-hex. plates, aggregates					M	impact metamorphism
Mon.	tabular, scaly, radiating, pseudomorphs	perfect basal	simple, polysynthetic	Z		I, M, S, O	pleochroic haloes

COLOURLESS MINERALS	Pleochroism	Relief	δ	2V Dispersion	Extinction	Orientation	Mineral	Composition
Biaxial +								
	W $\alpha=\beta>\gamma$	+L – +M	2	31° r<v	8° – 10°	cleavage length fast	CORUNDOPHILITE	$(Mg,Al,Fe)_{12}\{(Si,Al)_8O_{20}\}(OH)$
	W $\gamma>\beta>\alpha$	+M	1	50° r<v	parallel	length slow	CELESTINE	$SrSO_4$
		+M	1	63° – 64°	15°		RANKINITE	$Ca_3\{Si_2O_7\}$
	N – W	+M	1 – 2	48° – 68° r>v	parallel, symmetrical	cleavage fast	TOPAZ	$Al_2\{SiO_4\}(OH,F)_2$
	W $\gamma>\beta>\alpha$	+M	2	37° r<v	parallel	best cleavage slow	BARYTES	$BaSO_4$
	W	+M	2	67° – 70° r>v	33° – 40°	extinction nearest cleavage slow	JADEITE	$NaAl\{Si_2O_6\}$
		+M	2	$\simeq 80^\circ$	parallel	length fast	BOEHMITE	γ-AlO(OH)
	W $\gamma>\beta=\alpha$	+M	2 – 3	45° – 61° r>v	parallel, symmetrical	length slow	MULLITE	$3Al_2O_3 \cdot 2SiO_2$
	W	+M	2 – 3	58° – 68° r<v	22° – 26°	extinction nearest cleavage slow	SPODUMENE (HIDDENITE, KUNZITE)	$LiAl\{Si_2O_6\}$
	W $\gamma=\beta>\alpha$ $\gamma>\beta=\alpha$	+M	2 – 3	68° – 90° r$\lessgtr$v	parallel, symmetrical	length slow	ANTHOPHYLLITE (Fe-rich)	$(Mg,Fe^{+2})_7\{Si_8O_{22}\}(OH,F)_2$
	W $\gamma>\alpha$	+M	3	21° – 30° r>v	parallel	length slow	SILLIMANITE	Al_2SiO_5
		+M	3		parallel	length slow	FIBROLITE (SILLIMANITE)	Al_2SiO_5
	W	+M	3	50° – 62° r>v	38° – 48°	extinction nearest cleavage slow	DIOPSIDE (SALITE)	$Ca(Mg,Fe)\{Si_2O_6\}$

System	Form	Cleavage	Twinning	Zoning	Alteration	Occurrence	Remarks
Mon.	tabular, radiating	perfect basal				M	
Orth.	tabular, fibrous	three				S	cf. barytes
Mon.	granular					M	cf. tilleyite, spurrite, larnite, merwinite
Orth.	prismatic, aggregates	one perfect				I, M, S	cf. quartz, andalusite
Orth.	granular aggregates	three	polysynthetic			I, S	cf. celestite
Mon.	prismatic	two at 87°	simple, polysynthetic	Z	tremolite-actinolite	M	cf. nephrite, diopside, omphacite, fassaite
Orth.	tabular, fine-grained, oolitic	one good				S	cf. diaspore, gibbsite
Orth.	prismatic, needles, fibres, square x-sect.	one distinct				M	rare cf. sillimanite
Mon.	tabular	two at 87°	simple		muscovite, cymatolite, kaolinite, eucryptite	I	cf. aegirine-augite, diopside
Orth.	bladed, prismatic, fibrous, asbestiform	two at 54$\frac{1}{2}$$^{\circ}$			talc	M	cf. gedrite, tremolite, cummingtonite, holmquistite
Orth.	needles, slender prisms, fibrous, rhombic x-sect., faserkiesel	one, // long diagonal of x-sect.				M	cf. andalusite, mullite
Orth.	fibrous, felted, faserkiesel					M	stained brown in fibrous mats
Mon.	short prismatic	two at 87°	simple, polysynthetic	Z	tremolite-actinolite	M, I	cf. hedenbergite, tremolite, omphacite, wollastonite, epidote

COLOURLESS MINERALS	Pleochroism	Relief	δ	2V Dispersion	Extinction	Orientation	Mineral	Composition
Biaxial +								
		+M	3 (anom.)	$65° - 69°$ r>v	parallel, often wavy	cleavage length fast	PREHNITE	$Ca_2Al\{AlSi_3O_{10}\}(OH)_2$
	W $\alpha>\beta=\gamma$	+M	3	$65° - 84°$ r>v	parallel		HUMITE	$Mg(OH,F)\cdot 3Mg_2SiO_4$
	M - S $\gamma\geq\beta>\alpha$	+M	3	$67° - 90°$ r>v	$26°$	length slow	PARGASITE	$Na,Ca_2Mg_4Al_3Si_6O_{22}(OH)_2$
	W $\gamma>\beta=\alpha$	+M	3	$68° - 78°$ r>v	$28°$	length slow	ROSENBUSCHITE	$(Ca,Na,Mn)_3(Zr,Ti,Fe^{+3})$ $\{SiO_4\}_2(F,OH)$
	W $\alpha>\beta=\gamma$	+M	3	$71° - 85°$ r>v	$22° - 31°$	extinction nearest twin plane fast	CHONDRODITE	$Mg(OH,F)_2\cdot 2Mg_2SiO_4$
	W $\alpha>\gamma$	+M	3	$76° - 87°$ r>v	parallel, symmetrical	length slow, long diagonal of rhombic sect. slow	LAWSONITE	$CaAl_2(OH)_2\{Si_2O_7\}H_2O$
		+M	3 - 4	$35° - 63°$ r>v	$10° - 19°$	length slow	PECTOLITE (incl. SCHIZOLITE, SERANDITE)	$Ca_2NaH\{SiO_3\}_3$
	W $\gamma>\beta>\alpha$	+M	3 - 4	$65° - 90°$ Mg r<v Fe r>v	$15° - 21°$	length slow	CUMMINGTONITE	$(Mg,Fe^{+2})_7\{Si_8O_{22}\}(OH)_2$
		+M	4	$0° - 6°$	$\alpha : z$ $9°$		PSEUDOWOLLASTONITE	$\beta-CaSiO_3$
		+M	4	$82° - 90°$ r>v	parallel	cleavage length slow	FORSTERITE	Mg_2SiO_4
		+M	4	$85° - 89°$ r<v	$24° - 26°$		TILLEYITE	$Ca_3\{Si_2O_7\}\cdot 2CaCO_3$
	M	+M - H	1 anom.	$0° - 30°$ r>v	parallel	length fast	α - ZOISITE	$Ca_2Al\cdot Al_2O\cdot OH\{Si_2O_7\}\{SiO_4\}$

System	Form	Cleavage	Twinning	Zoning	Alteration	Occurrence	Remarks
Orth.	tabular, prismatic, sheaf-like aggregates, bow-tie structure	two	fine polysynthetic, in two directions at right angles	optical sectors, hour-glass		I, M	cf. lawsonite, topaz, wollastonite, datolite
Orth.	tabular, rounded	poor				M, O	cf. humite group, olivine
Mon.	prismatic	two at 56°, partings	simple, polysynthetic			M, I	cf. hornblende, hastingsite, cummingtonite
Tr.	prismatic, fibres, needles	one perfect, two poor				I	cf. låvenite
Mon.	rounded	poor	simple, polysynthetic, twin seams			M, O	cf. humite group, olivine, staurolite
Orth.	rectangular, rhombic	two perfect, one imperfect	simple, polysynthetic			M	cf. zoisite, prehnite, pumpellyite, andalusite, scapolite
Tr.	elongate, stellate, needles	two at 85°	rare		stevensite	M, I	cf. wollastonite
Mon.	prismatic, subradiating, fibrous, asbestiform	two at 55°	simple, polysynthetic			M, I	cf. tremolite, actinolite, grunerite, anthophyllite
Tr.	prismatic, anhedral	one	polysynthetic			M	rare
Orth.	anhedral - euhedral, rounded	uncommon, irregular fractures	uncommon	Z	chlorite, antigorite, serpentine, iddingsite, bowlingite	I, M	deformation lamellae cf. diopside, augite, humite group, epidote
Mon.	granular	one perfect, two poor	polysynthetic			M	cf. merwinite, rankinite, larnite, spurrite
Orth.	columnar aggregates, euhedral	one perfect, one imperfect	polysynthetic, rare	Z		M	cf. clinozoisite, epidote

COLOURLESS MINERALS	Pleochroism	Relief	δ	2V Dispersion	Extinction	Orientation	Mineral	Composition
Biaxial +								
	M	+M – H	1 normal	0° – 60° r<v	parallel	length fast/slow	β – ZOISITE (Fe-rich)	$Ca_2Al \cdot Al_2O \cdot OH\{Si_2O_7\}\{SiO_4\}$
		+M – H	1 – 2 anom.	14° – 90° r$\lessgtr$v	0° – 7°	length fast/slow	CLINOZOISITE	$Ca_2Al \cdot Al_2O \cdot OH\{Si_2O_7\}\{SiO_4\}$
	W β>α=γ	+M – H	2	26° – 85° r<v	4° – 32°	length slow	PUMPELLYITE	$Ca_4(Mg,Fe^{+2})(Al,Fe^{+3})_5O(OH)_3\{Si_2O_7\}_2\{SiO_4\}_2 \cdot 2H_2O$
		+M – H	2	55° – 90° r<v	parallel	length slow	ENSTATITE	$(Mg,Fe^{+2})\{SiO_3\}$
	N – W	+M – H	2 – 3	25° – 83° r>v	35° – 48°	extinction nearest cleavage fast	AUGITE-FERROAUGITE (incl. SALITE)	$(Ca,Na,Mg,Fe^{+2},Mn,Fe^{+3},Al,Ti)_2\{(Si,Al)_2O_6\}$
	W	+M – H	2 – 3	51° – 62°	35° – 48°	extinction nearest cleavage fast	FASSAITE	$(Ca,Na,Mg,Fe^{+2},Mn,Fe^{+3},Al,Ti)_2\{(Si,Al)_2O_6\}$
	W	+M – H	2 – 3	60° – 67°	39° – 41°	extinction nearest cleavage fast	OMPHACITE	$(Ca,Na,Mg,Fe^{+2},Mn,Fe^{+3},Al,Ti)_2\{(Si,Al)_2O_6\}$
	W γ=β>α γ>β=α	+M – H	2 – 3	68° – 90° r$\lessgtr$v	parallel, symmetrical	length slow	GEDRITE	$(Mg,Fe^{+2})_5Al_2\{Si_6Al_2O_{22}\}(OH,F)$
	N – M γ=α>β β>α=γ	+M – H	3	0° – 30° r$\lessgtr$v	37° – 44°	extinction nearest cleavage slow	PIGEONITE	$(Mg,Fe^{+2},Ca)(Mg,Fe^{+2})\{Si_2O_6\}$
	W	+M – VH	4	48° – 90° r$\lessgtr$v	parallel	cleavage length slow	OLIVINE	$(Mg,Fe)_2\{SiO_4\}$
	H		1 anom.	5° – 65° strong		length fast	VESUVIANITE (IDOCRASE) (incl. WILUITE)	$Ca_{10}(Mg,Fe)_2Al_4\{Si_2O_7\}_2\{SiO_4\}_5(OH,F)_4$

System	Form	Cleavage	Twinning	Zoning	Alteration	Occurrence	Remarks
Orth.	columnar aggregates, euhedral	one perfect, one imperfect	polysynthetic, rare	Z		M	cf. clinozoisite, epidote, diopside, augite
Mon.	columnar, six-sided, x-sect.	one perfect	polysynthetic, uncommon	Z		M	cf. epidote, zoisite, diopside, augite
Mon.	fibrous, needles, bladed, radial aggregates	two moderate	common	Z		M	cf. clinozoisite, zoisite, epidote, lawsonite
Orth.	prismatic, anhedral - euhedral, kelyphitic borders	two at 88°, parting	rare			I, M	often schillerized, Fe-enstatite = bronzite
Mon.	prismatic	two at 87°, parting	simple, polysynthetic, twin seams	Z	amphibole	I, M	herringbone structure, exsolution lamellae cf. pigeonite, diopside, epidote
Mon.	prismatic	two at 87°	simple, polysynthetic	Z	amphibole	I, M	cf. augite, diopside, omphacite, fassaite
Mon.	prismatic	two at 87°	simple, polysynthetic			M	inclusions cf. fassaite, diopside, jadeite
Orth.	bladed, prismatic, fibrous, asbestiform	two at $54\frac{1}{2}$°			talc	M	cf. Fe-anthophyllite, tremolite, cummingtonite, grunerite, zoisite
Mon.	prismatic, anhedral, overgrowths	two at 87°, parting	simple, polysynthetic	Z		I	exsolution lamellae cf. augite, olivine
Orth.	anhedral - euhedral, rounded	uncommon, one moderate, irregular fractures	uncommon, vicinal	Z	chlorite, antigorite, serpentine, iddingsite, bowlingite	I, M	deformation lamellae cf. diopside, augite, humite group, epidote
Tet.	variable, prismatic, fibrous, granular, radial	imperfect	sector	Z		M, I, S	cf. zoisite, clinozoisite, apatite, grossularite, melilite, andalusite

COLOURLESS MINERALS	Pleochroism	Relief	δ	2V Dispersion	Extinction	Orientation	Mineral	Composition
Biaxial +								
	W $\beta>\gamma>\alpha$	H	1	$50^\circ - 66^\circ$ r<v	$6^\circ - 9^\circ$	length slow	SAPPHIRINE	$(Mg,Fe)_2Al_4O_6\{SiO_4\}$
	W – M $\beta>\alpha>\gamma$	H	1 – 2 anom.	$45^\circ - 68^\circ$ r>v	$2^\circ - 30^\circ$	length fast	CHLORITOID (OTTRELITE)	$(Fe^{+2},Mg,Mn)_2(Al,Fe^{+3})Al_3O_2\{SiO_4\}_2(OH)_4$
		H	2	$35^\circ - 46^\circ$ r>v	64°		PYROXMANGITE	$(Mn,Fe)\{SiO_3\}$
	W $\alpha>\gamma$	H	2	$61^\circ - 76^\circ$ r<v	$5^\circ - 25^\circ$		RHODONITE	$(Mn,Ca,Fe)\{SiO_3\}$
	M $\gamma>\beta>\alpha$	H	2	$82^\circ - 90^\circ$ r>v	parallel, symmetrical	length slow	STAUROLITE	$(Fe^{+2},Mg)_2(Al,Fe^{+3})_9O_6\{SiO_4\}_4(O,OH)_2$
		H	2 – 3	$52^\circ - 76^\circ$ r>v	36°		MERWINITE	$Ca_3Mg\{Si_2O_8\}$
		H	3	moderate	$13^\circ - 14^\circ$		LARNITE	$Ca_2\{SiO_4\}$
		H	3	$55^\circ - 90^\circ$ r<v	parallel	length slow	ORTHOFERROSILITE	$Fe^{+2}\{SiO_3\}$
		H	3	$68^\circ - 70^\circ$ r>v	$46^\circ - 48^\circ$	extinction nearest cleavage slow	JOHANNSENITE	$Ca(Mn,Fe)\{Si_2O_6\}$
	W $\gamma>\beta>\alpha$	H	4	$84^\circ - 86^\circ$ r<v	parallel	length fast	DIASPORE	$\alpha-AlO(OH)$
	W $\beta>\alpha=\gamma$	H – VH	4 – 5	$6^\circ - 19^\circ$ r<v	$2^\circ - 7^\circ$	length fast/slow	MONAZITE	$(Ce,La,Th)PO_4$
	W – S	VH – E	5	$0^\circ - 38^\circ$ anom. strong	parallel, oblique to twin plane	length slow	CASSITERITE	SnO_2

System	Form	Cleavage	Twinning	Zoning	Alteration	Occurrence	Remarks
Mon.	tabular	poor	polysynthetic, uncommon			M	cf. corundum, cordierite, kyanite, zoisite, Na-amphibole
Mon., Tr.	tabular	one perfect, one imperfect, parting	polysynthetic	Z hour-glass		M, I	inclusions common cf. chlorite, clintonite, biotite, stilpnomelane
Tr.	prismatic	four, two at 92°	simple, polysynthetic			O, M	cf. rhodonite, bustamite
Tr.	prismatic, square x-sect., anhedral	three, two at $92\frac{1}{2}^{o}$	polysynthetic	Z	pyrolusite, rhodochrosite	O, M	rare, inclusions, exsolution lamellae cf. bustamite, pyroxmangite
Mon. (Psuedo-Orth.)	short, prismatic, six-sided x-sect., sieve texture	inconspicuous	interpenetrant	Z		M, S	inclusions cf. chondrodite, vesuvianite, melanite, Fe-olivine
Mon.	granular	one perfect	polysynthetic			M	cf. tilleyite, spurrite, larnite
Mon.	granular	one distinct, one imperfect at 90°	polysynthetic			M	cf. bredigite, merwinite, spurrite
Orth.	prismatic, anhedral	two at 88°	rare			M	
Mon.	prismatic	two at 87°	simple, polysynthetic			O	cf. pyroxene group
Orth.	tabular, aggregates, scales, acicular, stalactitic	one perfect				S, M, I	cf. anhydrite, boehmite, gibbsite, corundum, sillimanite
Mon.	small, euhedral	parting	rare		metamict	I, S	cf. sphene, zircon
Tet.	subhedral, veinlets, diamond-shaped x-sect.	prismatic	geniculate, cyclic, common	Z		O, I, S	cf. sphalerite, rutile, sphene

COLOURLESS MINERALS	Pleochroism	Relief	δ	2V Dispersion	Extinction	Orientation	Mineral	Composition
Biaxial +								
	W – M $\alpha>\beta>\gamma$	VH – E	5 anom.	17° – 40° r>v	40° symmetrical		SPHENE	$CaTi\{SiO_4\}(O,OH,F)$
	W $\gamma>\alpha$	E	1	90° r>v	$\simeq 45^{\circ}$		PEROVSKITE (incl. KNOPITE-LOPARITE-DYSANALYTE)	$Ca,Na,Fe^{+2},Ce)(Ti,Nb)O_3$
Biaxial –								
	–L	1	variable	variable	cleavage length fast		CLINOPTILITE	$(Na_2,K_2,Ca)\{Al_2Si_{10}O_{24}\}\cdot7H_2O$
	–L	1	0° – moderate	symmetrical			GMELINITE	$(Na_2,Ca)\{Al_2Si_4O_{12}\}\cdot6H_2O$
	–L	1 (anom.)	0° – 25° r>v	α : (001) 5° – 8°	optic plane $\perp$ (010)		SANIDINE	$(K,Na)\{AlSi_3O_8\}$
	–L	1	0° – 32°	symmetrical			CHABAZITE	$Ca\{Al_2Si_4O_{12}\}\cdot6H_2O$
	–L	1	0° – 63° r<v	α : (001) 5° – 11°	optic plane // (010)		HIGH SANIDINE	$(K,Na)\{AlSi_3O_8\}$
	–L	1	0° – 85°	parallel			ANALCITE	$Na\{AlSi_2O_6\}\cdot H_2O$
	–L	1	small-large	α : (001) 5° – 6°			ADULARIA	$(K,Na)\{AlSi_3O_8\}$
	–L	1	33° – 85° r>v	α : (001) 5° – 19°	(010) cleavage length fast		ORTHOCLASE	$(K,Na)\{AlSi_3O_8\}$

System	Form	Cleavage	Twinning	Zoning	Alteration	Occurrence	Remarks
Mon.	rhombs, irregular grains	parting	simple, polysynthetic		leucoxene	I, M, S	cf. monazite, rutile, cassiterite
Mon? Pseudo-Cub.	small cubes, skeletal	poor to distinct	polysynthetic, complex, interpenetrant	Z	leucoxene	I, M	cf. melanite, picotite, ilmenite
Pseudo-Mon.	tabular	one perfect				I	cf. heulandite
Trig.	tabular, prismatic, rhombohedral, approaching cube, radiating	one good, one imperfect, parting	interpenetrant			I	cf. chabazite
Mon.	clear, distinct crystals, tabular, microlites	two, parting	Carlsbad, Baveno, Manebach	Z		I, M	perthitic, cf. orthoclase, nepheline
Trig.	rhombohedral, approaching cube, anhedral, granular	one poor	interpenetrant	Z		I	cf. gmelinite, analcite
Mon.	clear distinct crystals, tabular, microlites	two, parting	Carlsbad, Baveno, Manebach	Z		I, M	perthitic cf. orthoclase, nepheline
Cub.	trapezohedral, rounded, radiating, irregular	poor	polysynthetic, complex, interpenetrant			I, S, M	cf. leucite, sodalite, wairakite
Mon.	minute crystals, rhombic x-sect.	three	albite, pericline	Z sector		O, S, M	cf. alkali feldspars
Mon.	phenocrysts, subhedral, anhedral, spherulitic aggregates	three	Carlsbad, Baveno, Manebach	Z	sericite, kaolinite	I, M	perthitic, inclusions cf. sanidine, nepheline

COLOURLESS MINERALS	Pleochroism	Relief	δ	2V Dispersion	Extinction	Orientation	Mineral	Composition
Biaxial −								
	−L	1	$36^{\circ} - 56^{\circ}$ r<v	18°	length fast		SCOLECITE	$Ca\{Al_2Si_3O_{10}\}\cdot 3H_2O$
	−L	1	$43^{\circ} - 60^{\circ}$ r>v	$\alpha : (001)$ $1^{\circ} - 6^{\circ}$			ANORTHOCLASE	$(K,Na)\{AlSi_3O_8\}$
	−L	1	45° r>v	inclined to twin lamellae			HIGH ALBITE	$Na\{AlSi_3O_8\}$
	−L	1	50°	parallel	length fast		GONNARDITE	$Na_2Ca\{(Al,Si)_5O_{10}\}_2\cdot 6H_2O$
	−L	1	$52^{\circ} - 73^{\circ}$ r>v	$0^{\circ} - 12^{\circ}$ (in albite twins)			HIGH OLIGOCLASE	$Na\{AlSi_3O_8\}-Ca\{Al_2Si_2O_8\}$
	−L	1	$70^{\circ} - 90^{\circ}$	parallel			WAIRAKITE	$CaAl_2Si_4O_{12}\cdot 2H_2O$
	−L	1	$76^{\circ} - 90^{\circ}$	parallel	length slow		MORDENITE	$(Na_2,K_2,Ca)\{Al_2Si_{10}O_{24}\}\cdot 7H_2O$
	−L	1 − 2	$66^{\circ} - 90^{\circ}$ r>v	$\alpha : (001)$ $15^{\circ} - 20^{\circ}$	cleavage length fast		MICROCLINE	$(K,Na)\{AlSi_3O_8\}$
	−L	2	$26^{\circ} - 44^{\circ}$ r<v	$8^{\circ} - 33^{\circ}$	length fast/slow		LEONHARDITE	$Ca\{Al_2Si_4O_{12}\}nH_2O$
	−L	2	$26^{\circ} - 47^{\circ}$ r<v	$8^{\circ} - 11^{\circ}$	length slow		LAUMONTITE	$Ca\{Al_2Si_4O_{12}\}\cdot 4H_2O$
	−L	2	$30^{\circ} - 49^{\circ}$ r<v	$0^{\circ} - 5^{\circ}$ wavy	cleavage fast/slow		STILBITE	$(Ca,Na_2,K_2)\{Al_2Si_7O_{18}\}\cdot 7H_2O$

System	Form	Cleavage	Twinning	Zoning	Alteration	Occurrence	Remarks
Mon. (Pseudo-Tet.)	columnar, fibrous	two at 88°	interpenetrant, common			I, M	cf. metascolecite, natrolite, zeolites
Mon.	phenocrysts, microlites	two	albite, Carlsbad, pericline, fine gridiron			I	perthitic cf. sanidine, orthoclase
Tr.	anhedral – euhedral, plates, laths	three	albite, Carlsbad, pericline, complex	Z	sericite, calcite, kaolinite, zeolites	I	
Orth. (Pseudo-Tet.)	fibrous, spherulitic					I	
Tr.	anhedral – euhedral, laths	three	albite, Carlsbad, pericline, complex	Z	sericite, calcite, albite, kaolinite	I	
Pseudo-Cub.	trapezohedral, anhedral	imperfect	simple			I	cf. analcite
Orth.	tabular, acicular, fibrous	one perfect				I	
Tr.	anhedral, subhedral, phenocrysts	two, parting	albite, Carlsbad, pericline, complex	Z	cloudy, sericite, kaolinite	I, M, S	perthitic cf. isomicrocline, orthoclase, albite, plagioclase
Mon.	prismatic, fibrous	two good				I	cf. laumontite
Mon.	prismatic, fibrous	two good			leonhardite	I, M	cf. leonhardite, phillipsite
Mon. (Pseudo-Hex.)	sheaf-like aggregates, spherulitic	one good	sector, interpenetrant, cruciform			I	cf. heulandite, phillipsite

COLOURLESS MINERALS	Pleochroism	Relief	δ	2V Dispersion	Extinction	Orientation	Mineral	Composition
Biaxial -								
		-L	2	44° r<v	10°	length slow	EPISTILBITE	$Ca\{Al_2Si_6O_{16}\}\cdot 5H_2O$
		-L	2 - 3	$40^{o} - 60^{o}$	$\simeq$ parallel	length slow	SEPIOLITE	$H_4Mg_2\{Si_3O_{10}\}$
		-L	3		mass extinction		PALYGORSKITE (ATTAPULGITE)	$2MgO\cdot 3SiO_2\cdot 4H_2O-$ $Al_2O_3\cdot 5SiO_2\cdot 6H_2O$
		-L	3	small			HECTORITE (SMECTITE)	$2MgO\cdot 3SiO_2\cdot nH_2O$
		-L - N	1 - 2	$48^{o} - 79^{o}$ r>v	α : x $0^{o} - 20^{o}$		HYALOPHANE	$(K,Na,Ba)\{(Al,Si)_4O_8\}$
		-L - +L	1	$76^{o} - 90^{o}$ r$\lessgtr$v	$0^{o} - 70^{o}$ (in albite twins)		PLAGIOCLASE	$Na\{AlSi_3O_8\}-Ca\{Al_2Si_2O_8\}$
	M	-L - +L	1 - 2	$65^{o} - 90^{o}$ r<v	parallel		CORDIERITE	$Al_3(Mg,Fe^{+2})_2\{Si_5AlO_{18}\}$
		-L - +L	2 - 3	moderate			SAPONITE (SMECTITE)	$(\tfrac{1}{2}Ca,Na)_{0.7}(Al,Mg,Fe)_4$ $\{(Si,Al)_8O_{20}\}(OH)_4\cdot nH_2O$
	W $\gamma>\beta>\alpha$	-L - +M	3	$0^{o} - 30^{o}$	small		MONTMORILLONITE-BEIDELLITE (SMECTITES)	$(\tfrac{1}{2}Ca,Na)_{0.7}(Al,Mg,Fe)_4$ $\{(Si,Al)_8O_{20}\}(OH)_4\cdot nH_2O$
		-L - +M	5	$4^{o} - 14^{o}$ (uniaxial)	symmetrical to cleavage		CALCITE	$CaCO_3$
	N		1 - 2	$15^{o} - 90^{o}$ r<v	small		GISMONDINE	$Ca\{Al_2Si_2O_8\}\cdot 4H_2O$

System	Form	Cleavage	Twinning	Zoning	Alteration	Occurrence	Remarks
Mon.	prismatic, sheaf-like aggregates, spherulitic	one good	sector, interpenetrant, cruciform			I	
Mon. (Orth.)	fibrous, aggregates, curved, matted					I, S	
Mon?	fine aggregates, clay					S	cf. montmorillonite
Mon.	shards, massive, scales, clay	perfect basal				I	
Mon.	tabular, prismatic	two	Carlsbad, Baveno, Manebach			O, M	cf. orthoclase
Tr.	anhedral - euhedral, laths	three	albite, Carlsbad, pericline, complex	Z	sericite, calcite, kaolinite, zeolites, saussurite	I, M, S	peristerite, schiller, exsolution lamellae cf. cordierite
Orth. (Pseudo-Hex.)	pseudo-hex., anhedral	moderate to absent	simple, polysynthetic, cyclic, sector, interpenetrant		pinite, talc	M, I	pleochroic haloes, inclusions (including opaque dust) common cf. quartz, plagioclase
Mon.	shards, massive, scales, lamellar, clay	perfect basal				I	cf. pyrophyllite, talc
Mon.	shards, scales, massive, microcrystalline	perfect basal				I, S	cf. nontronite
Trig.	anhedral	rhombohedral	polysynthetic, // long diagonal			M	twinkling cf. rhombohedral carbonate
Mon. (Pseudo-Tet.)	euhedral	distinct	complex, segmented			I, O	basal sections show four segments, opposite parts alike and extinction 5°

COLOURLESS MINERALS	Pleochroism	Relief	δ	2V Dispersion	Extinction	Orientation	Mineral	Composition
Biaxial -								
		N - +L	1	$86° - 90°$ r>v	$0° - 12°$ (in albite twins)		OLIGOCLASE	Na{AlSi$_3$O$_8$}-Ca{Al$_2$Si$_2$O$_8$}
		+L	1				LIZARDITE (SERPENTINE)	Mg$_3${Si$_2$O$_5$}(OH)$_4$
		+L	1		small	length slow	HALLOYSITE	Al$_4${Si$_4$O$_{10}$}(OH)$_8$·8H$_2$O
	W	+L	1	$0° - 17°$	parallel	long. sect. length fast, x-sect. length slow	BERYL	Be$_3$Al$_2${Si$_6$O$_{18}$}
	W $\gamma=\beta>\alpha$	+L	1 anom.	$0° - 40°$ r>v	$\simeq$ parallel	cleavage length slow	PENNINITE	(Mg,Al,Fe)$_{12}${(Si,Al)$_8$O$_{20}$}(OH)$_{16}$
	N - W $\gamma=\beta>\alpha$	+L	1	$24° - 50°$ r>v	$1° - 3\frac{1}{2}°$	length slow	KAOLINITE GROUP (KANDITES) (KAOLINITE, NACRITE)	Al$_4${Si$_4$O$_{10}$}(OH)$_8$
	$\gamma>\alpha$	+L	1 anom.	$37° - 61°$ r>v	parallel	length slow	ANTIGORITE (SERPENTINE)	Mg$_3${Si$_2$O$_5$}(OH)$_4$
		+L	1	$76° - 90°$ r<v	$13° - 27\frac{1}{2}°$ (in albite twins)		ANDESINE	Na{AlSi$_3$O$_8$}-Ca{Al$_2$Si$_2$O$_8$}
		+L	2	$30° - 32°$	parallel	length slow	CHRYSOTILE (SERPENTINE)	Mg$_3${Si$_2$O$_5$}(OH)$_4$
		+L	2	$54°$ r<v	parallel	length fast	EDINGTONITE	Ba{Al$_2$Si$_3$O$_{10}$}·4H$_2$O

System	Form	Cleavage	Twinning	Zoning	Alteration	Occurrence	Remarks
Tr.	anhedral - euhedral, laths	three	albite, Carlsbad, pericline, complex	Z	sericite, calcite, kaolinite, zeolites	I, M	peristerite, antiperthitic cf. cordierite, quartz
Mon.	tabular, fine grained aggregates	perfect basal				I, M	mesh, hour-glass structure cf. chlorite group
Tr. Mon(?)	fine aggregates, colloform, clay, fibrous	shatter cracks			sericite	S, I	cf. hydrohalloysite (endellite)
Orth. (Hex.)	prismatic, inclusions zoned	imperfect		Z	kaolin	I, M	liquid inclusions cf. apatite, quartz
Mon.	tabular, vermicular, radiating, pseudomorphs	perfect basal	simple, pennine law	Z		I, M	pleochroic haloes cf. clinochlore, prochlorite
Tr., Mon.	mosaic-like masses, veinlets, scales, fibrous, alteration, product of feldspar, clay	perfect basal	rare			S, I	cf. dickite, nacrite, phengite, sericite, montmorillonite
Mon.	anhedral, fibrolamellar, flakes, laths, plates	perfect basal	occasional			I, M	mesh, hour-glass structure cf. chrysotile, serpophite, chlorite, amphibole
Tr.	anhedral - euhedral, laths	three	albite, Carlsbad, pericline, complex	Z	sericite, calcite, kaolinite, saussurite	I, M	antiperthitic cf. cordierite
Mon.	fibrous, cross-fibre, veinlets	fibrous				I, M	mesh, hour-glass structure cf. antigorite, serpophite, chlorite, amphibole
Orth.	minute, fibrous	two at 90°				I	

COLOURLESS MINERALS	Pleochroism	Relief	δ	2V Dispersion	Extinction	Orientation	Mineral	Composition
Biaxial −								
		+L	2	$77^\circ - 79^\circ$ r>v	$51^\circ - 70^\circ$ (in albite twins)		ANORTHITE	$Ca\{Al_2Si_2O_8\}$
		+L	2	$79^\circ - 88^\circ$ r>v	$39^\circ - 51^\circ$ (in albite twins)		BYTOWNITE	$Na\{AlSi_3O_8\}-Ca\{Al_2Si_2O_8\}$
		+L	2	$83^\circ - 90^\circ$	$\alpha : z$ $3^\circ - 5^\circ$		CELSIAN	$Ba\{Al_2Si_2O_8\}$
	W $\gamma=\beta>\alpha$	+L	2 - 4	$0^\circ - 58^\circ$ $(30^\circ - 50^\circ)$ r>v	$0^\circ - 7^\circ$	cleavage length slow	LEPIDOLITE	$K_2(Li,Al)_{5-6}\{Si_{6-7}Al_{2-1}O_{20}\}(OH,F)_4$
	W $\gamma=\beta>\alpha$	+L	3	$0^\circ - 8^\circ$ r≤v	$1^\circ - 2^\circ$	cleavage length slow	VERMICULITE	$(Mg,Ca)_{0.7}(Mg,Fe^{+3},Al)_{6.0}\{(Al,Si)_8O_{20}\}(OH)_4 \cdot 8H_2O$
	W $\gamma>\beta>\alpha$	+L	3	$0^\circ - 40^\circ$ r>v	$0^\circ - 2^\circ$	cleavage length slow	ZINNWALDITE	$K_2(Fe^{+2}{}_{2-1},Li_{2-3},Al_2)\{Si_{6-7}Al_{2-1}O_{20}\}(F,OH)_4$
		+L	3	$< 10^\circ$	small		ILLITE GROUP (incl. ILLITE, BRAMMALLITE, HYDROMUSCOVITE)	$K_{1-1.5}Al_4\{Si_{7-6.5}Al_{1-1.5}O_{20}\}(OH)_4$
	W $\gamma=\beta>\alpha$	+L	3 - 4	small	$\simeq$ parallel	cleavage length slow	MINNESOTAITE	$(Fe^{+2},Mg)_3\{(Si,Al)_4O_{10}\}OH_2$
	W $\gamma=\beta>\alpha$	+L	4	$0^\circ - 30^\circ$ r>v	$\simeq$ parallel	cleavage length slow	TALC	$Mg_6\{Si_8O_{20}\}(OH)_4$
	M $\gamma \gtrsim \beta>\alpha$	+L	4	$32^\circ - 46^\circ$ r>v	$1^\circ - 3^\circ$	cleavage length slow	FUCHSITE	Cr- muscovite
	W	+L	4	$53^\circ - 62^\circ$ r>v	$\simeq$ parallel	cleavage length slow, elongate crystals length slow	PYROPHYLLITE	$Al_4\{Si_8O_{20}\}(OH)_4$

System	Form	Cleavage	Twinning	Zoning	Alteration	Occurrence	Remarks
Tr.	anhedral - euhedral, laths	three	albite, Carlsbad, pericline, complex	Z uncommon	calcite, albite, saussurite	I, M	antiperthitic cf. cordierite
Tr.	anhedral - euhedral, laths	three	albite, Carlsbad, pericline, complex	Z	calcite, albite, saussurite	I	antiperthitic cf. cordierite
Mon.	prismatic	three	Carlsbad, Baveno, Manebach			O, I	cf. orthoclase, hyalophane
Mon. (Trig.)	tabular, short prismatic, flakes, pseudo-hex.	perfect basal	simple			I	pleochroic haloes, inclusions cf. muscovite, zinnwaldite, phlogopite
Mon.	minute particles, plates	perfect basal				S, I, M	cf. biotite, smectites
Mon.	tabular, short prismatic, flakes, pseudo-hex.	perfect basal	simple			I	cf. lepidolite, biotite
Mon.	plates, minute flakes, clay aggregates	perfect basal, parting				S, I	cf. montmorillonite, kaolinite, muscovite
Mon.	minute plates, needles, radiating aggregates	perfect basal				M	cf. talc
Mon.	platy, tabular, fibrous, aggregates	perfect basal				M, I	cf. muscovite, pyrophyllite, brucite
Mon.	thin tablets, shreds	perfect basal	simple			M	birds-eye maple structure cf. muscovite
Mon.	fine-grained foliated lamellae, radiating, granular, spherulitic	perfect basal	ill-defined			M, O	rutile inclusions cf. talc, muscovite

COLOURLESS MINERALS	Pleochroism	Relief	δ	2V Dispersion	Extinction	Orientation	Mineral	Composition
Biaxial –								
	W – S $\gamma=\beta>\alpha$	+L – +M	1 – 2 (anom.)	0° – 20° r>v	0° – small	cleavage length slow	CHLORITE (Oxidised)	$(Mg,Al,Fe)_{12}\{(Si,Al)_8O_{20}\}(OH)_1$
	W – S $\gamma=\beta>\alpha$	+L – M	1 – 2 (anom.)	0° – 20° r>v	0° – small	cleavage length slow	CHLORITE (Unoxidised)	$(Mg,Al,Fe)_{12}\{(Si,Al)_8O_{20}\}(OH)_1$
	M $\gamma\geqslant\beta>\alpha$ $\alpha>\beta=\gamma$	+L – +M	3 – 4	0° – 15° r<v	0° – 5°	cleavage length slow	PHLOGOPITE	$K_2(Mg,Fe^{+2})_6\{Si_6Al_2O_{20}\}(OH,F)_4$
		+L – +M	3 – 4	0° – 40° r>v	$\simeq$ parallel	cleavage length slow	PARAGONITE	$Na_2Al_4\{Si_6Al_2O_{20}\}(OH)_4$
		+L – +M	3 – 4	small			SAUCONITE (SMECTITE)	$(\tfrac{1}{2}Ca,Na)_{0.7}Zn_{4-5}(Mg,Al,Fe^{+3})_{2-}\{(Si,Al)_8O_{20}\}(OH)_4\cdot nH_2O$
	W $\gamma=\beta>\alpha$	+L – +M	3 – 4	25° – 70°	indistinct	length slow	NONTRONITE (SMECTITE)	$(\tfrac{1}{2}Ca,Na)_{0.7}(Al,Mg,Fe)_4\{(Si,Al)_8O_{20}\}(OH)_4\cdot nH_2O$
	W	+L – +M	4	30° – 47° r>v	1° – 3°	cleavage length slow	MUSCOVITE (incl. PHENGITE, SERICITE)	$K_2Al_4\{Si_6Al_2O_{20}\}(OH,F)_4$
	W $\gamma=\beta>\alpha$	+M	1 anom.	0° – small r>v	$\simeq$ parallel	cleavage length slow	RIPIDOLITE (PROCHLORITE)	$(Mg,Al,Fe)_{12}\{(Si,Al)_8O_{20}\}(OH)_1$
	W $\gamma=\beta>\alpha$	+M	1 anom.	0° – 20°	small	cleavage length slow	DAPHNITE	$(Mg,Al,Fe)_{12}\{(Si,Al)_8O_{20}\}(OH)_1$
	W $\varepsilon>\omega$	+M	1	0° – 36° (uniaxial)		length fast, tabular length slow	APATITE (incl. DAHLLITE, FRANCOLITE)	$Ca_5(PO_4)_3(OH,F,Cl)$
	S $\gamma>\beta>\alpha$ $\beta>\gamma>\alpha$	+M	1 – 2 anom.	2° – 65° r<v	12° – 30°	length slow	CROSSITE	$Na_2(Mg_3Fe_3^{+2},Fe_2^{+3},Al_2)\{Si_8O_{22}\}(OH)_2$

System	Form	Cleavage	Twinning	Zoning	Alteration	Occurrence	Remarks
Mon. (Pseudo-Hex.)	tabular, scaly, radiating, pseudomorphs	perfect basal	simple, polysynthetic	Z		I, M, S, O	pleochroic haloes
Mon.	tabular, scaly, radiating, pseudomorphs	perfect basal	simple, polysynthetic	Z		I, M, S, O	pleochroic haloes
Mon.	tabular, flakes, plates, pseudo-hex.	perfect basal	inconspicuous	Z colour zoning		I, M	inclusions common, birds-eye maple structure cf. biotite, muscovite, lepidolite, rutile, tourmaline
Mon.	scaly aggregates	perfect basal				M, S	cf. muscovite
Mon.	shards, massive, scales	perfect basal				I, S	
Mon.	shards, scales, fibrous, massive	perfect basal				I, M	cf. montmorillonite, kaolinite
Mon.	tabular, flakes, scales, aggregates	perfect basal	simple			M, I, S	birds-eye maple structure cf. talc, pyrophyllite, mica group
Mon.	tabular, scaly, vermicular, fan-shaped aggregates	perfect basal				M, I, O	cf. clinochlore, penninite
Mon.	concentric aggregates, fibrous plates	perfect basal				O	
Hex.	small, prismatic	poor basal				I, S, M, O	cf. beryl, topaz
Mon.	prismatic, columnar aggregates	two at 58°	simple, polysynthetic	Z		M	cf. glaucophane, riebeckite

COLOURLESS MINERALS	Pleochroism	Relief	δ	2V Dispersion	Extinction	Orientation	Mineral	Composition
Biaxial −								
	S $\gamma>\beta>\alpha$	+M	1 − 3 anom.	0° − 50° r>v	4° − 14°	length slow		GLAUCOPHANE	$Na_2Mg_3Al_2\{Si_8O_{22}\}(OH)_2$
	M $\gamma=\beta>\alpha$	+M	2	0° − 23° r<v	parallel	cleavage length slow	XANTHOPHYLLITE	$Ca_2(Mg,Fe)_{4\cdot6}Al_{1\cdot4}\{Si_{2\cdot5}Al_{5\cdot5}O_{20}\}(OH)_4$
	W	+M	2	0° − 20°	small	cleavage length slow	THURINGITE	$(Mg,Al,Fe)_{12}\{(Si,Al)_8O_{20}\}(OH)_1$
	S patchy $\alpha>\beta\geqslant\gamma$	+M	2	20° − 40° r<v	parallel	length fast	DUMORTIERITE	$HBAl_8Si_3O_{20}$
	M $\gamma=\beta>\alpha$	+M	2	32° r<v	parallel	cleavage length slow	CLINTONITE	$Ca_2(Mg,Fe)_{4\cdot6}Al_{1\cdot4}\{Si_{2\cdot5}Al_{5\cdot5}O_{20}\}(OH)_4$
		+M	2	38° − 60° r>v	α:z 30°-44° ≃ parallel	length fast/slow	WOLLASTONITE	$Ca\{SiO_3\}$
	N − W	+M	2	40° − 67° r<v	6° − 8°	cleavage length slow	MARGARITE	$Ca_2Al_4\{Si_4Al_4O_{20}\}(OH)_4$
		+M	2	44°	α:z 38° β:γ 0°	length fast/slow	PARAWOLLASTONITE	$Ca\{SiO_3\}$
	W	+M	2	63° − 80° r<v	inclined		AXINITE	$(Ca,Mn,Fe^{+2})_3Al_2BO_3\{Si_4O_{12}\}OH$
		+M	2	72° − 82° r>v	parallel	cleavage length slow	MONTICELLITE	$CaMg\{SiO_4\}$
	W patchy $\alpha>\beta=\gamma$	+M	2	73° − 86° r<v	parallel	length fast	ANDALUSITE (incl. CHIASTOLITE)	Al_2SiO_5
	M $\beta>\gamma>\alpha$	+M	2 − 3 anom.	66° − 87° r<v	15° − 40°	length slow	RICHTERITE-FERRORICHTERITE	$Na_2Ca(Mg,Fe^{+3},Fe^{+2},Mn)_5\{Si_8O_{22}\}\{OH,F\}_2$

System	Form	Cleavage	Twinning	Zoning	Alteration	Occurrence	Remarks
Mon.	prismatic, columnar aggregates	two at 58°	simple, polysynthetic	Z		M	resembles holmquistite in pleochroism cf. arfvedsonite, eckermannite, riebeckite
Mon.	tabular, short prismatic	perfect basal	simple			M	cf. clintonite, chlorite, chloritoid
Mon. (Pseudo-Hex.)	tabular, radiating	perfect basal				O	
Orth.	prismatic, acicular	imperfect, cross fractures	interpenetrant, trillings		sericite	M, I	cf. tourmaline, sillimanite
Mon.	tabular, short prismatic	perfect basal	simple			M	cf. xanthophyllite, chlorite, chloritoid
Tr.	subhedral - euhedral, columnar, fibrous	three	polysynthetic	Z	pectolite, calcite	M, I	cf. tremolite, pectolite
Mon.	tabular, flakes, pseudo-hex.	perfect basal	polysynthetic		vermiculite	M	cf. muscovite, talc, chlorite, chloritoid
Tr.	columnar, fibrous aggregates	three	polysynthetic			M	cf. wollastonite
Tr.	bladed, wedge-shaped, clusters	imperfect in several directions		Z		I, M	inclusions common cf. quartz
Orth.	aggregates, euhedral, rounded	poor			idocrase, fassaite, serpentine	M, I	cf. olivine
Orth.	columnar, square x-sect., inclusions, in shape of a cross (chiastolite)	two at 89°			sericite, kyanite, sillimanite	M, S	variety chiastolite, cf. sillimanite, hypersthene, viridine
Mon.	long prismatic, fibrous	two at 56°, partings	simple, polysynthetic	Z		M, I	

COLOURLESS MINERALS	Pleochroism	Relief	δ	2V Dispersion	Extinction	Orientation	Mineral	Composition
Biaxial –								
	W $\gamma > \beta > \alpha$	+M	2 – 3	68° – 90° $r \lessgtr v$	parallel, symmetrical	length slow	ANTHOPHYLLITE (Mg-rich)	$(Mg,Fe^{+2})_7\{Si_8O_{22}\}(OH,F)_2$
	W $\gamma > \beta > \alpha$	+M	2 – 3	73° – 86° $r < v$	10° – 17°	length slow	ACTINOLITE-FERROACTINOLITE	$Ca_2(Mg,Fe^{+2})_5\{Si_8O_{22}\}(OH,F)_2$
		+M	3	65° – 86° $r < v$	15° – 21°	length slow	TREMOLITE	$Ca_2(Mg,Fe^{+2})_5\{Si_8O_{22}\}(OH,F)_2$
		+M	4	35° – 41° $r > v$	0° – 33°		SPURRITE	$2Ca_2\{SiO_4\}\cdot CaCO_3$
		+M	4	72° – 75° $r > v$	1° – 4°		DATOLITE	$CaB\{SiO_4\}(OH)$
		+M	5	7° – 10° $r < v$	parallel		STRONTIANITE	$SrCO_3$
		+M	5	16° $r > v$	parallel		WITHERITE	$BaCO_3$
		+M	5	18° – $18\frac{1}{2}^\circ$ $r < v$	parallel	length fast	ARAGONITE	$CaCO_3$
	N – S $\gamma > \beta > \alpha$	+M – H	1 – 2	50° – 90° $r \lessgtr v$	parallel	length slow	HYPERSTHENE (incl. BRONZITE-EULITE)	$(Mg,Fe^{+2})\{SiO_3\}$
	W	+M – H	2	30° – 44° $r < v$	15° – 35°		BUSTAMITE	$(Mn,Ca,Fe)\{SiO_3\}$
	W $\gamma = \beta > \alpha$ $\gamma > \beta = \alpha$	+M – H	2 – 3	68° – 90° $r \lessgtr v$	parallel, symmetrical	length slow	FERROGEDRITE	$Fe_5Al_4Si_6O_{22}(OH)_2$
	W $\gamma > \beta = \alpha$	+M – H	3 – 4	84° – 90° $r > v$	10° – 15°	length slow	GRUNERITE	$(Fe^{+2},Mg)_7\{Si_8O_{22}\}(OH)_2$

System	Form	Cleavage	Twinning	Zoning	Alteration	Occurrence	Remarks
Orth.	bladed, prismatic, fibrous, asbestiform	two at $54\frac{1}{2}^{o}$			talc	M	cf. gedrite, tremolite, zoisite, cummingtonite, grunerite, holmquistite
Mon.	long prismatic, fibrous	two at 56°, parting	simple, polysynthetic	Z		M, I	cf. tremolite, orthoamphibole, hornblende
Mon.	long prismatic, fibrous	two at 56°, partings	simple, polysynthetic			M, I	cf. actinolite, orthoamphibole, cummingtonite, wollastonite
Mon.	granular masses	one distinct, one poor	polysynthetic, in two directions at 57°			M	cf. merwinite, rankinite, larnite
Mon.	aggregates, glassy					I, M	cf. danbourite, topaz
Orth.	fibrous masses	one good	simple, repeated, polysynthetic			S, I, O	twinkling cf. aragonite, witherite
Orth.	pyramidal, veins, pseudo-hex.	one distinct, two poor	repeated, cyclic, always present			S, I	twinkling cf. aragonite, strontianite
Orth.	columnar, fibrous, acicular, pseudo-hex.	imperfect, // length	repeated, polysynthetic, cyclic, interpenetrant		calcite	S, I, M	twinkling cf. calcite
Orth.	prismatic, anhedral - euhedral	two at 88°	polysynthetic, twin seams	Z		I, M	exsolution lamellae, schiller inclusions, cf. andalusite
Tr.	prismatic, fibrous	three, two at 95°	simple			M, O	cf. rhodonite, pyroxmangite
Orth.	prismatic, asbestiform	two at $54\frac{1}{2}^{o}$				M	cf. anthophyllite, zoisite, cummingtonite, grunerite
Mon.	prismatic, subradiating, fibrous, asbestiform	two at 55°, cross fractures	simple, polysynthetic		limonite	M	cf. tremolite, actinolite, cummingtonite, anthophyllite

COLOURLESS MINERALS	Pleochroism	Relief	δ	2V Dispersion	Extinction	Orientation	Mineral	Composition
Biaxial −								
	W	+M − VH	4	48° − 90° r$\lessgtr$v	parallel	cleavage length slow	OLIVINE	$(Mg,Fe)_2\{SiO_4\}$
		M − VH	4	52° − 90° r>v	parallel	cleavage length slow	CHRYSOLITE- HYALOSIDERITE- HORTONOLITE- FERROHORTONOLITE	$(Mg,Fe)_2\{SiO_4\}$
	W $\omega>\epsilon$	H	1	0° − 30° (uniaxial) moderate	parallel	tabular length slow, prismatic length fast	CORUNDUM	$\alpha-Al_2O_3$
		H	1 anom.	5° − 65° strong		length fast	VESUVIANITE (IDOCRASE)	$Ca_{10}(Mg,Fe)_2Al_4\{Si_2O_7\}_2\{SiO_4\}_5(OH,F)_4$
	W $\beta>\gamma>\alpha$	H	1	50° − 66° r<v	6° − 9°	length slow	SAPPHIRINE	$(Mg,Fe)_2Al_4O_6\{SiO_4\}$
	W − M $\beta>\alpha>\gamma$	H	1 − 2	45° − 68° r>v	2° − 30°	length fast	CHLORITOID (OTTRELITE)	$(Fe^{+2},Mg,Mn)_2(Al,Fe^{+3})Al_3O_2\{SiO_4\}_2(OH)_4$
		H	2	35° − 46° r>v	64°		PYROXMANGITE	$(Mn,Fe)\{SiO_3\}$
	W $\gamma>\beta>\alpha$	H	2	82° − 83° r>v	0° − 32°	length slow	KYANITE	Al_2SiO_5
	W $\beta>\gamma>\alpha$	H	3 − 4 anom.	74° − 90° r>v	0° − 15° parallel in elong. sect.	length fast/slow	EPIDOTE	$Ca_2Fe^{+3}Al_2O\cdot OH\{Si_2O_7\}\{SiO_4\}$
	W $\gamma>\beta>\alpha$	H	4	73° − 85° r<v	40° − 41°	length fast	LÅVENITE	$(Na,Ca,Mn,Fe^{+2})_3(Zr,Nb,Ti)\{Si_2O_7\}(OH,F)$
	W $\gamma>\beta>\alpha$	VH	4	44° − 70° r>v	parallel	length fast/slow	KNEBELITE	$(Mn,Fe)_2\{SiO_4\}$

System	Form	Cleavage	Twinning	Zoning	Alteration	Occurrence	Remarks
Orth.	anhedral - euhedral, rounded	uncommon, one moderate, irregular fractures	uncommon, vicinal	Z	chlorite, antigorite, serpentine, iddingsite, bowlingite	I, M	deformation lamellae cf. diopside, augite, humite group, epidote
Orth.	anhedral	moderate, irregular fractures		Z	chlorite, antigorite, serpentine, iddingsite, bowlingite	I, M	cf. olivine
Trig.	tabular, prismatic, six-sided x-sect.	parting	simple, lamellar seams	Z colour banding		M, I	inclusions cf. sapphirine
Tet.	variable, prismatic, fibrous, granular, radial	imperfect	sector	Z		M, I, S	cf. zoisite, clinozoisite, apatite, grossularite, melilite, andalusite
Mon.	tabular	poor	polysynthetic, uncommon			M	cf. corundum, cordierite, kyanite, zoisite, Na-amphibole
Mon., Tr.	tabular	one perfect, one imperfect, parting	polysynthetic	Z hour-glass		M, I	inclusions common cf. chlorite, clintonite, biotite, stilpnomelane
Tr.	prismatic	four, two at 92°	simple, polysynthetic			O, M	cf. rhodonite, bustamite
Tr.	bladed, prismatic	two, parting	simple, polysynthetic			M	cf. sillimanite, pyroxene
Mon.	distinct crystals, columnar aggregates, six-sided x-sect.	one perfect	uncommon	Z		M, I, S	cf. zoisite, clinozoisite, diopside, augite, sillimanite
Mon.	prismatic	one good	polysynthetic			I	
Orth.	anhedral - euhedral, rounded	two moderate, imperfect	uncommon			O, M	cf. tephroite, olivine

COLOURLESS MINERALS	Pleochroism	Relief	δ	2V Dispersion	Extinction	Orientation	Mineral	Composition
Biaxial –								
	W $\gamma > \beta > \alpha$	VH	4	$44^{\circ} - 70^{\circ}$ r>v	parallel	length fast/slow	**TEPHROITE**	$Mn_2\{SiO_4\}$
	W	VH	4	$48^{\circ} - 52^{\circ}$ r>v	parallel	cleavage length slow	**FAYALITE**	Fe_2SiO_4

System	Form	Cleavage	Twinning	Zoning	Alteration	Occurrence	Remarks
Orth.	anhedral - euhedral, rounded	two moderate, imperfect	uncommon			O, M	cf. knebelite, olivine
Orth.	anhedral - euhedral, rounded	one moderate, irregular fractures	uncommon, vicinal		grunerite, serpentine	I, M	cf. olivine, knebelite, pyroxene

PINK MINERALS	Pleochroism	Relief	δ	2V Dispersion	Extinction	Orientation	Mineral	Composition
Isometric								
		H					DANALITE	$Fe_4\{Be_3Si_3O_{12}\}S$
		H					GENTHELVITE	$Zn_4\{Be_3Si_3O_{12}\}S$
		H					PYROPE	$Mg_3Al_2Si_3O_{12}$
		-L	(weak)				SODALITE (incl. HACKMANITE)	$Na_8\{Al_6Si_6O_{24}\}Cl_2$
	W $\varepsilon>\omega$	-L - +M	1 (aniso-tropic)				EUDIALYTE-EUCOLITE	$(Na,Ca,Fe)_6Zr\{(Si_3O_9)_2\}$ (OH,F,Cl)
		H					ALMANDINE	$Fe_3^{+2}Al_2Si_3O_{12}$
		H - VH	(anom.)				GARNET	$(Mg,Fe^{+2},Mn,Ca)_3$ $(Fe^{+3},Ti,Cr,Al)_2\{Si_3O_{12}\}$
		VH	(anom.)				SPESSARTITE	$Mn_3Al_2Si_3O_{12}$
Uniaxial +								
	W $\varepsilon>\omega$	+L - +M	1 (iso-tropic)				EUDIALYTE	$(Na,Ca,Fe)_6Zr\{(Si_3O_9)_2\}$ (OH,F,Cl)
	M - S $\alpha=\beta>\gamma$	+L - +M	1	$0^{\circ} - 60^{\circ}$ $r<v$	0° - small	length fast	Mn and Cr-CHLORITES	$(Mg,Mn,Al,Cr,Fe)_{12}$ $\{(Si,Al)_8O_{20}\}(OH)_{16}$

System	Form	Cleavage	Twinning	Zoning	Alteration	Occurrence	Remarks
Cub.	tetrahedral, triangular x-sect., granular	one poor				I, M	cf. garnet, helvite
Cub.	tetrahedral, triangular x-sect., granular	one poor				I, M	cf. garnet
Cub.	polygonal x-sect., aggregates	parting, irregular fractures				M, I	inclusions
Cub.	hexagonal x-sect., anhedral aggregates	poor	simple		zeolites, diaspore, gibbsite	I	cf. fluorite, leucite
Trig.	rhombohedral, aggregates	one good, one poor		Z		I	usually anisotropic cf. catapleite, låvenite, rosenbuschite, garnet
Cub.	polygonal x-sect., aggregates	parting, irregular fractures			chlorite	M, I	inclusions
Cub.	four, six, eight-sided, polygonal x-sect., aggregates	parting, irregular fractures	complex, sector	Z	chlorite	M, I, S	inclusions cf. spinel
Cub.	polygonal x-sect., aggregates	parting, irregular fractures				M	inclusions cf. almandine
Trig.	rhombohedral, aggregates	one good, one poor		Z		I	cf. catapleite, låvenite, rosenbuschite, garnet, eucolite
Mon.	tabular, radiating	perfect basal				O, I, M	

PINK MINERALS	Pleochroism	Relief	δ	2V Dispersion	Extinction	Orientation	Mineral	Composition
Uniaxial +								
	N – M $\gamma=\alpha>\beta$ $\beta>\alpha=\gamma$	+M – H	3	$0°$ – $30°$	$37°$ – $44°$ extinction nearest cleavage slow		PIGEONITE	$(Mg,Fe^{+2},Ca)(Mg,Fe^{+2})\{Si_2O_6\}$
	W	H	5		straight		XENOTIME	YPO_4
Uniaxial –								
	W $\epsilon>\omega$	+L – +M	1 (iso-tropic)				EUCOLITE	$(Na,Ca,Fe)_6Zr\{(Si_3O_9)_2\}$ (OH,F,Cl)
	M – S $\gamma=\beta>\alpha$	+L – +M	1	$0°$ – $20°$ r>v	$0°$ – small	length slow	Mn and Cr-CHLORITES	$(Mg,Mn,Al,Cr,Fe)_{12}$ $\{(Si,Al)_8O_{20}\}(OH)_{16}$
	W $\omega>\epsilon$	N – VH	5		symmetrical to cleavage		RHODOCHROSITE	$MnCO_3$
	S $\omega>\epsilon$	+M	2 – 3		parallel	length fast	ELBAITE (TOURMALINE)	$Na(Li,Al)_3Al_6B_3Si_6O_{27}(OH,F)_4$
	W $\omega>\epsilon$	H	1	$0°$ – $30°$ anom.	parallel, symmetrical	tabular length slow, prismatic length fast	CORUNDUM	$\alpha-Al_2O_3$
Biaxial +								
	M – S $\alpha=\beta>\gamma$	+L – +M	1	$0°$ – $60°$ r<v	$0°$ – small	length fast	Mn and Cr-CHLORITES	$(Mg,Mn,Al,Cr,Fe)_{12}$ $\{(Si,Al)_8O_{20}\}(OH)_{16}$
	N – W	+M	1 – 2	$48°$ – $68°$ r>v	parallel, symmetrical	cleavage fast	TOPAZ	$Al_2\{SiO_4\}(OH,F)_2$
	W $\gamma>\beta=\alpha$	+M	2 – 3	$45°$ – $61°$ r>v	parallel, symmetrical	length slow	MULLITE	$3Al_2O_3\cdot2SiO_2$

System	Form	Cleavage	Twinning	Zoning	Alteration	Occurrence	Remarks
Mon.	prismatic, anhedral, overgrowths	two at 87°, parting	simple, polysynthetic	Z		I	exsolution lamellae cf. augite, olivine
Tet.	small, prismatic					I, M, S	inclusions, pleochroic haloes around grains cf. zircon, sphene, monazite
Trig.	aggregates	one good, one poor		Z		I	some zones isotropic cf. catapleite, rosenbuschite, lavenite, garnet, eudialyte
Mon.	tabular, radiating	perfect basal				O, I, M	
Trig.	rhombohedral, aggregates, bands	rhombohedral	polysynthetic, rare	Z	manganese oxide	O, I, M	twinkling cf. rhombohedral carbonates
Trig.	prismatic, radiating	fractures at 90°	rare	Z		M	cf. tourmaline group
Trig.	tabular, prismatic, six-sided x-sect.	parting	simple, polysynthetic, twin seams	Z		M, I	inclusions cf. sapphirine
Mon.	tabular, radiating	perfect basal				O, I, M	
Orth.	prismatic, aggregates	one perfect				I, M, S	usually colourless cf. quartz, andalusite
Orth.	prismatic, needles, fibres, square x-sect.	one distinct				M	rare cf. sillimanite

PINK MINERALS	Pleochroism	Relief	δ	2V Dispersion	Extinction	Orientation	Mineral	Composition
Biaxial +								
	M	+M - H	1 - 3	0° - 60° r>v	parallel	length fast/slow	THULITE	$Ca_2Mn^{+3}Al_2O \cdot OH\{Si_2O_7\}\{SiO_4\}$
	M - S	+M - H	2 - 3 (anom.)	25° - 50° r>v	35° - 48°	extinction nearest cleavage fast	TITANAUGITE	$(Ca,Na,Mg,Fe^{+2},Mn,Fe^{+3},Al,Ti)_2\{(Si,Al)_2O_6\}$
	W $\alpha>\gamma$	H	2	61° - 76° r<v	5° - 25°		RHODONITE	$(Mn,Ca,Fe)\{SiO_3\}$
	W $\gamma>\beta>\alpha$	H	4	84° - 86° r<v	parallel	length fast	DIASPORE	$\alpha \cdot AlO(OH)$
	S $\gamma>\alpha>\beta$ $\gamma>\beta>\alpha$	H - VH	3 - 5	64° - 85° r$\lessgtr$v	2° - 9°	length fast/slow	PIEMONTITE	$Ca_2(Mn,Fe^{+3},Al)_2AlO \cdot OH$ $\{Si_2O_7\}\{SiO_4\}$
Biaxial -								
	W $\gamma>\alpha$	+L	1	small			KÄMMERERITE	$(Mg,Al,Cr,Fe)_{12}$ $\{(Si,Al)_8O_{20}\}(OH)_{16}$
	M - S $\gamma=\beta>\alpha$	+L - +M	1	0° - 20° r>v	0° - small	length slow	Mn and Cr-CHLORITES	$(Mg,Mn,Al,Cr,Fe)_{12}$ $\{(Si,Al)_8O_{20}\}(OH)_{16}$
	S patchy $\alpha>\beta>\gamma$	+M	2	20° - 40° r$\lessgtr$v	parallel	length fast	DUMORTIERITE	$HBAl_8Si_3O_{20}$
	W patchy $\alpha>\gamma$	+M	2	73° - 86° r<v	parallel	length fast	ANDALUSITE	Al_2SiO_5
	S $\omega>\epsilon$	+M	2 - 3		parallel	length fast	ELBAITE (TOURMALINE)	$Na(Li,Al)_3Al_6B_3Si_6O_{27}(OH,F)_4$
	N - S $\gamma>\beta>\alpha$	+M - H	1 - 2	50° - 90° r$\lessgtr$v	parallel	length slow	HYPERSTHENE (incl. BRONZITE-EULITE)	$(Mg,Fe^{+2})\{SiO_3\}$

System	Form	Cleavage	Twinning	Zoning	Alteration	Occurrence	Remarks
Orth.	columnar aggregates	one perfect, one imperfect	polysynthetic, rare			M	cf. zoisite, clinozoisite, vesuvianite, sillimanite
Mon.	prismatic	two at 93°	polysynthetic, twin seams	Z hour-glass		I	cf. hypersthene
Tr.	prismatic, square x-sect., anhedral	three, two at 92½°	polysynthetic	Z	pyrolusite, rhodochrosite	O, M	rare, inclusions, exsolution lamellae cf. bustamite, pyroxmangite
Orth.	tabular aggregates, scales, acicular, stalactitic	one perfect				S, M, I	cf. anhydrite, boehmite, gibbsite, corundum, sillimanite
Mon.	columnar, six-sided x-sect.	one perfect	polysynthetic, uncommon			M, I	cf. thulite, titanaugite, dumortierite
Mon.	tabular	perfect				O	cf. chlorite group
Mon.	tabular, radiating	perfect basal				O, I, M	
Orth.	prismatic, acicular	imperfect, cross fractures	interpenetrant, trillings		sericite	M, I	cf. tourmaline, sillimanite
Orth.	columnar, square x-sect.	two at 89°	rare		sericite, kyanite, sillimanite	M	variety chiastolite, inclusions in shape of a cross cf. hypersthene, viridine
Trig.	prismatic, radiating	fractures at 90°	rare	Z		M	cf. tourmaline group
Orth.	prismatic, anhedral - euhedral	two at 88°	polysynthetic, twin seams	Z		I, M	exsolution lamellae, schiller inclusions cf. andalusite, titanaugite

PINK MINERALS	Pleochroism	Relief	δ	2V Dispersion	Extinction	Orientation	Mineral	Composition
Biaxial –								
	W	+M – H	2	$30^{\circ} - 44^{\circ}$ $r<v$	$15^{\circ} - 35^{\circ}$		BUSTAMITE	$(Mn,Ca,Fe)\{SiO_3\}$
	W $\omega>\varepsilon$	H	1	$0^{\circ} - 30^{\circ}$ (uniaxial) moderate	parallel, symmetrical	tabular length slow, prismatic length fast	CORUNDUM	$\alpha-Al_2O_3$

System	Form	Cleavage	Twinning	Zoning	Alteration	Occurrence	Remarks
Tr.	prismatic, fibrous	three, two at 95°	simple			M, O	cf. rhodonite, pyroxmangite
Trig.	tabular, prismatic, six-sided x-sect.	parting	simple, polysynthetic twin seams	Z		M, I	inclusions cf. sapphirine

RED MINERALS	Pleochroism	Relief	δ	2V Dispersion	Extinction	Orientation	Mineral	Composition
Isometric								
		−L	(weak)				SODALITE (incl. HACKMANITE)	$Na_8\{Al_6Si_6O_{24}\}Cl_2$
		+L					CLIACHITE	$Al_2O_3(H_2O)x$
		H					PYROPE	$Mg_3Al_2Si_3O_{12}$
		H	(anom.)				GROSSULARITE (HESSONITE)	$Ca_3Al_2Si_3O_{12}$
		H − VH					ALLANITE (ORTHITE)	$(Ca,Ce)_2(Fe^{+2},Fe^{+3})Al_2O\cdot OH\{Si_2O_7\}\{SiO_4\}$
		H − VH	(anom.)				GARNET	$(Mg,Fe^{+2},Mn,Ca)_3(Fe^{+3},Ti,Cr,Al)_2\{Si_3O_{12}\}$
		H − E					SPINEL (MAGNESIOFERRITE, GALAXITE, FRANKLINITE)	$(Mg,Mn,Zn)(Fe_2^{+3},Al_2)O_4$
		VH					ALMANDINE	$Fe_3^{+2}Al_2Si_3O_{12}$
		VH	(anom.)				SPESSARTITE	$Mn_3Al_2Si_3O_{12}$
		VH	(anom.)				MELANITE	$Ca_3(Fe^{+3},Ti)_2Si_3O_{12}$
		VH	(anom.)				SCHORLOMITE	$Ca_3(Fe^{+3},Ti)_2Si_3O_{12}$

System	Form	Cleavage	Twinning	Zoning	Alteration	Occurrence	Remarks
Cub.	hexagonal x-sect., anhedral aggregates	poor			zeolites, diaspore, gibbsite	I	cf. fluorite, leucite
M'loid.	pisolitic, massive	contraction cracks				S	often with gibbsite and siderite
Cub.	four, six, eight-sided, polygonal x-sect., aggregates	parting, irregular fractures				M, I	inclusions cf. garnet group
Cub.	four, six, eight-sided, polygonal x-sect., aggregates	parting, irregular fractures	sector	Z		M	cf. garnet group, periclase, vesuvianite
Mon.	distinct crystals, columnar, six-sided x-sect., irregular	two imperfect	uncommon	Z	metamict	I, M	anastomosing cracks, isotropic in metamict state cf. epidote, melanite
Cub.	four, six, eight-sided, polygonal x-sect., grains, aggregates	parting, irregular fractures	complex, sector	Z	chlorite	M, I, S	inclusions cf. spinel
Cub.	small grains, cubes, octahedra, rhombic x-sect.	parting		Z		M, I, O	
Cub.	four, six, eight-sided, polygonal x-sect., aggregates	parting, irregular fractures			chlorite	M, I	inclusions cf. garnet group
Cub.	four, six, eight-sided, polygonal x-sect., aggregates	parting, irregular fractures				M	inclusions cf. almandine
Cub.	four, six, eight-sided, polygonal x-sect., aggregates	parting, irregular fractures	complex, sector	Z		I, M	cf. garnet group
Cub.	four, six, eight-sided, polygonal x-sect., aggregates	parting, irregular fractures	complex, sector	Z		I, M	cf. garnet group

RED MINERALS	Pleochroism	Relief	δ	2V Dispersion	Extinction	Orientation	Mineral	Composition
Isometric								
		VH	(anom.)	90° rare			ANDRADITE	$Ca_3(Fe^{+3},Ti)_2Si_3O_{12}$
	W $\gamma>\alpha$	E	(weak)				PEROVSKITE (KNOPITE-LOPARITE-DYSANALYTE)	$(Ca,Na,Fe^{+2},Ce)(Ti,Nb)O_3$
		E	(weak)				LIMONITE	$FeO \cdot OH \cdot nH_2O$
Uniaxial +								
	M – S $\alpha=\beta>\gamma$	+L – +M	1	0° – 60° r<v	0° – small	cleavage length fast	Mn and Cr-CHLORITES	$(Mg,Mn,Al,Cr,Fe)_{12}$ $\{(Si,Al)_8O_{20}\}(OH)_{16}$
	W – S $\varepsilon>\omega$	VH – E	5		parallel, oblique to twin plane	length slow	CASSITERITE	SnO_2
	W – M $\varepsilon>\omega$	E	5		parallel		RUTILE	TiO_2
Uniaxial –								
	M – S $\gamma=\beta>\alpha$	+L – +M	1	0° – 20° r>v	0° – small	cleavage length slow	Mn and Cr-CHLORITES (incl. DELESSITE)	$(Mg,Mn,Al,Cr,Fe)_{12}$ $\{(Si,Al)_8O_{20}\}(OH)_{16}$
	S $\gamma\gtrsim\beta>\alpha$ $\beta>\gamma>\alpha$	+M	4 – 5	0° – 25° Mg r$\lessgtr$v Fe r$>$v	0° – 9° wavy	cleavage length slow	BIOTITE	$K_2(Mg,Fe^{+2})_{6-4}(Fe^{+3},Al,Ti)_{0-2}$ $\{Si_{6-5}Al_{2-3}O_{20}\}(OH,F)_4$

System	Form	Cleavage	Twinning	Zoning	Alteration	Occurrence	Remarks
Cub.	four, six, eight-sided, polygonal x-sect., aggregates	parting, irregular fractures	complex, sector	Z		M	cf. garnet group
Mon? Pseudo-Cub.	small cubes, skeletal	poor to distinct	polysynthetic, complex, interpenetrant	Z	leucoxene	I, M	cf. melanite, picotite, ilmenite
M'loid.	stain or border to other minerals, pseudomorphs					I, M, S, O	opaque to translucent cf. goethite
Mon.	tabular, radiating	perfect basal				O, I, M	
Tet.	subhedral veinlets, diamond-shaped x-sect.	prismatic	geniculate, cyclic, common	Z		O, I, S	cf. sphalerite, rutile
Tet.	prismatic, acicular, reticulate network, inclusions	one good	geniculate			M, I, S	often opaque cf. brookite, anatase, sphene, baddeleyite, cassiterite
Mon.	tabular, radiating	perfect basal				O, I, M	
Mon. (Pseudo-Hex.)	tabular, flakes, plates, pseudo-hex.	perfect basal	simple	Z	chlorite, vermiculite, prehnite	M, I, S	pleochroic haloes, inclusions common, birds-eye maple structure cf. stilpnomelane, phlogopite

RED MINERALS	Pleochroism	Relief	δ	2V Dispersion	Extinction	Orientation	Mineral	Composition
Uniaxial +								
	W $\omega>\epsilon$	H	1	$0^{o} - 30^{o}$	parallel, symmetrical	tabular length slow, prismatic length fast	CORUNDUM	$\alpha-Al_2O_3$
	W	E	5				HAEMATITE	$\alpha-Fe_2O_3$
Biaxial +								
	M − S $\alpha=\beta>\gamma$	+L − +M	1	$0^{o} - 60^{o}$ r<v	0° − small	cleavage length fast	**Mn and Cr-CHLORITES**	$(Mg,Mn,Al,Cr,Fe)_{12}$ $\{(Si,Al)_8O_{20}\}(OH)_{16}$
	M − S	+M − H	2 − 3 (anom.)	$25^{o} - 50^{o}$ r>v	$35^{o} - 48^{o}$	extinction nearest cleavage fast	TITANAUGITE	$(Ca,Na,Mg,Fe^{+2},Mn,Fe^{+3},Al,Ti)_2$ $\{(Si,Al)_2O_6\}$
	W $\alpha>\gamma$	H	2	$61^{o} - 76^{o}$ r<v	$5^{o} - 25^{o}$		RHODONITE	$(Mn,Ca,Fe)\{SiO_3\}$
	W − M $\gamma>\beta>\alpha$	H	4	$25^{o} - 75^{o}$ r<v	parallel		IDDINGSITE	$Mg,Fe^{+3}\{Si_3O_{10}\}4H_2O$
	W $\gamma>\alpha$	H − VH	2 − 3	$40^{o} - 90^{o}$ r$\lessgtr$v	$1^{o} - 42^{o}$ straight, // elongation	length fast/slow	ALLANITE (ORTHITE)	$(Ca,Ce)_2(Fe^{+2},Fe^{+3})Al_2O\cdot OH$ $\{Si_2O_7\}\{SiO_4\}$
	S $\gamma>\alpha>\beta$ $\gamma>\beta>\alpha$	H − VH	3 − 5	$64^{o} - 85^{o}$ r$\lessgtr$v	$2^{o} - 9^{o}$	length fast/slow	PIEMONTITE	$Ca_2(Mn,Fe^{+3},Al)_2AlO\cdot OH$ $\{Si_2O_7\}\{SiO_4\}$
	W − S	VH − E	5	$0^{o} - 38^{o}$ anom. strong	parallel, oblique to twin plane	length slow	CASSITERITE	SnO_2
	W $\gamma>\alpha$	E	1	$\simeq 90^{o}$ r>v	$\simeq 45^{o}$		PEROVSKITE (KNOPITE-LOPARITE-DYSANALYTE)	$(Ca,Na,Fe^{+2},Ce)(Ti,Nb)O_3$

System	Form	Cleavage	Twinning	Zoning	Alteration	Occurrence	Remarks
Trig.	tabular, prismatic, six-sided x-sect.	parting	simple, lamellar seams	Z colour banding		M, I	inclusions
Trig.	scales, flakes, grains, irregular masses	parting	polysynthetic			M, I, O, S	cf. goethite, limonite
Mon.	tabular, radiating	perfect basal				O, I, M	
Mon.	prismatic	two at 93°	polysynthetic, twin seams	Z hour-glass		I	cf. hypersthene
Tr.	prismatic, square x-sect., anhedral	three, two at 92½^{o}	polysynthetic	Z	pyrolusite, rhodochrosite	O, M	usually pink, rare, inclusions, exsolution lamellae cf. bustamite, pyroxmangite
Orth.	tabular, pseudomorphs	reported			limonite	I, M	alteration of olivine
Mon.	distinct crystals, columnar, six-sided x-sect., irregular	two imperfect	uncommon	Z	metamict	I, M	anastomosing cracks, isotropic in metamict state cf. epidote, melanite
Mon.	columnar, six-sided x-sect.	one perfect	polysynthetic, uncommon			M, I	cf. thulite, titanaugite, dumortierite
Tet.	subhedral, veinlets, diamond-shaped x-sect.	prismatic	geniculate, cyclic, common	Z		O, I, S	cf. sphalerite, rutile
Mon? Pseudo-Cub.	small cubes, skeletal	poor to distinct	polysynthetic, complex, interpenetrant	Z	leucoxene	I, M	cf. melanite, picotite, ilmenite

RED MINERALS	Pleochroism	Relief	δ	2V Dispersion	Extinction	Orientation	Mineral	Composition
Biaxial +								
	W – M $\varepsilon>\omega$	E	5	anom. (uniaxial)	parallel		RUTILE	TiO_2
Biaxial –								
	M – S $\gamma=\beta>\alpha$	+L – +M	1	$0°$ – $20°$ r>v	$0°$ – small	cleavage length slow	Mn and Cr-CHLORITES (incl. DELESSITE)	$(Mg,Mn,Al,Cr,Fe)_{12}\{(Si,Al)_8O_{20}\}(OH)_{16}$
	M $\gamma\gtrsim\beta>\alpha$ $\alpha>\beta=\gamma$	+L – +M	3 – 4	$0°$ – $15°$ r<v	$0°$ – $5°$	cleavage length slow	PHLOGOPITE	$K_2(Mg,Fe^{+2})_6\{Si_6Al_2O_{20}\}(OH,F)_4$
	M $\gamma=\beta>\alpha$	+M	2	$0°$ – $23°$ r<v	parallel	cleavage length slow	XANTHOPHYLLITE	$Ca_2(Mg,Fe)_{4.6}Al_{1.4}\{Si_{2.5}Al_{5.5}O_{20}\}(OH)_4$
	S patchy $\alpha>\beta\geq\gamma$	+M	2	$20°$ – $40°$ r$\lesssim$v	parallel	length fast	DUMORTIERITE	$HBAl_8Si_3O_{20}$
	W patchy $\alpha>\gamma$	+M	2	$73°$ – $86°$ r<v	parallel	length fast	ANDALUSITE	Al_2SiO_5
	M $\beta>\gamma>\alpha$	+M	2 – 3 anom.	$66°$ – $87°$ r<v	$15°$ – $40°$	length slow	RICHTERITE-FERRORICHTERITE	$Na_2Ca(Mg,Fe^{+3},Fe^{+2},Mn)_5\{Si_8O_{22}\}(OH,F)_2$
	S $\gamma\geq\beta>\alpha$ $\beta>\gamma>\alpha$	+M	4 – 5	$0°$ – $25°$ Mg r$\lesssim$v Fe r$\gtrsim$v	$0°$ – $9°$ wavy	cleavage length slow	BIOTITE	$K_2(Mg,Fe^{+2})_{6-4}(Fe^{+3},Al,Ti)_{0-2}\{Si_{6-5}Al_{2-3}O_{20}\}(OH,F)_4$
	W – S $\gamma>\beta>\alpha$	+M – H	1 – 2	$50°$ – $90°$ r$\lesssim$v	parallel	length slow	HYPERSTHENE (incl. BRONZITE-EULITE)	$(Mg,Fe^{+2})\{SiO_3\}$
	S $\gamma>\beta>\alpha$	+M – H	2 – 5	$60°$ – $82°$ r<v	$0°$ – $18°$	length slow	BASALTIC HORNBLENDE (LAMPROBOLITE, OXYHORNBLENDE)	$(Ca,Na)_{2-3}(Mg,Fe^{+2})_{3-2}(Fe^{+3},Al)_{2-3}O_2\{Si_6Al_2O_{22}\}$

System	Form	Cleavage	Twinning	Zoning	Alteration	Occurrence	Remarks
Tet.	prismatic, acicular, reticulate network, inclusions	one good	geniculate			M, I, S	often opaque cf. brookite, anatase, sphene, baddeleyite, cassiterite
Mon.	tabular, radiating	perfect basal				O, I, M	
Mon.	tabular, flakes, plates, pseudo-hex.	perfect basal	inconspicuous	Z colour zoning		I, M	inclusions common, birds-eye maple structure cf. biotite, muscovite, lepidolite, rutile, tourmaline
Mon.	tabular, short prismatic	perfect basal	simple			M	cf. clintonite, chlorite, chloritoid
Orth.	prismatic, acicular	imperfect, cross fractures	interpenetrant, trillings		sericite	M, I	cf. tourmaline, sillimanite
Orth.	columnar, square x-sect.	distinct, two at 89°	rare		sericite, kyanite, sillimanite	M	variety chiastolite, inclusions in shape of a cross cf. hypersthene, viridine
Mon.	long prismatic, fibrous	two at 56°, partings	simple, polysynthetic	Z		M, I	
Mon. (Pseudo-Hex.)	tabular, flakes, plates, pseudo-hex.	perfect basal	simple	Z	chlorite, vermiculite, prehnite	M, I, S	pleochroic haloes, inclusions common, birds-eye maple structure cf. stilpnomelane, phlogopite
Orth.	prismatic, anhedral - euhedral	two at 88°	polysynthetic, twin seams	Z		I, M	exsolution lamellae, schiller inclusions cf. andalusite, titanaugite
Mon.	short prismatic	two at 56°	simple, polysynthetic	Z		I	cf. kaersutite, barkevikite, katophorite, cossyrite

RED MINERALS	Pleochroism	Relief	δ	2V Dispersion	Extinction	Orientation	Mineral	Composition
Biaxial –								
	W $\omega>\varepsilon$	H	1	$0^{\circ} - 30^{\circ}$ (uniaxial) moderate	parallel, symmetrical	tabular length slow, prismatic length fast	CORUNDUM	$\alpha-Al_2O_3$
	M $\alpha>\beta>\alpha$	H	5	$70^{\circ} - 88^{\circ}$ r>v	13°	length slow	ASTROPHYLLITE (KUPLETSKITE)	$(K,Na)_3(Fe,Mn)_7Ti_2\{Si_4O_{12}\}_2(O,OH,F)_7$
	W $\gamma>\alpha$	H – VH	2 – 3	$40^{\circ} - 90^{\circ}$ $r \lessgtr v$	$1^{\circ} - 42^{\circ}$ straight, // elongation	length fast/slow	ALLANITE (ORTHITE)	$(Ca,Ce)_2(Fe^{+2},Fe^{+3})Al_2O \cdot OH\{Si_2O_7\}\{SiO_4\}$
	W $\gamma>\beta>\alpha$	VH	4	$44^{\circ} - 70^{\circ}$ r>v	parallel	length slow	TEPHROITE (incl. ROEPPERITE)	$Mn_2\{SiO_4\}$
	M $\alpha>\beta>\gamma$	E	5	$0^{\circ} - 27^{\circ}$ extreme	parallel	length fast/slow	GOETHITE	$\alpha \cdot FeO \cdot OH$
	S $\gamma>\beta>\alpha$	E	5	83° slight	parallel		LEPIDOCROCITE	$\gamma \cdot FeO \cdot OH$

System	Form	Cleavage	Twinning	Zoning	Alteration	Occurrence	Remarks
Trig.	tabular, prismatic, six-sided x-sect.	parting	simple, lamellar seams	Z colour banding		M, I	inclusions
Tr.	tabular plates	perfect				I	cf. biotite, staurolite, ottrelite
Mon.	distinct crystals, columnar, six-sided x-sect., irregular	two imperfect	uncommon	Z	metamict	I, M	anastomosing cracks, isotropic in metamict state cf. epidote group, melanite
Orth.		two moderate, imperfect				O, M	cf. olivine, knebelite
Orth.	pseudomorphs, fibrous, oolitic, concretions	one perfect				O, S, I, M	weathering product cf. bowlingite, limonite
Orth.	tabular	two perfect, moderate				O, S	

BROWN MINERALS	Pleochroism	Relief	δ	2V Dispersion	Extinction	Orientation	Mineral	Composition
Isometric								
		-M	irregular				OPAL	SiO_2
		-L	(weak)				NOSEAN	$Na_8\{Al_6Si_6O_{24}\}SO_4$
		-L	1 (weak)				ANALCITE	$Na\{AlSi_2O_6\} \cdot H_2O$
		-L - +M	(weak)				PALAGONITE	Altered glass
		-L - +M	(weak)				VOLCANIC GLASS	
		+L					CLIACHITE	$Al_2O_3(H_2O)x$
		+M					GREENALITE	$Fe_6^{+2}Si_4O_{10}(OH)_8$
		+M	(aniso-tropic)				MELILITE	$(Ca,Na)_2\{(Mg,Fe^{+2},Al,Si)_3O_7\}$
		H					HELVITE	$Mn_4\{Be_3Si_3O_{12}\}S$
		H					PYROPE	$Mg_3Al_2Si_3O_{12}$
		H	(anom.)				GROSSULARITE (HESSONITE)	$Ca_3Al_2Si_3O_{12}$
		H - VH					ALLANITE (ORTHITE)	$(Ca,Ce)_2(Fe^{+2},Fe^{+3})Al_2O \cdot OH$ $\{Si_2O_7\}\{SiO_4\}$

System	Form	Cleavage	Twinning	Zoning	Alteration	Occurrence	Remarks
M'loid.	colloform, veinlets, cavity fillings	irregular fractures				I, S	cf. lechatelierite
Cub.	hexagonal x-sect., anhedral aggregates	imperfect	simple	Z	zeolites, diaspore, gibbsite, limonite	I	clouded with inclusions
Cub.	trapezohedral, rounded, radiating, irregular	poor				I, S, M	cf. leucite, sodalite, wairakite
M'loid.	amorphous, oolitic				chlorite	I	cf. volcanic glass, opal, collophane
M'loid.	amorphous, massive	perlitic parting			frequent devit-rification	I	often with crystallites and phenocrysts cf. tachylyte, lechatelierite
M'loid.	pisolitic, massive	contraction cracks				S	often with gibbsite and siderite
M'loid.	massive					O	restricted to Iron Formation, Mesabi Range, Minnesota
Tet.	tabular, peg structure	moderate, single crack		Z	zeolites, carbonate	I	cf. zoisite, vesuvianite, apatite, nepheline
Cub.	tetrahedral, triangular x-sect., granular	one poor	simple		ochre, manganese oxide	I, M	cf. garnet, danalite
Cub.	four, six, eight-sided, polygonal x-sect.	parting, irregular fractures				M, I	inclusions cf. garnet group
Cub.	four, six, eight-sided, polygonal x-sect.	parting, irregular fractures	sector	Z		M	cf. garnet group, periclase, vesuvianite
Mon.	distinct crystals, columnar, tabular, six-sided x-sect., irregular	two imperfect	uncommon	Z	metamict	I, M	anastomosing cracks, isotropic in metamict state cf. epidote, melanite

BROWN MINERALS	Pleochroism	Relief	δ	2V Dispersion	Extinction	Orientation	Mineral	Composition
Isometric								
		H – VH	(anom.)				GARNET	$(Mg,Fe^{+2},Mn,Ca)_3$ $(Fe^{+3},Ti,Cr,Al)_2\{Si_3O_{12}\}$
		VH	(anom.)	$90°$ rare			ANDRADITE	$Ca(Fe^{+3},Ti)_2Si_3O_{12}$
		VH	(anom.)				MELANITE	$Ca_3(Fe^{+3},Ti)_2Si_3O_{12}$
		VH	(anom.)				SCHORLOMITE	$Ca_3(Fe^{+3},Ti)_2Si_3O_{12}$
		VH					ALMANDINE	$Fe_3^{+2}Al_2Si_3O_{12}$
		H – E					SPINEL (PICOTITE)	$(Mg,Fe)\{(Al,Cr)_2O_4\}$
		E	(weak)				SPHALERITE	ZnS
	W $\gamma>\alpha$	E	(weak)				PEROVSKITE (KNOPITE-LOPARITE-DYSANALYTE)	$(Ca,Na,Fe^{+2},Ce)(Ti,Nb)O_3$
		E	(weak)				LIMONITE	$FeO\cdot OH\cdot nH_2O$

System	Form	Cleavage	Twinning	Zoning	Alteration	Occurrence	Remarks
Cub.	four, six, eight-sided, polygonal x-sect., grains, aggregates	parting, irregular fractures		Z	chlorite	M, I, S	inclusions cf. spinel
Cub.	four, six, eight-sided, polygonal x-sect.	parting, irregular fractures	complex, sector	Z		M	cf. garnet group
Cub.	four, six, eight-sided, polygonal x-sect.	parting, irregular fractures	complex, sector	Z		I, M	cf. garnet group
Cub.	four, six, eight-sided, polygonal x-sect.	parting, irregular fractures	complex, sector	Z		I, M	cf. garnet group
Cub.	four, six, eight-sided, polygonal x-sect.	parting, irregular fractures			chlorite	M, I	inclusions cf. garnet group
Cub.	small grains, cubes, rhombic x-sect.	parting				M, I, O	cf. garnet group
Cub.	irregular, anhedral, curved surfaces	six perfect	polysynthetic, lamellar intergrowths	Z		O	colour variable, (uniaxial) cf. cassiterite
Mon? Pseudo-Cub.	small cubes, skeletal	poor to distinct	polysynthetic complex, interpenetrant	Z	leucoxene	I, M	cf. melanite, picotite, ilmenite
M'loid.	stain or border to other minerals, pseudomorphs					I, M, S, O	opaque to translucent cf. goethite

BROWN MINERALS	Pleochroism	Relief	δ	2V Dispersion	Extinction	Orientation	Mineral	Composition
Uniaxial +								
	W $\omega>\varepsilon$	+M	1 anom. (iso-tropic)		parallel	length slow	MELILITE	$(Ca,Na)_2\{(Mg,Fe^{+2},Al,Si)_3O_7\}$
	N – M $\gamma=\alpha>\beta$ $\beta>\alpha=\gamma$	+M – H	3	$0^o - 30^o$ $r\lessgtr v$	$37^o - 44^o$	extinction nearest cleavage slow	PIGEONITE	$(Mg,Fe^{+2},Ca)(Mg,Fe^{+2})\{Si_2O_6\}$
		+H	1 anom.		parallel	length fast	VESUVIANITE (IDOCRASE) (incl. WILUITE)	$Ca_{10}(Mg,Fe)_2Al_4\{Si_2O_7\}_2$ $\{SiO_4\}_5(OH,F)_4$
	W	H	5		straight		XENOTIME	YPO_4
	W $\varepsilon>\omega$		4 – 5		parallel	length slow	ZIRCON	$Zr\{SiO_4\}$
	W – S $\varepsilon>\omega$	VH – E	5	$0^o - 38^o$ anom. strong	parallel, oblique to twin plane	length slow	CASSITERITE	SnO_2
		E	2				WURTZITE	ZnS
	W – M $\varepsilon>\omega$	E	5		parallel		RUTILE	TiO_2
Uniaxial –								
	W $\omega>\varepsilon$ rare	-L – H	5		symmetrical to cleavage	subhedral microcrystalline	MAGNESITE	$MgCO_3$
	S $\gamma=\beta>\alpha$	+L – H	3 – 5	$\simeq 0^o$	$\simeq$ parallel	length slow	STILPNOMELANE	$(K,Na,Ca)_{0-1.4}$ $(Fe^{+3},Fe^{+2},Mg,Al,Mn)_{5.9-8.2}$ $\{Si_8O_{20}\}(OH)_4(O,OH,H_2O)_{3.6-8.5}$

System	Form	Cleavage	Twinning	Zoning	Alteration	Occurrence	Remarks
Tet.	tabular, peg structure	moderate, single crack		Z	zeolites, carbonate	I	cf. zoisite, vesuvianite, apatite, nepheline
Mon.	prismatic, anhedral, overgrowths	two at 87°, parting	simple, polysynthetic	Z		I	exsolution lamellae cf. augite, olivine
Tet.	variable, prismatic, fibrous, granular, radial	imperfect		Z		M, I, S	cf. zoisite, clinozoisite, apatite, grossularite, melilite, andalusite
Tet.	small, prismatic					I, M, S	inclusions, pleochroic haloes around grains cf. zircon, sphene, monazite
Tet.	minute prisms	poor, absent		Z	metamict	I, S, M	cf. apatite
Tet.	subhedral, veinlets, diamond-shaped x-sect.	prismatic	geniculate, cyclic, common	Z		O, I, S	cf. sphalerite, rutile, sphene
Hex.	irregular					O	
Tet.	prismatic, acicular, reticulate network, inclusions	one good	geniculate			M, I, S	often opaque cf. brookite, anatase, sphene baddeleyite, cassiterite
Trig.	rhombohedral				huntite	M, S	cf. rhombohedral carbonates
Mon.	plates, pseudo-hex. micaceous masses	two at 90°, perfect basal	polysynthetic	Z		I, M, O	cf. biotite, chlorite, chloritoid, clintonite

BROWN MINERALS	Pleochroism	Relief	δ	2V Dispersion	Extinction	Orientation	Mineral	Composition
Uniaxial −								
		+L − VH	5		symmetrical to cleavage		SIDERITE	$FeCO_3$
	W $\omega > \varepsilon$	+M	1 anom. (iso-tropic)			length fast	MELILITE	$(Ca,Na)_2\{(Mg,Fe^{+2},Al,Si)_3O_7\}$
	S $\omega > \varepsilon$	+M	3		parallel	length fast	DRAVITE (TOURMALINE)	$NaMg_3Al_6B_3Si_6O_{27}(OH,F)_4$
	S $\gamma \geqslant \beta > \alpha$ $\beta > \gamma > \alpha$	+M	4 − 5	$0° - 25°$ Mg $r \lessgtr v$ Fe $r > v$	$0° - 9°$ wavy	cleavage length slow	BIOTITE	$K_2(Mg,Fe^{+2})_{6-4}(Fe^{+3},Al,Ti)_{0-2}$ $\{Si_{6-5}Al_{2-3}O_{20}\}(OH,F)_4$
		+H	1 anom.		parallel	length fast	VESUVIANITE (IDOCRASE)	$Ca_{10}(Mg,Fe)_2Al_4\{Si_2O_7\}_2$ $\{SiO_4\}_5(OH,F)_4$
	W $\omega > \varepsilon, \varepsilon > \omega$	E	5	$0°$ − small			ANATASE	TiO_2
Biaxial +								
		+L	3	$0° - 40°$ $r > v$	$21°$	elongate twinned crystals, length slow	GIBBSITE	$Al(OH)_3$
	W $\beta > \alpha = \gamma$	+M − H	2	$26° - 85°$ $r < v$	$4° - 32°$	length slow	PUMPELLYITE	$Ca_4(Mg,Fe^{+2})(Al,Fe^{+3})_5O(OH)_3$ $\{Si_2O_7\}_2\{SiO_4\}_2 \cdot 2H_2O$
	W $\gamma = \beta > \alpha$ $\gamma > \beta = \alpha$	+M	2 − 3	$68° - 90°$ $r \lessgtr v$	parallel, symmetrical	length slow	ANTHOPHYLLITE (Fe-rich)	$(Mg,Fe^{+2})_7\{Si_8O_{22}\}(OH,F)_2$

System	Form	Cleavage	Twinning	Zoning	Alteration	Occurrence	Remarks
Trig.	anhedral, aggregates, oolitic, spherulitic, colloform	rhombohedral	polysynthetic, // long diagonal, uncommon		brown spots	I, S, O	cf. rhombohedral carbonates
Tet.	tabular, peg structure	moderate, single crack		Z	zeolites, carbonate	I	twinkling, brown stain around borders and along cleavage cracks cf. zoisite, vesuvianite, apatite
Trig.	large, prismatic	fractures at 90°	rare	Z		M	cf. tourmaline, chondrodite
Mon. (Pseudo-Hex.)	tabular, flakes, plates, pseudo-hex.	perfect basal	simple	Z	chlorite, vermiculite, prehnite,	M, I, S	pleochroic haloes, inclusions common, birds-eye maple structure cf. stilpnomelane, phlogopite
Tet.	variable, prismatic, fibrous, granular, radial	imperfect	sector	Z		M, I, S	cf. zoisite, clinozoisite, apatite, grossularite, melilite, andalusite
Tet.	small, prismatic, acicular	two perfect		Z	leucoxene	S, I, M	usually yellow to blue cf. rutile, brookite
Mon.	aggregates, tabular, very small, stalactitic	one perfect	polysynthetic			S M, I	cf. chalcedony, dahllite, muscovite, kaolinite, boehmite, diaspore
Mon.	fibrous, needles, bladed, radial aggregates	two moderate	common	Z		M	cf. clinozoisite, zoisite, epidote, lawsonite
Orth.	bladed, prismatic, fibrous, asbestiform	two at $54\frac{1}{2}^{\circ}$				M	cf. gedrite, cummingtonite

BROWN MINERALS	Pleochroism	Relief	δ	2V Dispersion	Extinction	Orientation	Mineral	Composition
Biaxial +								
	W $\gamma>\alpha$	+M	3	$21^{\circ} - 30^{\circ}$ r>v	parallel	length slow	SILLIMANITE	Al_2SiO_5
		+M	3		parallel		FIBROLITE (SILLIMANITE)	Al_2SiO_5
	M - S $\gamma\geq\beta>\alpha$	+M	3	$67^{\circ} - 90^{\circ}$ r>v	26°	length slow	PARGASITE	$NaCa_2Mg_4Al_3Si_6O_{22}(OH)_2$
	M - S	+M - H	2 - 3 (anom.)	$25^{\circ} - 50^{\circ}$ r>v	$35^{\circ} - 48^{\circ}$	extinction nearest cleavage fast	TITANAUGITE	$(Ca,Na,Mg,Fe^{+2},Mn,Fe^{+3},Al,Ti)\{(SiAl)_2O_6\}$
	N - W	+M - H	2 - 3	$25^{\circ} - 83^{\circ}$ r>v	$35^{\circ} - 48^{\circ}$	extinction nearest cleavage fast	AUGITE-FERROAUGITE (incl. SALITE)	$(Ca,Na,Mg,Fe^{+2},Mn,Fe^{+3},Al,Ti)\{(Si,Al)_2O_6\}$
	W $\gamma=\beta>\alpha$ $\gamma>\beta=\alpha$	+M - H	2 - 3	$68^{\circ} - 90^{\circ}$ r<v	parallel, symmetrical	length slow	GEDRITE	$(Mg,Fe^{+2})_5Al_2\{Si_6Al_2O_{22}\}(OH,F$
	N - M $\gamma=\alpha>\beta$ $\beta>\alpha=\gamma$	+M - H	3	$0^{\circ} - 30^{\circ}$ r$\lessgtr$v	$37^{\circ} - 44^{\circ}$	extinction nearest cleavage slow	PIGEONITE	$(Mg,Fe^{+2},Ca)(Mg,Fe^{+2})\{Si_2O_6\}$
	W $\gamma>\beta>\alpha$	+M - H	3	$50^{\circ} - 62^{\circ}$ r>v	$38^{\circ} - 48^{\circ}$	cleavage fast	HEDENBERGITE (incl. FERROSALITE)	$Ca(Fe,Mg)\{Si_2O_6\}$
	W $\alpha>\gamma$	+M - VH	4	$48^{\circ} - 90^{\circ}$ r$\lessgtr$v	parallel	cleavage length slow	OLIVINE	$(Mg,Fe)_2\{SiO_4\}$
		H	3 - 4	$70^{\circ} - 90^{\circ}$ r>v	$\gamma : z$ $70^{\circ} - 90^{\circ}$	extinction nearest cleavage fast	AEGIRINE-AUGITE	$(Na,Ca)Fe^{+3},Fe^{+2},Mg)\{Si_2O_6\}$
	W - M $\gamma>\beta>\alpha$	H	4	$25^{\circ} - 75^{\circ}$ r<v	parallel		IDDINGSITE	$Mg,Fe_2^{+3}\{Si_3O_{10}\}4H_2O$

System	Form	Cleavage	Twinning	Zoning	Alteration	Occurrence	Remarks
Orth.	needles, slender prisms, fibrous, rhombic x-sect., faserkiesel	one, // long diagonal of rhombic x-sect.				M	cf. andalusite, mullite
Orth.	fibrous, felted, faserkiesel					M	stained brown in fibrous mats
Mon.	prismatic	two at 56°, partings	simple, polysynthetic			M, I	cf. hornblende, hastingsite, cummingtonite
Mon.	prismatic	two at 93°	polysynthetic, twin seams	Z hour-glass		I	cf. hypersthene
Mon.	prismatic	two at 87°, parting	simple, polysynthetic, twin seams	Z	amphibole	I, M	herringbone structure, exsolution lamellae cf. pigeonite, diopside, epidote
Orth.	bladed, prismatic, fibrous, asbestiform	two at 54½°			talc	M	cf. Fe-anthophyllite, tremolite, cummingtonite, grunerite, zoisite
Mon.	prismatic, anhedral, overgrowths	two at 87°, parting	simple, polysynthetic	Z		I	exsolution lamellae cf. augite, olivine
Mon.	prismatic	two at 87°	simple, polysynthetic			M, I	cf. diopside, aegirine-augite, augite
Orth.	anhedral - euhedral, rounded	uncommon, one moderate, irregular fractures	uncommon, vicinal	Z	chlorite, antigorite, serpentine, iddingsite, bowlingite	I, M	deformation lamellae cf. diopside, augite, humite group, epidote
Mon.	short prismatic, needles, felted aggregates	two at 87°, parting	simple, polysynthetic, twin seams	Z hour-glass		I	cf. aegirine, acmite, Na-amphibole
Orth.	tabular, pseudomorphs	reported			limonite	I, M	alteration of olivine

BROWN MINERALS	Pleochroism	Relief	δ	2V Dispersion	Extinction	Orientation	Mineral	Composition
	M $\alpha>\beta>\gamma$	H	5	70° - 88° r>v	13°	length slow	ASTROPHYLLITE (KUPLETSKITE)	$(K,Na)_3(Fe,Mn)_7Ti_2\{Si_4O_{12}\}_2(O,OH,F)_7$
	W $\gamma>\alpha$	H - VH	2 - 3	40° - 90° r≶v	1° - 42° straight, // elongation	length fast/slow	ALLANITE (ORTHITE)	$(Ca,Ce)_2(Fe^{+2},Fe^{+3})Al_2O{\cdot}OH\{Si_2O_7\}\{SiO_4\}$
	M - S $\gamma>\alpha$	VH	5	32° r<v	4° - 45°		COSSYRITE (AENIGMATITE)	$Na_2Fe_5^{+2}TiSi_6O_{20}$
	W - S	VH - E	5	0° - 38° anom. strong	parallel, oblique to twin plane	length slow	CASSITERITE	SnO_2
	W - M $\alpha>\beta>\gamma$	VH - E	5 anom.	17° - 40° r>v	40° symmetrical		SPHENE	$CaTi\{SiO_4\}(O,OH,F)$
	W $\gamma>\alpha$	E	1	90° r>v	≃ 45°		PEROVSKITE (KNOPITE-LOPARITE-DYSANALYTE)	$(Ca,Na,Fe^{+2},Ce)(Ti,Nb)O_3$
	W - M $\varepsilon>\omega$	E	5	anom. (uniaxial)		parallel	RUTILE	TiO_2
	W $\gamma>\alpha$	E	5 anom.	0° - 30° strong	parallel	length slow	BROOKITE	TiO_2
Biaxial -								
		-L	1	0° - 85°			ANALCITE	$Na\{AlSi_2O_6\}{\cdot}H_2O$
	W $\gamma=\beta>\alpha$	+L	3	0° - 8° r≶v	1° - 2°	cleavage length slow	VERMICULITE	$(Mg,Ca)_{0.7}(Mg,Fe^{+3}Al)_{6.0}\{(Al,Si)_8O_{20}\}(OH)_4{\cdot}8H_2O$
	W $\gamma>\beta>\alpha$	+L	3	0° - 40° r>v	0° - 2°	cleavage length slow	ZINNWALDITE	$K_2(Fe^{+2}_{2-1},Li_{2-3},Al_2)\{Si_{6-7}Al_{2-1}O_{20}\}(F,OH)_4$

System	Form	Cleavage	Twinning	Zoning	Alteration	Occurrence	Remarks
Tr.	tabular plates	perfect				I	cf. biotite, staurolite, ottrelite
Mon.	distinct crystals, columnar six-sided x-sect., irregular	two imperfect	uncommon	Z	metamict	I, M	anastomosing cracks, isotropic in metamict state cf. epidote, melanite
Tr.	small prismatic, aggregates	two at 66°	simple, repeated			I	cf. katophorite, kaersutite, basaltic hornblende
Tet.	subhedral veinlets, diamond-shaped x-sect.	prismatic	geniculate, cyclic, common	Z		O, I, S	cf. sphalerite, rutile, sphene
Mon.	rhombs, irregular grains	parting	simple, polysynthetic		leucoxene	I, M, S	cf. monazite, rutile, cassiterite
Mon? Pseudo-Cub.	small cubes, skeletal	poor to distinct	polysynthetic, complex, interpenetrant	Z	leucoxene	I, M	cf. melanite, picotite, ilmenite
Tet.	prismatic, acicular, reticulate network, inclusions	one good	geniculate			M, I, S	often opaque cf. brookite, anatase, sphene, baddeleyite, cassiterite
Orth.	prismatic	poor				I, M, S	cf. rutile, pseudobrookite
Cub.	trapezohedral, rounded, radiating, irregular	poor	polysynthetic, complex, interpenetrant			I, S, M	cf. leucite, sodalite, wairakite
Mon.	minute particles, plates	perfect basal				S, I, M	cf. biotite, smectites
Mon.	tabular, short prismatic, flakes, pseudo-hex.	perfect basal	simple			I	cf. lepidolite, biotite

BROWN MINERALS	Pleochroism	Relief	δ	2V Dispersion	Extinction	Orientation	Mineral	Composition
Biaxial $-$								
	M $\gamma \geqslant \beta > \alpha$ $\alpha > \beta = \gamma$	+L $-$ +M	3 $-$ 4	$0° - 15°$ $r < v$	$0° - 5°$	cleavage length slow	PHLOGOPITE	$K_2(Mg,Fe^{+2})_6\{Si_6Al_2O_{20}\}(OH,F)$
	S $\gamma = \beta > \alpha$	+L $-$ H	3 $-$ 5	$\simeq 0°$	$\simeq$ parallel	length slow	STILPNOMELANE	$(K,Na,Ca)_{0-1.4}$ $(Fe^{+3},Fe^{+2},Mg,Al,Mn)_{5.9-8.2}$ $\{Si_8O_{20}\}(OH)_4(O,OH,H_2O)_{3.6-8.}$
	N $-$ W	+M	1 anom.	small	small	cleavage length slow	CHAMOSITE	$(Mg,Al,Fe)_{12}\{(Si,Al)_8O_{20}\}(OH)$
	M $-$ S $\gamma > \beta > \alpha$ $\gamma < \beta > \alpha$	+M	1 $-$ 3	$0° - 50°$ $r < v$	$36° - 70°$	length fast	KATOPHORITE- MAGNESIOKATOPHORITE	$Na_2Ca(Mg,Fe^{+2})_4Fe^{+3}$ $\{Si_7AlO_{22}\}(OH,F)_2$
	M $\gamma = \beta > \alpha$	+M	2	$0° - 23°$ $r < v$	parallel	cleavage length slow	XANTHOPHYLLITE	$Ca_2(Mg,Fe)_{4.6}Al_{1.4}$ $\{Si_{2.5}Al_{5.5}O_{20}\}(OH)_4$
	M $\gamma = \beta > \alpha$	+M	2	$32°$ $r < v$	parallel	cleavage length slow	CLINTONITE	$Ca_2(Mg,Fe)_{4.6}Al_{1.4}$ $\{Si_{2.5}Al_{5.5}O_{20}\}(OH)_4$
	S $\gamma > \beta > \alpha$	+M	2	$40° - 50°$ $r > v$	$11° - 18°$	length slow	BARKEVIKITE	$Ca_2(Na,K)(Fe^{+2},Mg,Fe^{+3},Mn)_5$ $\{Si_{6.5}Al_{1.5}O_{22}\}(OH)_2$
	W	+M	2	$63° - 80°$ $r < v$	inclined		AXINITE	$(Ca,Mn,Fe^{+2})_3Al_2BO_3$ $\{Si_4O_{12}\}OH$
	W $\gamma > \beta > \alpha$	+M	2 $-$ 3	$68° - 90°$ $r \lessgtr v$	parallel, symmetrical	length slow	ANTHOPHYLLITE (Mg-rich)	$(Mg,Fe^{+2})_7\{Si_8O_{22}\}(OH,F)_2$
	S $\gamma \geqslant \beta > \alpha$ $\beta > \gamma > \alpha$	+M	4 $-$ 5	$0° - 25°$ Mg $r \lessgtr v$ Fe $r > v$	$0° - 9°$ wavy	cleavage length slow	BIOTITE	$K_2(Mg,Fe^{+2})_{6-4}(Fe^{+3},Al,Ti)_{0-2}$ $\{Si_{6-5}Al_{2-3}O_{20}\}(OH,F)_4$
	N $-$ S $\gamma > \beta > \alpha$	+M $-$ H	1 $-$ 2	$50° - 90°$ $r \lessgtr v$	parallel	length slow	HYPERSTHENE (incl. BRONZITE- EULITE)	$(Mg,Fe^{+2})\{SiO_3\}$

248

System	Form	Cleavage	Twinning	Zoning	Alteration	Occurrence	Remarks
Mon.	tabular, flakes, plates, pseudo-hex.	perfect basal	inconspicuous	Z colour zoning		I, M	inclusions common, birds-eye maple structure cf. biotite, muscovite, lepidolite, rutile, tourmaline
Mon.	plates, pseudo-hex., micaceous masses	two at 90°, perfect basal	polysynthetic	Z		I, M, O	cf. biotite, chlorite, chloritoid, clintonite
Mon.	pseudospherulitic, concentric, tabular, massive	one good, concentric parting				S, O	cf. glauconite, collophane, greenalite, thuringite
Mon.	prismatic	two at 56°, parting	simple	Z		I	cf. barkevikite, kaersutite, basaltic hornblende, arfvedsonite, cossyrite
Mon.	tabular, short prismatic	perfect basal	simple			M	cf. clintonite, chlorite, chloritoid
Mon.	tabular, short prismatic	perfect basal	simple			M	cf. xanthophyllite, chlorite, chloritoid
Mon.	prismatic	two at $\simeq 56^o$	simple	Z		I	cf. hastingsite, kaersutite, basaltic hornblende
Tr.	bladed, wedge-shaped, clusters	imperfect in several directions		Z		I, M	inclusions common cf. quartz
Orth.	bladed, prismatic, fibrous, asbestiform	two at $54\frac{1}{2}^o$			talc	M	cf. gedrite, tremolite, zoisite, cummingtonite, grunerite, holmquistite
Mon.	tabular, flakes, plates, pseudo-hex.	perfect basal	simple	Z	chlorite, vermiculite, prehnite	M, I, S	pleochroic haloes, inclusions common, birds-eye maple structure cf. stilpnomelane, phlogopite
Orth.	prismatic, anhedral - euhedral	two at 88°	polysynthetic, twin seams	Z		I, M	exsolution lamellae, schiller inclusions cf. andalusite, titanaugite

BROWN MINERALS	Pleochroism	Relief	δ	2V Dispersion	Extinction	Orientation	Mineral	Composition
Biaxial −								
	S $\gamma>\beta>\alpha$ $\beta>\gamma>\alpha$	+M – H	2 – 3	$66^\circ - 85^\circ$ r>v	$13^\circ - 34^\circ$	length slow	HORNBLENDE	$(Na,K)_{0-1}Ca_2(Mg,Fe^{+2},Fe^{+3},Al)_5\{Si_{6-7}Al_{2-1}O_{22}\}(OH)F_2$
	W $\gamma=\beta>\alpha$ $\gamma>\beta=\alpha$	+M – H	2 – 3	$68^\circ - 90^\circ$	parallel, symmetrical	length slow	FERROGEDRITE	$Fe_5Al_4Si_6O_{22}(OH)_2$
	S $\gamma>\beta>\alpha$	+M – H	2 – 5	$60^\circ - 82^\circ$ r<v	$0^\circ - 18^\circ$	length slow	BASALTIC HORNBLENDE (LAMPROBOLITE, OXYHORNBLENDE)	$(Ca,Na)_{2-3}(Mg,Fe^{+2})_{3-2}(Fe^{+3},Al)_{2-3}O_2\{Si_6Al_2O_{22}\}$
	S $\beta\gtrless\gamma>\alpha$ $\gamma\gtrless\beta>\alpha$	+M – H	3	$10^\circ - 90^\circ$ r≶v	$9^\circ - 40^\circ$	length slow	HASTINGSITE-FERROHASTINGSITE	$NaCa_2(Mg,Fe^{+2})_4(Al,Fe^{+3})Al_2Si_6O_{22}(OH,F)_2$
	W $\gamma>\beta=\alpha$	+M – H	3 – 4	$84^\circ - 90^\circ$ r>v	$10^\circ - 15^\circ$	length slow	GRUNERITE	$(Fe^{+2},Mg)_7\{Si_8O_{22}\}(OH)_2$
	S $\gamma>\beta>\alpha$	+M – H	3 – 5	$66^\circ - 82^\circ$ r>v	$0^\circ - 19^\circ$	length slow	KAERSUTITE	$Ca_2(Na,K)(Mg,Fe^{+2},Fe^{+3})_4Ti\{Si_6Al_2O_{22}\}(O,OH,F)_2$
	W	+M – VH	4	$48^\circ - 90^\circ$ r≶v	parallel	cleavage length slow	OLIVINE	$(Mg,Fe)_2\{SiO_4\}$
	S $\alpha>\gamma$	H	3 – 4	$70^\circ - 90^\circ$ r>v	$\gamma : z$ $70^\circ - 90^\circ$	extinction nearest cleavage fast	AEGIRINE-AUGITE	$(Na,Ca)(Fe^{+3},Fe^{+2},Mg)\{Si_2O_6\}$
	W – M $\gamma>\beta>\alpha$	+H	4	$25^\circ - 75^\circ$ r<v	parallel		IDDINGSITE (BOWLINGITE)	$Mg,Fe_2^{+3}\{Si_3O_{10}\}4H_2O$
	W $\gamma>\alpha$	H – VH	2 – 3	$40^\circ - 90^\circ$ r≶v	$1^\circ - 42^\circ$ straight, // elongation	length fast/slow	ALLANITE (ORTHITE)	$(Ca,Ce)_2(Fe^{+2},Fe^{+3})Al_2O\cdot OH\{Si_2O_7\}\{SiO_4\}$

System	Form	Cleavage	Twinning	Zoning	Alteration	Occurrence	Remarks
Mon.	prismatic	two at 56°, parting	simple, polysynthetic	Z	mica, chlorite	M, I	cf. edenite, biotite, pargasite ferrohastingsite, basaltic hornblende
Orth.	prismatic, asbestiform	two at 54½°				M	cf. gedrite
Mon.	short prismatic	two at 56°	simple, polysynthetic	Z		I	cf. kaersutite, barkevikite, katophorite, cossyrite
Mon.	prismatic	two at 56°, parting	simple, polysynthetic	Z		I, M	cf. arfvedsonite, hornblende
Mon.	prismatic, subradiating, fibrous, asbestiform	two at 55°, cross fractures	simple, polysynthetic		limonite	M	cf. tremolite, actinolite, cummingtonite, anthophyllite
Mon.	prismatic	two at 56°, parting	simple, polysynthetic	Z		I	cf. barkevikite, katophorite, basaltic hornblende, cossyrite, titanaugite
Orth.	anhedral - euhedral, rounded	uncommon, one moderate, irregular fractures	uncommon, vicinal	Z	chlorite, antigorite, serpentine, iddingsite, bowlingite	I, M	deformation lamellae cf. diopside, augite, humite group, epidote
Mon.	short prismatic, needles, felted aggregates	two at 87°, parting	simple, polysynthetic, twin seams	Z hour-glass		I	cf. aegirine, acmite, Na-amphibole
Orth.	tabular, pseudomorphs	reported			limonite	I, M	alteration of olivine
Mon.	distinct crystals, columnar, six-sided x-sect., irregular	two imperfect	uncommon	Z	metamict	I, M	anastomosing cracks, isotropic in metamict state cf. epidote, melanite

BROWN MINERALS	Pleochroism	Relief	δ	2V Dispersion	Extinction	Orientation	Mineral	Composition
Biaxial –								
	W $\alpha > \beta \gtrless \gamma$	H – VH	4 – 5	$60° - 70°$ r>v	$0° - 10°$	extinction nearest cleavage fast	**ACMITE**	$NaFe^{+3}\{Si_2O_6\}$
	M $\alpha > \beta > \gamma$	VH	3	$\simeq 0°$			**CRONSTEDTITE**	$(Fe_4^{+2}Fe_2^{+3})(Si_2Fe_2^{+3})O_{10}(OH)_8$
	W	VH	3				**DEERITE**	$(Fe,Mn)_{13}(Fe,Al)_7Si_{13}O_{44}(OH)_{11}$
	W $\omega > \varepsilon,\ \varepsilon > \omega$	E	5	(uniaxial)			**ANATASE**	TiO_2
	W – M $\alpha > \beta > \gamma$	E	5 anom.	$30°$ r>v strong	$12°$		**BADDELEYITE**	ZrO_2

System	Form	Cleavage	Twinning	Zoning	Alteration	Occurrence	Remarks
Mon.	short, prismatic, needles, pointed terminations	two at 87°, parting	simple, polysynthetic, twin seams	Z		I	cf. aegirine, aegirine-augite, Na-amphibole
Mon. (Pseudo-Hex.)						O	cf. septochlorites
Mon.	acicular, amphibole-like	one good	simple			M	
Tet.	small, prismatic, acicular	two perfect		Z	leucoxene	S, I, M	usually yellow to blue cf. rutile, brookite
Mon.	tabular	one	simple, polysynthetic			I, S	cf. zircon, rutile, brookite, anatase

ORANGE MINERALS	Pleochroism	Relief	δ	2V Dispersion	Extinction	Orientation	Mineral	Composition
Isometric								
		$-L - +M$					CHLOROPHAEITE	Chloritic mineral
	H – VH	(anom.)					GARNET	$(Mg,Fe^{+2},Mn,Ca)_3$ $(Fe^{+3},Ti,Cr,Al)_2\{Si_3O_{12}\}$
Uniaxial +								
	M – S $\alpha=\beta>\gamma$	$+L - +M$	1	$0^{\circ} - 60^{\circ}$ r<v	0° – small	cleavage length fast	Mn and Cr-CHLORITES	$(Mg,Mn,Al,Cr,Fe)_{12}$ $\{(Si,Al)_8O_{20}\}(OH)_{16}$
	W – M $\varepsilon>\omega$ often opaque	E	5			parallel	RUTILE	TiO_2
Uniaxial –								
	M – S $\gamma=\beta>\alpha$	$+L - +M$	1	$0^{\circ} - 20^{\circ}$ r>v	0° – small	cleavage length slow	Mn and Cr-CHLORITES	$(Mg,Mn,Al,Cr,Fe)_{12}$ $\{(Si,Al)_8O_{20}\}(OH)_{16}$
	S $\gamma\gtrless\beta>\alpha$ $\beta>\gamma>\alpha$	$+M$	4 – 5	$0^{\circ} - 25^{\circ}$ Mg r$\lessgtr$v Fe r<v	0° – 9°	cleavage length slow	BIOTITE	$K_2(Mg,Fe^{+2})_{6-4}(Fe^{+3},Al,Ti)_{0-2}$ $\{Si_{6-5}Al_{2-3}O_{20}\}(OH,F)_4$
	W $\omega>\varepsilon,\ \varepsilon>\omega$	E	5 anom.	0° – small			ANATASE	TiO_2
Biaxial +								
	M – S $\alpha=\beta>\gamma$	$+L - +M$	1	$0^{\circ} - 60^{\circ}$ r<v	0° – small	cleavage length fast	Mn and Cr-CHLORITES	$(Mg,Mn,Al,Cr,Fe)_{12}$ $\{(Si,Al)_8O_{20}\}(OH)_{16}$
	W $\alpha>\gamma$	$+M$	3 – 4 (anom.)	$73^{\circ} - 76^{\circ}$ r>v	9° – 15°		CLINOHUMITE (TITANOCLINOHUMITE)	$Mg(OH,F)_2\cdot 4Mg_2SiO_4$

254

System	Form	Cleavage	Twinning	Zoning	Alteration	Occurrence	Remarks
	massive, cryptocrystalline					I	pseudomorphs after olivine cf. bowlingite, iddingsite
Cub.	four, six, eight-sided, polygonal x-sect., aggregates	parting, irregular fractures	complex, sector	Z		M, I, S	inclusions cf. spinel
Mon.	tabular, radiating	perfect basal				O, I, M	
Tet.	prismatic, acicular, reticulate network, inclusions	one good	geniculate			M, I, S	rarely biaxial cf. brookite, anatase, baddeleyite, cassiterite
Mon.	tabular, radiating	perfect basal				O, I, M	
Mon.	tabular, flakes, plates, pseudo-hex.	perfect basal	simple	Z	chlorite, vermiculite, prehnite	M, I, S	birds-eye maple structure, pleochroic haloes
Tet.	small prismatic, acicular	two perfect		Z	leucoxene	S, I, M	usually yellow to blue cf. rutile, brookite
Mon.	tabular, radiating	perfect basal				O, I, M	
Mon.	rounded	poor	simple, polysynthetic			M, O	cf. humite group, olivine, staurolite

ORANGE MINERALS	Pleochroism	Relief	δ	2V Dispersion	Extinction	Orientation	Mineral	Composition
Biaxial +								
	M $\gamma>\beta>\alpha$	H	2	82° – 90° r>v	parallel, symmetrical	length slow	STAUROLITE	$(Fe^{+2},Mg)_2(Al,Fe^{+3})_9O_6$ $\{SiO_4\}_4(O,OH)_2$
	W – M $\alpha>\beta>\gamma$	VH – E	5 anom.	17° – 40° r>v	40° symmetrical		SPHENE	$CaTi\{SiO_4\}(O,OH,F)$
	W $\gamma>\alpha$	E	5 anom.	0° – 30° strong	parallel	length slow	BROOKITE	TiO_2
Biaxial –								
	W $\gamma>\alpha$	+L	1	small			KOCHUBEITE	Cr-chlorite
	M – S $\gamma=\beta>\alpha$	+L – +M	1	0° – 20° r>v	0° – small	cleavage length slow	Mn and Cr- CHLORITES	$(Mg,Mn,Al,Cr,Fe)_{12}$ $\{(Si,Al)_8O_{20}\}(OH)_{16}$
	M $\gamma\geqslant\beta>\alpha$ $\alpha>\beta=\gamma$	+L – +M	3 – 4	0° – 15° r<v	0° – 5°	cleavage length slow	PHLOGOPITE	$K_2(Mg,Fe^{+2})_6\{Si_6Al_2O_{20}\}(OH,F)_4$
	M $\gamma=\beta>\alpha$	+M	2	0° – 23° r<v	parallel	cleavage length slow	XANTHOPHYLLITE	$Ca_2(Mg,Fe)_{4.6}Al_{1.4}$ $\{Si_{2.5}Al_{5.5}O_{20}\}(OH)_4$
	$\gamma=\beta>\alpha$	+M	2	32° r<v	parallel	cleavage length slow	CLINTONITE	$Ca_2(Mg,Fe)_{4.6}Al_{1.4}$ $\{Si_{2.5}Al_{5.5}O_{20}\}(OH)_4$
	S $\gamma>\beta>\alpha$	+M	2	40° – 50° r>v	11° – 18°	length slow	BARKEVIKITE	$Ca_2(Na,K)(Fe^{+2},Mg,Fe^{+3},Mn)_5$ $\{Si_{6.5}Al_{1.5}O_{22}\}(OH)_2$
	M $\beta>\gamma>\alpha$	+M	2 – 3 anom.	66° – 87° r<v	15° – 40°	length slow	RICHTERITE- FERRORICHTERITE	$Na_2Ca(Mg,Fe^{+3},Fe^{+2},Mn)_5$ $\{Si_8O_{22}\}(OH,F)_2$
	S $\gamma\geqslant\beta>\alpha$ $\beta>\gamma>\alpha$	+M	4 – 5	0° – 25° Mg r$\lessgtr$v Fe r<v	0° – 9° wavy	cleavage length slow	BIOTITE	$K_2(Mg,Fe^{+2})_{6-4}(Fe^{+3},Al,Ti)_{0-2}$ $\{Si_{6-5}Al_{2-3}O_{20}\}(OH,F)_4$

System	Form	Cleavage	Twinning	Zoning	Alteration	Occurrence	Remarks
Mon. (Pseudo-Orth.)	short prismatic, six-sided x-sect., sieve texture	inconspicuous	interpenetrant	Z		M, S	inclusions cf. chondrodite, vesuvianite, melanite, Fe-olivine
Mon.	rhombs, irregular grains	parting	simple, polysynthetic		leucoxene	I, M, S	cf. monazite, rutile
Orth.	prismatic	poor				I, M, S	cf. rutile, pseudo-brookite
Mon.	tabular	perfect				O	cf. chlorite
Mon.	tabular, radiating	perfect basal				O, I, M	
Mon.	tabular, flakes, plates, pseudo-hex.	perfect basal	inconscipuous	Z colour zoning		I, M	inclusions common, birds-eye maple structure cf. biotite, muscovite, lepidolite, rutile, tourmaline
Mon.	tabular, short prismatic	perfect basal	simple			M	cf. clintonite, chlorite, chloritoid
Mon.	tabular, short prismatic	perfect basal	simple			M	cf. xanthophyllite, chlorite, chloritoid
Mon.	prismatic	two at $\simeq 56^{o}$	simple	Z		I	cf. hastingsite, kaersutite, basaltic hornblende
Mon.	long prismatic, fibrous	two at 56°, parting	simple, polysynthetic	Z		M, I	
Mon. (Pseudo-Hex.)	tabular, flakes, plates, pseudo-hex.	perfect basal	simple	Z	chlorite, vermiculite, prehnite	M, I, S	pleochroic haloes, inclusions common, birds-eye maple structure cf. stilpnomelane, phlogopite

ORANGE MINERALS	Pleochroism	Relief	δ	2V Dispersion	Extinction	Orientation	Mineral	Composition
Biaxial −								
	W	+M − H	2	30° − 44° r<v	15° − 35°		BUSTAMITE	$(Mn,Ca,Fe)\{SiO_3\}$
	S $\gamma>\beta>\alpha$	+M − H	3 − 5	66° − 82° r>v	0° − 19°	length slow	KAERSUTITE	$Ca_2(Na,K)(Mg,Fe^{+2},Fe^{+3})_4Ti\{Si_6Al_2O_{22}\}(O,OH,F)_2$
	W $\beta>\gamma>\alpha$	H	1	50° − 66° r<v	6° − 9°	length slow	SAPPHIRINE	$(Mg,Fe)_2Al_4O_6\{SiO_4\}$
	W $\alpha>\beta>\gamma$	E	5	0° − 27° extreme	parallel	length fast/slow	GOETHITE	$\alpha\cdot FeO\cdot OH$
	S $\gamma>\beta>\alpha$	E	5	83° slight	parallel		LEPIDOCROCITE	$\gamma-FeO\cdot OH$

System	Form	Cleavage	Twinning	Zoning	Alteration	Occurrence	Remarks
Tr.	prismatic, fibrous	three, two at 95°	simple			M, O	cf. rhodonite, pyroxmangite
Mon.	prismatic	two at 56°, parting	simple, polysynthetic	Z		I	cf. barkevikite, katophorite, basaltic hornblende, cossyrite, titanaugite
Mon.	tabular	poor	polysynthetic, uncommon			M	cf. corundum, cordierite, kyanite, zoisite, Na-amphibole
Orth.	pseudomorphs, fibrous, oolitic, concretions	one perfect				O, S, I, M	weathering product cf. bowlingite, limonite
Orth.	tabular	two perfect, moderate				O, S	

YELLOW MINERALS	Pleochroism	Relief	δ	2V Dispersion	Extinction	Orientation	Mineral	Composition
Isometric								
		−L	(weak)				SODALITE (incl. HACKMANITE)	$Na_8\{Al_6Si_6O_{24}\}Cl_2$
	W $\varepsilon>\omega$	+L − +M	1 (aniso-tropic)				EUDIALYTE-EUCOLITE	$(Na,Ca,Fe)_6Zr\{(Si_3O_9)_2\}$ $\{(Si_3O_9)_2\}(OH,F,Cl)$
		+M					GREENALITE	$Fe_6^{+2}Si_4O_{10}(OH)_8$
		+M	(aniso-tropic)				MELILITE	$(Ca,Na)_2\{(Mg,Fe^{+2},Al,Si)_3O_7\}$
		H					HELVITE	$Mn_4\{Be_3Si_3O_{12}\}S$
		H					PYROPE	$Mg_3Al_2Si_3O_{12}$
		H	(anom.)				GROSSULARITE (HESSONITE)	$Ca_3Al_2Si_3O_{12}$
		H − VH					ALLANITE (ORTHITE)	$(Ca,Ce)_2(Fe^{+2},Fe^{+3})Al_2O\cdot OH$ $\{Si_2O_7\}\{SiO_4\}$
		H − VH	(anom.)				GARNET	$(Mg,Fe^{+2},Mn,Ca)_3$ $(Fe^{+3},Ti,Cr,Al)_2\{Si_3O_{12}\}$
		VH	(anom.)	$90°$ rare			ANDRADITE	$Ca_3(Fe^{+3},Ti)_2Si_3O_{12}$
	W $\gamma>\alpha$	E	(weak)				PEROVSKITE (incl. KNOPITE-LOPARITE-DYSANALYTE)	$(Ca,Na,Fe^{+2},Ce)(Ti,Nb)O_3$

System	Form	Cleavage	Twinning	Zoning	Alteration	Occurrence	Remarks
Cub.	hexagonal x-sect., anhedral aggregates	poor	simple		zeolites, diaspore, gibbsite	I	cf. fluorite, leucite
Trig.	rhombohedral aggregates	one good, one poor		Z		I	cf. catapleite, låvenite, rosenbuschite, garnet
M'loid.	massive					O	restricted to Iron Formation, Mesabi Range, Minnesota
Tet.	tabular, peg structure	moderate, single crack		Z	zeolites, carbonate	I	usually colourless cf. zoisite, vesuvianite, apatite, nepheline
Cub.	tetrahedral, triangular x-sect., granular	one poor	simple		ochre, manganese oxide	I, M	cf. garnet, danalite
Cub.	four, six, eight-sided, polygonal x-sect., aggregates	parting, irregular fractures		Z		M, I	inclusions cf. garnet group
Cub.	four, six, eight-sided, polygonal x-sect., aggregates	parting, irregular fractures	sector	Z		M	cf. garnet group, periclase, vesuvianite
Mon.	distinct crystals, columnar, six-sided x-sect., irregular	two imperfect	uncommon	Z	metamict	I, M	anastomosing cracks, isotropic in metamict state cf. epidote, melanite
Cub.	four, six, eight-sided, polygonal x-sect., aggregates	parting, irregular fractures	complex, sector	Z	chlorite	M, I, S	inclusions cf. spinel
Cub.	four, six, eight-sided, polygonal x-sect., aggregates	parting, irregular fractures	complex, sector	Z		M	cf. garnet group
Mon? Pseudo-Cub.	small cubes, skeletal	poor to distinct	polysynthetic, complex, interpenetrant	Z	leucoxene	I, M	cf. melanite, picotite, ilmenite

YELLOW MINERALS	Pleochroism	Relief	δ	2V Dispersion	Extinction	Orientation	Mineral	Composition
Isometric								
		E	(weak)				SPHALERITE	ZnS
		+E	(weak)				LIMONITE	$FeO \cdot OH \cdot nH_2O$
Uniaxial +								
	W $\varepsilon > \omega$	$-L - +M$	1 (iso-tropic)				EUDIALYTE	$(Na,Ca,Fe)_6 Zr\{(Si_3O_9)_2\}$ (OH,F,Cl)
	W $\alpha = \beta > \gamma$	+L	1 anom.	$0^o - 20^o$ $r<v$	$\simeq$ parallel	cleavage length fast	PENNINITE	$(Mg,Al,Fe)_{12}\{(Si,Al)_8O_{20}\}(OH)_1$
	W $\alpha = \beta > \gamma$	+L	1 - 2	$0^o - 40^o$ $r<v$	$0^o - 9^o$	cleavage length fast	CLINOCHLORE	$(Mg,Al,Fe)_{12}\{(Si,Al)_8O_{20}\}(OH)_1$
	W - S $\alpha = \beta > \gamma$	$+L - +M$	1 - 2 (anom.)	$0^o - 60^o$ $r<v$	$0^o - 9^o$	cleavage length fast	CHLORITE (Unoxidised)	$(Mg,Al,Fe)_{12}\{(Si,Al)_8O_{20}\}(OH)_1$
	W $\varepsilon > \omega$	+M	1 anom. (iso-tropic)		parallel	length slow	MELILITE	$(Ca,Na)_2\{(Mg,Fe^{+2},Al,Si)_3O_7\}$
	N - M $\gamma = \alpha > \beta$ $\beta > \alpha = \gamma$	$+M - H$	3	$0^o - 30^o$ $r \lessgtr v$	$37^o - 44^o$	extinction nearest cleavage slow	PIGEONITE	$(Mg,Fe^{+2},Ca)(Mg,Fe^{+2})\{Si_2O_6\}$
	H	1 anom.			parallel	length fast	VESUVIANITE (IDOCRASE) (incl. WILUITE)	$Ca_{10}(Mg,Fe)_2Al_4\{Si_2O_7\}_2$ $\{SiO_4\}_5(OH,F)_4$
	W	H	5		straight		XENOTIME	YPO_4

System	Form	Cleavage	Twinning	Zoning	Alteration	Occurrence	Remarks
Cub.	irregular, anhedral, curved surfaces	six perfect	polysynthetic, lamellar intergrowths	Z		O	colour variable, (uniaxial) cf. cassiterite
M'loid.	stain or border to other minerals, pseudomorphs					I, M, S, O	opaque to translucent cf. goethite
Trig.	rhombohedral, aggregates	one good, one poor		Z		I	some zones isotropic cf. catapleite, lavenite, rosenbuschite, garnet, eucolite
Mon.	tabular, vermicular, radiating, pseudomorphs	perfect basal	simple, pennine law	Z		I, M	pleochroic haloes cf. clinochlore, prochlorite
Mon.	tabular, fibrous, pseudo-hex.	perfect basal	polysynthetic			M	pleochroic haloes, cf. penninite, prochlorite, leuchtenbergite, katschubeite
Mon. (Pseudo-Hex.)	tabular, scaly, radiating, pseudomorphs	perfect basal	simple, polysynthetic	Z		I, M, S, O	pleochroic haloes
Tet.	tabular, peg structure	moderate, single crack		Z	zeolites, carbonate	I	cf. zoisite, vesuvianite, apatite, nepheline
Mon.	prismatic, anhedral, overgrowths	two at 87°, parting	simple, polysynthetic	Z		I	exsolution lamellae cf. augite, olivine
Tet.	variable, prismatic, fibrous, granular, radial	imperfect	sector	Z		M, I, S	cf. zoisite, clinozoisite, apatite, grossularite, melilite, andalusite
Tet.	small, prismatic					I, M, S	biaxial varieties known, inclusions, pleochroic haloes around grains cf. zircon, sphene, monazite

YELLOW MINERALS	Pleochroism	Relief	δ	2V Dispersion	Extinction	Orientation	Mineral	Composition
Uniaxial +								
	W $\varepsilon>\omega$	VH	4 - 5		parallel	length slow	ZIRCON	$Zr\{SiO_4\}$
	W - S $\varepsilon>\omega$	VH - E	5	0° - 38° anom. strong	parallel, oblique to twin plane	length slow	CASSITERITE	SnO_2
	W - M $\varepsilon>\omega$	E	5		parallel		RUTILE	TiO_2
Uniaxial -								
		-L	3	anom.	parallel	length fast	CANCRINITE	$(Na,Ca,K)_{6-8}\{Al_6Si_6O_{24}\}$ $(CO_3,SO_4,Cl)_{1-2} \cdot 1-5H_2O$
	W $\varepsilon>\omega$	+L - +M	1 (iso-tropic)				EUCOLITE	$(Na,Ca,Fe)_6Zr\{(Si_3O_9)_2\}$ (OH,F,Cl)
	W - S $\gamma=\beta>\alpha$	+L - +M	1 - 2 anom.	0° - 20° r>v	0° - small	cleavage length slow	CHLORITE (Unoxidised)	$(Mg,Al,Fe)_{12}\{(Si,Al)_8O_{20}\}(OH)_1$
	S $\gamma=\beta>\alpha$	+L - H	3 - 5	$\simeq 0^{\circ}$	$\simeq$ parallel	length slow	STILPNOMELANE	$(K,Na,Ca)_{0-1\cdot4}$ $(Fe^{+3},Fe^{+2},Mg,Al,Mn)_{5\cdot9-8\cdot2}$ $\{Si_8O_{20}\}(OH)_4(O,OH,H_2O)_{3\cdot6-8\cdot5}$
	W $\omega>\varepsilon$	+M	1 anom. (iso-tropic)			length slow	MELILITE	$(Ca,Na)_2\{(Mg,Fe^{+2},Al,Si)_3O_7\}$
	S $\omega>\varepsilon$	+M	3		parallel	length fast	DRAVITE (TOURMALINE)	$NaMg_3Al_6B_3Si_6O_{27}(OH,F)_4$
	S $\omega>\varepsilon$	+M	3		parallel	length fast	SCHORL (TOURMALINE)	$Na(Fe,Mn)_3Al_6B_3Si_6O_{27}(OH,F)_4$

System	Form	Cleavage	Twinning	Zoning	Alteration	Occurrence	Remarks
Tet.	minute prisms	poor, absent		Z	metamict	I, S, M	cf. apatite
Tet.	subhedral, veinlets, diamond-shaped x-sect.	prismatic	geniculate, cyclic, common	Z		O, I, S	cf. sphalerite, rutile, sphene
Tet.	prismatic, acicular, reticulate network, inclusions	one good	geniculate			M, I, S	often opaque, rarely biaxial cf. brookite, anatase, sphene, baddeleyite, cassiterite
Hex.	anhedral	one perfect, one poor	polysynthetic, rare			I	cf. vishnevite
Trig.	aggregates	one good, one poor		Z		I	some zones isotropic cf. catapleite, lavenite, rosenbuschite, garnet, eudialyte
Mon. (Pseudo-Hex.)	tabular, scaly, radiating, pseudomorphs	perfect basal	simple, polysynthetic	Z		I, M, S, O	pleochroic haloes
Mon.	plates, pseudo-hex., micaceous masses	two at 90°, perfect basal	polysynthetic	Z		I, M, O	cf. biotite, chlorite, chloritoid, clintonite
Tet.	tabular, peg structure	moderate, single crack		Z		I	usually colourless cf. zoisite, vesuvianite, apatite
Trig.	prismatic	fractures at 90°		Z		M	cf. tourmaline group, chondrodite
Trig.	hexagonal, rounded, suns, triangular x-sect.	fractures		Z		I, M, S	cf. tourmaline group, biotite

YELLOW MINERALS	Pleochroism	Relief	δ	2V Dispersion	Extinction	Orientation	Mineral	Composition
Uniaxial −								
	S $\gamma\geq\beta>\alpha$ $\beta>\gamma>\alpha$	+M	4 − 5	$0°$ − $.25°$ Mg $r<v$ Fe $r>v$	$0°$ − $9°$ wavy	cleavage length slow	BIOTITE	$K_2(Mg,Fe^{+2})_{6-4}(Fe^{+3},Al,Ti)_{0-}$ $\{Si_{6-5}Al_{2-3}O_{20}\}(OH,F)_4$
	S $\gamma>\alpha$	+M − H	5		parallel or symmetrical		JAROSITE	$KFe_3^{+3}(OH)_6(SO_4)_2$
		H	1 anom.		parallel	length fast	VESUVIANITE (IDOCRASE)	$Ca_{10}(Mg,Fe)_2Al_4\{Si_2O_7\}_2$ $\{SiO_4\}_5(OH,F)_4$
	W $\omega>\varepsilon$	H	1	$0°$ − $30°$	parallel, symmetrical	tabular length slow, prismatic length fast	CORUNDUM	$\alpha\text{-}Al_2O_3$
	W $\omega>\varepsilon$, $\varepsilon>\omega$	E	5 anom.	$0°$ − small			ANATASE	TiO_2
Biaxial +								
	M	−L − +L	1 − 2	$65°$ − $90°$ $r<v$	parallel		CORDIERITE	$Al_3(Mg,Fe^{+2})_2\{Si_5AlO_{18}\}$
	W $\alpha=\beta>\gamma$	+L	1 anom.	$0°$ − $20°$ $r<v$	$\simeq$ parallel	cleavage length fast	PENNINITE	$(Mg,Al,Fe)_{12}\{(Si,Al)_8O_{20}\}(OH)$
	W $\alpha=\beta>\gamma$	+L	1 − 2	$0°$ − $40°$ $r<v$	$0°$ − $9°$	cleavage length fast	CLINOCHLORE	$(Mg,Al,Fe)_{12}\{(Si,Al)_8O_{20}\}(OH)$
	W $\alpha=\beta>\gamma$	+L	3	$44°$ − $50°$ $r>v$	parallel		NORBERGITE	$Mg(OH,F)_2Mg_2SiO_4$
	W − S $\alpha=\beta>\gamma$	+L − +M	1 − 2 (anom.)	$0°$ − $60°$ $r<v$	$0°$ − $9°$	cleavage length fast	CHLORITE (Unoxidised)	$(Mg,Al,Fe)_{12}\{(Si,Al)_8O_{20}\}(OH)$

System	Form	Cleavage	Twinning	Zoning	Alteration	Occurrence	Remarks
Mon. (Pseudo-Hex.)	tabular, flakes, bent plates, pseudo-hex.	perfect basal	simple	Z	chlorite, vermiculite, prehnite	M, I,S	pleochroic haloes, inclusions common, birds-eye maple structure cf. stilpnomelane, phlogopite
Hex.	tabular, aggregates				limonite	I, O	cf. alunite, natrojarosite
Tet.	variable, prismatic, fibrous, granular, radial	imperfect		Z		M, I	usually colourless, cf. zoisite, clinozoisite, grossularite
Trig.	tabular, prismatic, six-sided x-sect.	parting	simple, polysynthetic	Z		M, I	inclusions
Tet.	small, prismatic, acicular	two perfect		Z	leucoxene	S, I, M	cf. rutile, brookite
Orth. (Pseudo-Hex.)	pseudo-hex., anhedral	moderate to absent	simple, polysynthetic, cyclic, sector, interpenetrant		pinite, talc	M, I	pleochroic haloes, inclusions (including opaque dust) common cf. quartz, plagioclase
Mon.	tabular, vermicular, radiating, pseudomorphs	perfect basal	simple, pennine law	Z		I, M	pleochroic haloes cf. clinochlore, prochlorite
Mon.	tabular, fibrous, pseudo-hex.	perfect basal	polysynthetic			M	pleochroic haloes cf. penninite, prochlorite, leuchtenbergite, katschubeite
Orth.	massive					M, O	cf. humite group
Mon. (Pseudo-Hex.)	tabular, scaly, radiating, pseudomorphs	perfect basal	simple, polysynthetic	Z		I, M, S, O	pleochroic haloes

YELLOW MINERALS	Pleochroism	Relief	δ	2V Dispersion	Extinction	Orientation	Mineral	Composition
Biaxial +								
	W	+M	1 - 2	$48^{\circ} - 68^{\circ}$ r>v	parallel, symmetrical	cleavage fast	TOPAZ	$Al_2\{SiO_4\}(OH,F)_2$
	W $\gamma=\beta>\alpha$ $\gamma>\beta=\alpha$	+M	2	$68^{\circ} - 90^{\circ}$ r$\lessgtr$v	parallel, symmetrical	length slow	ANTHOPHYLLITE (Fe-rich)	$(Mg,Fe^{+2})_7\{Si_8O_{22}\}(OH,F)_2$
	W $\gamma>\alpha$	+M	3	$21^{\circ} - 30^{\circ}$ r>v	parallel	length slow	SILLIMANITE	Al_2SiO_5
	W $\alpha>\beta=\gamma$	+M	3	$65^{\circ} - 84^{\circ}$ r>v	parallel		HUMITE	$Mg(OH,F) \cdot 3Mg_2SiO_4$
	M - S $\gamma\geq\beta>\alpha$	+M	3	$67^{\circ} - 90^{\circ}$ r>v	26°	length slow	PARGASITE	$NaCa_2Mg_4Al_3Si_6O_{22}(OH)_2$
	W $\gamma>\beta=\alpha$	+M	3	$68^{\circ} - 78^{\circ}$ r>v	28°	length slow	ROSENBUSCHITE	$(Ca,Na,Mn)_3(Zr,Ti,Fe^{+3})\{SiO_4\}_2(F,OH)$
	W $\alpha>\beta=\gamma$	+M	3	$71^{\circ} - 85^{\circ}$ r>v	$22^{\circ} - 31^{\circ}$	extinction nearest twin plane fast	CHRONDRODITE	$Mg(OH,F)_2 \cdot 2Mg_2SiO_4$
	W $\alpha>\gamma$	+M	3	$76^{\circ} - 87^{\circ}$ r>v	parallel, symmetrical	length fast, long diagonal of rhombic sect. slow	LAWSONITE	$CaAl_2(OH)_2\{Si_2O_7\}H_2O$
	W $\alpha>\gamma$	+M	3 - 4 (anom.)	$73^{\circ} - 76^{\circ}$ r>v	$9^{\circ} - 15^{\circ}$		CLINOHUMITE	$Mg(OH,F)_2 \cdot 4Mg_2SiO_4$
	M	+M - H	1 - 3	$0^{\circ} - 60^{\circ}$ r>v	parallel	length fast/slow	THULITE	$Ca_2Mn^{+3}Al_2O \cdot OH\{Si_2O_7\}\{SiO_4\}$
	W $\beta>\alpha=\gamma$	+M - H	2	$26^{\circ} - 85^{\circ}$ r<v	$4^{\circ} - 32^{\circ}$	length slow	PUMPELLYITE	$Ca_4(Mg,Fe^{+2})(Al,Fe^{+3})_5O(OH)\{Si_2O_7\}_2\{SiO_4\}_2 \cdot 2H_2O$
	W $\gamma=\beta>\alpha$ $\gamma>\beta=\alpha$	+M - H	2 - 3	$68^{\circ} - 90^{\circ}$ r$\lessgtr$v	parallel	length slow	GEDRITE	$(Mg,Fe^{+2})_5Al_2\{Si_6Al_2O_{22}\}(OH,$

System	Form	Cleavage	Twinning	Zoning	Alteration	Occurrence	Remarks
Orth.	prismatic, aggregates	one perfect				I, M, S	usually colourless cf. quartz, andalusite
Orth.	bladed, prismatic, fibrous, asbestiform	two at $54\frac{1}{2}^{o}$			talc	M	cf. gedrite, tremolite, cummingtonite, holmquistite
Orth.	needles, slender prisms, fibrous, rhombic x-sect., faserkiesel	one, // long diagonal of x-sect.				M, O	cf. andalusite, mullite
Orth.	tabular, rounded	poor				M	cf. humite group, olivine, staurolite
Mon.	prismatic	two at 56°, parting	simple, polysynthetic			M, I	cf. hornblende, hastingsite, cummingtonite
Tr.	prismatic, fibres, needles	one perfect, two poor				I	cf. låvenite
Mon.	rounded	poor	simple, polysynthetic, twin seams			M, O	cf. humite group, olivine, staurolite
Orth.	rectangular, rhombic	two perfect	simple, polysynthetic			M	cf. zoisite, prehnite, pumpellyite, andalusite, scapolite
Mon.	rounded	poor	simple, polysynthetic			M, O	cf. humite group, olivine, staurolite
Orth.	columnar aggregates	one perfect, one imperfect	polysynthetic, rare			M	cf. zoisite, clinozoisite, vesuvianite, sillimanite
Mon.	fibrous, needles, bladed, radial aggregates	two moderate	common	Z		M	cf. clinozoisite, zoisite, epidote, lawsonite
Orth.	bladed, prismatic, fibrous, asbestiform	two at $54\frac{1}{2}^{o}$			talc	M	cf. Fe-anthophyllite, holmquistite, cummingtonite, grunerite, zoisite, tremolite

YELLOW MINERALS	Pleochroism	Relief	δ	2V Dispersion	Extinction	Orientation	Mineral	Composition
Biaxial +								
	N - M $\gamma=\alpha>\beta$ $\beta>\alpha=\gamma$	+M - H	3	$0°$ - $30°$ $r \lesseqgtr v$	$37°$ - $44°$	extinction nearest cleavage slow	PIGEONITE	$(Mg,Fe^{+2},Ca)(Mg,Fe^{+2})\{SiO_6\}$
	W	+M - VH	4	$48°$ - $90°$ $r \lesseqgtr v$	parallel	cleavage length slow	OLIVINE	$(Mg,Fe)_2\{SiO_4\}$
	W - M $\beta>\alpha>\gamma$	H	1 - 2 (anom.)	$45°$ - $68°$ $r>v$	$2°$ - $30°$	length fast	CHLORITOID (OTTRELITE)	$(Fe^{+2},Mg,Mn)_2(Al,Fe^{+3})Al_3O_2$ $\{SiO_4\}_2(OH)_4$
	M $\gamma>\beta>\alpha$	H	2	$82°$ - $90°$ $r>v$	parallel, symmetrical	length slow	STAUROLITE	$(Fe^{+2},Mg)_2(Al,Fe^{+3})_9O_6$ $\{SiO_4\}_4(O,OH)_2$
	M $\alpha>\beta>\gamma$	H	5	$70°$ - $88°$ $r>v$	$13°$	length slow	ASTROPHYLLITE	$(K,Na)_3(Fe,Mn)_7Ti_2$ $\{Si_4O_{12}\}_2(O,OH,F)_7$
	W $\gamma>\alpha$	H - VH	2 - 3	$40°$ - $90°$ $r \lesseqgtr v$	$1°$ - $42°$ straight, // elongation	length fast/slow	ALLANITE (ORTHITE)	$(Ca,Ce)_2(Fe^{+2},Fe^{+3})Al_2O \cdot OH$ $\{Si_2O_7\}\{SiO_4\}$
	S $\gamma>\alpha>\beta$ $\gamma>\beta>\alpha$	H - VH	3 - 5	$64°$ - $85°$ $r \lesseqgtr v$	$2°$ - $9°$	length fast/slow	PIEMONTITE	$Ca_2(Mn,Fe^{+3},Al)_2AlO \cdot OH$ $\{Si_2O_7\}\{SiO_4\}$
	W $\beta>\alpha=\gamma$	H - VH	4 - 5	$6°$ - $19°$ $r<v$	$2°$ - $7°$	length fast/slow	MONAZITE	$(Ce,La,Th)PO_4$
	W - S	VH - E	5	$0°$ - $38°$ anom. strong	parallel, oblique to twin plane	length slow	CASSITERITE	SnO_2
	W - M $\alpha>\beta>\gamma$	VH - E	5 anom.	$17°$ - $40°$ $r>v$	$40°$ symmetrical		SPHENE	$CaTi\{SiO_4\}(O,OH,F)$
	W $\gamma>\alpha$	E	5 anom.	$0°$ - $30°$ strong	parallel	length slow	BROOKITE	TiO_2

System	Form	Cleavage	Twinning	Zoning	Alteration	Occurrence	Remarks
Mon.	prismatic, anhedral, overgrowths	two at 87°, parting	simple, polysynthetic	Z		I	exsolution lamellae cf. augite, olivine
Orth.	anhedral – euhedral, rounded	uncommon, one moderate, irregular fractures	uncommon, vicinal	Z	chlorite, antigorite, serpentine, iddingsite, bowlingite	I, M	deformation lamellae cf. diopside, augite, humite group, epidote
Mon., Tr.	tabular	one perfect, one imperfect, parting	polysynthetic	Z hourglass		M, I	inclusions common cf. chlorite, clintonite, biotite, stilpnomelane
Mon. (Pseudo-Orth.)	short prismatic, six-sided x-sect., sieve texture	inconspicuous	interpenetrant	Z		M, S	inclusions cf. chondrodite, vesuvianite, melanite, Fe-olivine
Tr.	tabular, plates	perfect				I	cf. biotite, staurolite, ottrelite
Mon.	distinct crystals, columnar aggregates, six-sided x-sect.	two imperfect	uncommon	Z	metamict	I, M	anastomosing cracks, isotropic in metamict state cf. epidote, melanite
Mon.	columnar	one perfect	polysynthetic, uncommon			M, I	cf. thulite, dumortierite
Mon.	small, euhedral	parting	rare		metamict	I, S	cf. sphene, zircon
Tet.	subhedral, veinlets, diamond-shaped x-sect.	prismatic	geniculate, cyclic, common	Z		O, S	cf. sphalerite, rutile, sphene
Mon.	rhombs, irregular grains	parting	simple, polysynthetic		leucoxene	I, M, S	cf. monazite, rutile, cassiterite
Orth.	prismatic	poor				I, M, S	cf. rutile, pseudobrookite

YELLOW MINERALS	Pleochroism	Relief	δ	2V Dispersion	Extinction	Orientation	Mineral	Composition
Biaxial −								
		−L	2 − 3	40° − 60°	≃ parallel	length slow	SEPIOLITE	$H_4Mg_2\{Si_3O_{10}\}$
	M	−L − +L	1 − 2	65° − 90° r<v	parallel		CORDIERITE	$Al_3(Mg,Fe^{+2})_2\{Si_5AlO_{18}\}$
	W γ>β>α	−L − +M	3	0° − 30°	small		MONTMORILLONITE (SMECTITE)	$(\frac{1}{2}Ca,Na)_{0.7}(Al,Mg,Fe)_4\{(Si,Al)_8O_{20}\}(OH)_4\cdot nH_2O$
	W γ=β>α	+L	1 anom.	0° − 40° r>v	≃ parallel	cleavage length slow	PENNINITE	$(Mg,Al,Fe)_{12}\{(Si,Al)_8O_{20}\}(OH)_{16}$
	W γ=β>α	+L	3 − 4	small	≃ parallel	cleavage length slow	MINNESOTAITE	$(Fe^{+2},Mg)_3\{(Si,Al)_4O_{10}\}OH_2$
	W − S γ=β>α	+L − +M	1 − 2 (anom.)	0° − 20° r>v	0° − small	cleavage length slow	CHLORITE (Oxidised)	$(Mg,Al,Fe)_{12}\{(Si,Al)_8O_{20}\}(OH)_{16}$
	W − S γ=β>α	+L − +M	1 − 2 (anom.)	0° − 20° r>v	0° − small	cleavage length slow	CHLORITE (Unoxidised)	$(Mg,Al,Fe)_{12}\{(Si,Al)_8O_{20}\}(OH)_{16}$
	M γ≥β>α α>β=γ	+L − +M	3 − 4	0° − 15° r<v	0° − 5°	cleavage length slow	PHLOGOPITE	$K_2(Mg,Fe^{+2})_6\{Si_6Al_2O_{20}\}(OH,F)_4$
	W γ=β>α	+L − +M	3 − 4	25° − 70°	indistinct	length slow	NONTRONITE (SMECTITE)	$(\frac{1}{2}Ca,Na)_{0.7}(Al,Mg,Fe)_4\{(Si,Al)_8O_{20}\}(OH)_4\cdot nH_2O$
	S γ=β>α	+L − H	3 − 5	≃ 0°	≃ parallel	length slow	STILPNOMELANE	$(K,Na,Ca)_{0-1.4}(Fe^{+3},Fe^{+2},Mg,Al,Mn)_{5.9-8.2}\{Si_8O_{20}\}(OH)_4(O,OH,H_2O)_{3.6-8.5}$
	S γ>β>α β>γ>α	+M	1 − 2 anom.	12° − 65° r<v	12° − 30°	length slow	CROSSITE	$Na_2(Mg_3,Fe_3^{+2},Fe_2^{+3},Al_2)\{Si_8O_{22}\}(OH)_2$
	M − S γ>β>α γ<β>α	+M	1 − 3	0° − 50° r<v	36° − 70°	length fast	KATOPHORITE-MAGNESIOKATOPHORITE	$Na_2Ca(Mg,Fe^{+2})_4Fe^{+3}\{Si_7AlO_{22}\}(OH,F)_2$

ystem	Form	Cleavage	Twinning	Zoning	Alteration	Occurrence	Remarks
on. (Orth.)	fibrous aggregates, curved, matted					I, S	
rth. Pseudo-ex.)	pseudo-hex., anhedral	moderate to absent	simple, polysynthetic, cyclic, sector, interpenetrant		pinite, talc	M, I	pleochroic haloes, inclusions (including opaque dust) common cf. quartz, plagioclase
on.	shards, scales, massive, microcrystalline	perfect basal				I, S	cf. nontronite
on.	tabular, vermicular, radiating, pseudomorphs	perfect basal	simple, pennine law	Z		I, M	pleochroic haloes cf. clinochlore, prochlorite
on.	minute plates, needles, radiating, aggregates	perfect basal				M	cf. talc
on. Pseudo-ex.)	tabular, scaly, radiating, pseudomorphs	perfect basal	simple, polysynthetic	Z		I, M, S, O	pleochroic haloes
on. Pseudo-ex.)	tabular, scaly, radiating, pseudomorphs	perfect basal	simple, polysynthetic	Z		I, M, S, O	pleochroic haloes
on.	tabular, flakes, plates, pseudo-hex.	perfect basal	inconspicuous	Z colour zoning		I, M	inclusions common, birds-eye maple structure cf. biotite, muscovite, lepidolite, rutile, tourmaline
on.	shards, scales, massive, fibrous	perfect basal				I, M	cf. montmorillonite, kaolinite
on.	plates, pseudo-hex., micaceous masses	two at 90°, perfect basal	polysynthetic	Z		I, M, O	cf. biotite, chlorite, chloritoid, clintonite
on.	prismatic, columnar aggregates	two at 58°	simple, polysynthetic	Z		M	cf. glaucophane, riebeckite
on.	prismatic	two at 56°, parting	simple	Z		I	cf. barkevikite, kaersutite, basaltic hornblende, arfvedsonite, cossyrite

YELLOW MINERALS	Pleochroism	Relief	δ	2V Dispersion	Extinction	Orientation	Mineral	Composition
Biaxial –								
	M $\gamma=\beta>\alpha$	+M	2	$0°-23°$ r<v	parallel	cleavage length slow	XANTHOPHYLLITE	$Ca_2(Mg,Fe)_{4.6}Al_{1.4}\{Si_{2.5}Al_{5.5}O_{20}\}(OH)_4$
	M $\gamma=\beta>\alpha$	+M	2	$32°$ r<v	parallel	cleavage length slow	CLINTONITE	$Ca_2(Mg,Fe)_{4.6}Al_{1.4}\{Si_{2.5}Al_{5.5}O_{20}\}(OH)_4$
	S $\gamma>\beta>\alpha$	+M	2	$40°-50°$ r>v	$11°-18°$	length slow	BARKEVIKITE	$Ca_2(Na,K)(Fe^{+2},Mg,Fe^{+3},Mn)_5\{Si_{6.5}Al_{1.5}O_{22}\}(OH)_2$
	W	+M	2	$63°-80°$ r<v	inclined		AXINITE	$(Ca,Mn,Fe^{+2})_3Al_2BO_3\{Si_4O_{12}\}($
	W patchy $\alpha>\gamma$	+M	2	$73°-86°$ r<v	parallel	length fast	ANDALUSITE	Al_2SiO_5
	M $\gamma=\beta>\alpha$	+M	2 – 3	$0°-20°$ r<v	$0°-3°$	cleavage length slow	GLAUCONITE (incl. CELADONITE)	$(K,Na,Ca)_{1.2-2.0}(Fe^{+3},Al,Fe^{+2},Mg)_{4.0}\{Si_{7-7.6}Al_{1-0.4}O_{20}\}(OH)_4\cdot nH_2$
	M $\beta>\gamma>\alpha$	+M	2 – 3 anom.	$66°-87°$ r<v	$15°-40°$	length slow	RICHTERITE-FERRORICHTERITE	$Na_2Ca(Mg,Fe^{+3},Fe^{+2},Mn)_5\{Si_8O_{22}\}(OH,F)_2$
	W $\gamma=\beta>\alpha$ $\gamma>\beta=\alpha$	+M	2 – 3	$68°-90°$ r≲v	parallel, symmetrical	length slow	ANTHOPHYLLITE (Fe-rich)	$(Mg,Fe^{+2})_7\{Si_8O_{22}\}(OH,F)_2$
	W $\gamma>\beta>\alpha$	+M	2 – 3	$73°-86°$ r<v	$10°-17°$	length slow	ACTINOLITE-FERROACTINOLITE	$Ca_2(Mg,Fe^{+2})_5\{Si_8O_{22}\}(OH,F)_2$
	S	+M	3	$51°$ r>v	parallel	length slow	HOLMQUISTITE	$Li_2(Mg,Fe^{+2})_3(Al,Fe^{+3})_2\{Si_8O_{22}\}(OH)_2$
	S $\gamma\geqslant\beta>\alpha$ $\beta>\gamma>\alpha$	+M	4 – 5	$0°-25°$ Mg r≲v Fe r>v	$0°-9°$ wavy	cleavage length slow	BIOTITE	$K_2(Mg,Fe^{+2})_{6-4}(Fe^{+3},Al,Ti)_{0-}\{Si_{6-5}Al_{2-3}O_{20}\}(OH,F)_4$
	S $\alpha>\beta>\gamma$	+M – H	1 – 2	$0°-50°$ r<v	$0°-30°$ anom.	length fast	ARFVEDSONITE	$Na_3(Mg,Fe^{+2})_4Al\{Si_8O_{22}\}(OH,F$

System	Form	Cleavage	Twinning	Zoning	Alteration	Occurrence	Remarks
on.	tabular, short prismatic	perfect basal	simple			M	cf. clintonite, chlorite chloritoid
on.	tabular, short prismatic	perfect basal	simple			M	cf. xanthophyllite, chlorite, chloritoid
on.	prismatic	two at $\simeq 56^{\circ}$	simple	Z		I	cf. hastingsite, kaersutite, basaltic hornblende
r.	bladed, wedge-shaped clusters	imperfect in several directions		Z		I, M	inclusions common cf. quartz
rth.	columnar, square x-sect., inclusions in shape of a cross (chiastolite)	two at 89°			sericite, kyanite, sillimanite	M, S	cf. sillimanite
on.	grains, pellets, plates, pseudomorphs	perfect basal			limonite, goethite	S	cf. chamosite, biotite, chlorites
on.	long prismatic, fibrous	two at 56°, parting	simple, polysynthetic	Z		M, I	
rth.	bladed, prismatic, fibrous, asbestiform	two at $54\frac{1}{2}^{\circ}$			talc	M	cf. gedrite, tremolite, cummingtonite, holmquistite
on.	long prismatic, fibrous	two at 56°, parting	simple, polysynthetic	Z		M, I	cf. tremolite, orthoamphibole, hornblende
rth.	fibrous, bladed, prismatic	two at $54\frac{1}{2}^{\circ}$				M, I	cf. anthophyllite, glaucophane
on. (Pseudo-hex.)	tabular, flakes, bent plates, pseudo-hex.	perfect basal	simple	Z	chlorite, vermiculite, prehnite	M, I, S	pleochroic haloes, inclusions common, birds-eye maple structure cf. stilpnomelane, phlogopite
on.	prismatic	two at 56°, parting	simple, polysynthetic	Z		I	cf. riebeckite, katophorite, glaucophane, tourmaline

YELLOW MINERALS	Pleochroism	Relief	δ	2V Dispersion	Extinction	Orientation	Mineral	Composition
Biaxial -								
	W $\gamma=\beta>\alpha$ $\gamma>\beta=\alpha$	+M – H	2 – 3	$68^{\circ} - 90^{\circ}$ $r \lessgtr v$	parallel, symmetrical	length slow	FERROGEDRITE	$Fe_5Al_4Si_6O_{22}(OH)_2$
	S $\gamma>\beta>\alpha$	+M – H	2 – 5	$60^{\circ} - 82^{\circ}$ $r<v$	$0^{\circ} - 18^{\circ}$	length slow	BASALTIC HORNBLENDE (LAMPROBOLITE, OXYHORNBLENDE)	$(Ca,Na)_{2-3}(Mg,Fe^{+2})_{3-2}$ $(Fe^{+3},Al)_{2-3}O_2\{Si_6Al_2O_{22}\}$
	S $\beta\gtrless\gamma>\alpha$ $\gamma>\beta>\alpha$	+M – H	3	$10^{\circ} - 90^{\circ}$ $r\lessgtr v$	$9^{\circ} - 40^{\circ}$	length slow	HASTINGSITE- FERROHASTINGSITE	$NaCa_2(Mg,Fe^{+2})_4$ $(Al,Fe^{+3})Al_2Si_6O_{22}(OH,F)_2$
	W $\gamma>\beta=\alpha$	+M – H	3 – 4	$84^{\circ} - 90^{\circ}$ $r>v$	$10^{\circ} - 15^{\circ}$	length slow	GRUNERITE	$(Fe^{+2},Mg)_7\{Si_8O_{22}\}(OH)_2$
	S $\gamma>\beta>\alpha$	+M – H	3 – 5	$66^{\circ} - 82^{\circ}$ $r>v$	$0^{\circ} - 19^{\circ}$	length slow	KAERSUTITE	$Ca_2(Na,K)(Mg,Fe^{+2},Fe^{+3})_4Ti$ $\{Si_6Al_2O_{22}\}(O,OH,F)_2$
	W	+M – VH	4	$48^{\circ} - 90^{\circ}$ $r\lessgtr v$	parallel	cleavage length slow	OLIVINE	$(Mg,Fe)_2\{SiO_4\}$
	W – M $\beta>\alpha>\gamma$	H	1 – 2 anom.	$45^{\circ} - 68^{\circ}$ $r>v$	$20^{\circ} - 30^{\circ}$	length fast	CHLORITOID (OTTRELITE)	$(Fe^{+2},Mg,Mn)_2(Al,Fe^{+3})Al_3O_2$ $\{SiO_4\}_2(OH)_4$
	S	H	3	65° $r<v$			HOWIEITE	$Na(Fe,Mn)_{10}$ $(Fe,Al)_2Si_{12}O_{31}(OH)_{13}$
	W $\beta>\gamma>\alpha$	H	3 – 4 anom.	$74^{\circ} - 90^{\circ}$ $r>v$	$0^{\circ} - 15^{\circ}$ parallel in elong. sect.	length fast/slow	EPIDOTE	$Ca_2Fe^{+3}Al_2O \cdot OH\{Si_2O_7\}\{SiO_4\}$
	W $\gamma>\beta>\alpha$	H	4	$73^{\circ} - 85^{\circ}$ $r<v$	$40^{\circ} - 41^{\circ}$	length fast	LÅVENITE	$(Na,Ca,Mn,Fe^{+2})_3(Zr,Nb,Ti)$ $\{Si_2O_7\}(OH,F)$
	W $\gamma>\alpha$	H – VH	2 – 3	$40^{\circ} - 90^{\circ}$ $r\lessgtr v$	$1^{\circ} - 42^{\circ}$ straight, // elongation	length fast/slow	ALLANITE (ORTHITE)	$(Ca,Ce)_2(Fe^{+2},Fe^{+3})Al_2O \cdot OH$ $\{Si_2O_7\}\{SiO_4\}$

System	Form	Cleavage	Twinning	Zoning	Alteration	Occurrence	Remarks
Orth.	prismatic, asbestiform	two at $54\frac{1}{2}^{\circ}$				M	cf. anthophyllite, zoisite, cummingtonite, grunerite
Mon.	short prismatic	two at 56°	simple, polysynthetic	Z		I	cf. kaersutite, barkevikite, katophorite, cossyrite
Mon.	prismatic	two at 56°, parting	simple, polysynthetic	Z		I, M	cf. arfvedsonite, hornblende
Mon.	prismatic, subradiating, fibrous, asbestiform	two at 55°, cross fractures	simple, polysynthetic		limonite	M	cf. tremolite, actinolite, cummingtonite, anthophyllite
Mon.	prismatic	two at 56°, parting	simple, polysynthetic	Z		I	cf. barkevikite, katophorite, basaltic hornblende, cossyrite, titanaugite
Orth.	anhedral - euhedral, rounded	uncommon, one moderate, irregular fractures	uncommon, vicinal	Z	chlorite, antigorite, serpentine, iddingsite, bowlingite	I, M	deformation lamellae cf. diopside, augite, humite group, epidote
Mon., Tr.	tabular	one perfect, one imperfect, parting	polysynthetic	Z hour-glass		M, I	usually positive cf. chlorite, clintonite, biotite, stilpnomelane
Tr.	bladed	three, one good				M	
Mon.	distinct crystals, columnar aggregates, six-sided x-sect.	one perfect	uncommon	Z		M, I, S	cf. zoisite, clinozoisite, diopside, augite, sillimanite
Mon.	prismatic	one good	polysynthetic			I	
Mon.	distinct crystals, columnar, six-sided x-sect., irregular	two imperfect	uncommon	Z	metamict	I, M	anastomosing cracks, isotropic in metamict state cf. epidote, melanite

YELLOW MINERALS	Pleochroism	Relief	δ	2V Dispersion	Extinction	Orientation	Mineral	Composition
Biaxial –								
	W $\alpha>\beta\geqslant\gamma$	H – VH	4 – 5	$60^{\circ} - 70^{\circ}$ r>v	$0^{\circ} - 10^{\circ}$	extinction nearest cleavage fast	ACMITE	$NaFe^{+3}\{Si_2O_6\}$
	W $\gamma>\beta>\alpha$	VH	4	$44^{\circ} - 70^{\circ}$ r>v	parallel	length fast/slow	TEPHROITE	$Mn_2\{SiO_4\}$
	W	VH	4	$44^{\circ} - 70^{\circ}$ r>v	parallel	length fast/slow	KNEBELITE	$(Mn,Fe_2)\{SiO_4\}$
	W	VH	4	$48^{\circ} - 52^{\circ}$ r>v	parallel	cleavage length slow	FAYALITE	Fe_2SiO_4
	W $\omega>\epsilon, \epsilon>\omega$	E	5 anom.	(uniaxial)			ANATASE	TiO_2
	$\alpha>\gamma>\beta$	E	5	$0^{\circ} - 27^{\circ}$ extreme	parallel	length fast/slow	GOETHITE	$\alpha-FeO\cdot OH$
	W – M $\alpha>\beta>\gamma$	E	5	30° r>v strong	12°		BADDELEYITE	ZrO_2
	S $\gamma>\beta>\alpha$	E	5	83° slight	parallel		LEPIDOCROCITE	$\gamma-FeO\cdot OH$

System	Form	Cleavage	Twinning	Zoning	Alteration	Occurrence	Remarks
Mon.	short prismatic, needles, pointed terminations	two at 87°, parting	simple, polysynthetic, twin seams	Z		I	cf. aegirine, aegirine-augite, Na-amphibole
Orth.	anhedral - euhedral, rounded	two moderate, imperfect	uncommon			O	cf. knebelite, olivine
Orth.	anhedral - euhedral, rounded	two moderate, imperfect	uncommon			O, M	cf. tephroite, olivine
Orth.	anhedral - euhedral, rounded	one moderate, irregular fractures	uncommon, vicinal		grunerite, serpentine	I, M	cf. olivine, knebelite, pyroxene
Tet.	small, prismatic, acicular	two perfect		Z	leucoxene	S, I, M	cf. rutile, brookite
Orth.	pseudomorphs, fibrous, oolitic concretions	one perfect				O, S, I, M	weathering product cf. bowlingite, limonite
Mon.	tabular	one	simple, polysynthetic			I, S	cf. zircon, rutile, brookite, anatase
Orth.	tabular	two perfect, moderate				O, S	

GREEN MINERALS	Pleochroism	Relief	δ	2V Dispersion	Extinction	Orientation	Mineral	Composition
Isometric								
		$-M$					FLUORITE	CaF_2
		$-L - +M$					CHLOROPHAEITE	Chloritic mineral
		$-L - +M$ (weak)					PALAGONITE	Altered glass
		$+M$					GREENALITE	$Fe_6^{+2}Si_4O_{10}(OH)_8$
		H (anom.)					GROSSULARITE	$Ca_3Al_2Si_3O_{12}$
		H - VH (anom.)					GARNET (GROSSULARITE, UVAROVITE)	$Ca_3(Al,Cr)_2Si_3O_{12}$
		H - E					SPINEL (HERCYNITE, GAHNITE, PLEONASTE, CEYLONITE)	$(Mg,Fe^{+2},Zn)Al_2O_4$
		VH (anom.)					UVAROVITE	$Ca_3Cr_2Si_3O_{12}$
	W $\gamma>\alpha$	$+E$	(weak)				PEROVSKITE (DYSANALYTE)	$(Ca,Na,Fe^{+2},Ce)(Ti,Nb)O_3$
Uniaxial +								
	W - S $\alpha=\beta>\gamma$	$+L - +M$ (anom.)	1 - 2	$0° - 60°$ r<v	$0° - 9°$	cleavage length fast	CHLORITE (Unoxidised)	$(Mg,Al,Fe)_{12}\{(Si,Al)_8O_{20}\}(OH)$
	N - M $\gamma=\alpha>\beta$ $\beta>\alpha=\gamma$	$+M - H$	3	$0° - 30°$ r$\lessgtr$v	$37° - 44°$	extinction nearest cleavage slow	PIGEONITE	$(Mg,Fe^{+2},Ca)(Mg,Fe^{+2})\{Si_2O_6\}$

System	Form	Cleavage	Twinning	Zoning	Alteration	Occurrence	Remarks
Cub.	anhedral, hexagonal x-sect.	two or three perfect	interpenetrant	Z		I, S, O, M	colour spots cf. cryolite, halite
	massive, cryptocrystalline					I	pseudomorphs after olivine cf. bowlingite, iddingsite
M'loid.	amorphous, oolitic				chlorite	I	cf. volcanic glass, opal, collophane
M'loid.	massive					O	restricted to Iron Formation, Mesabi Range, Minnesota
Cub.	four, six, eight-sided, polygonal x-sect., aggregates	parting, irregular fractures	sector	Z		M	cf. uvarovite
Cub.	four, six or eight-sided x-sect., grains, aggregates	parting, irregular fractures	complex, sector	Z		M, I, S	inclusions cf. spinel
Cub.	small grains, cubes, octahedra, rhombic x-sect.	parting		Z		M, I	colour variable according to composition cf. garnet
Cub.	four, six, eight-sided, polygonal x-sect., aggregates	parting, irregular fractures	complex, sector	Z		M	cf. grossularite
Mon? Pseudo-Cub.	small cubes, skeletal	one poor to distinct	polysynthetic, complex, interpenetrant	Z	leucoxene	I, M	cf. melanite, picotite, ilmenite
Mon. (Pseudo-Hex.)	tabular, scaly, radiating, pseudomorphs	one perfect	simple, polysynthetic	Z		I, S, M, O	pleochroic haloes cf. bowlingite
Mon.	prismatic, anhedral, overgrowths	two at 87°, parting	simple, polysynthetic	Z		I	exsolution lamellae cf. augite, olivine

GREEN MINERALS	Pleochroism	Relief	δ	2V Dispersion	Extinction	Orientation	Mineral	Composition
Uniaxial +								
		H	1 anom.		parallel	length fast	**VESUVIANITE** (IDOCRASE)	$Ca_{10}(Mg,Fe)_2Al_4\{Si_2O_7\}_2$ $\{SiO_4\}_5(OH,F)_4$
	W - M $\beta>\alpha>\gamma$	H	1 - 2 anom.	$45°-68°$	$2°-30°$	length fast	**CHLORITOID** (OTTRELITE)	$(Fe^{+2},Mg,Mn)_2(Al,Fe^{+3})Al_3O_2$ $\{SiO_4\}_2(OH)_4$
	W	H	5		straight		**XENOTIME**	YPO_4
	W - M $\epsilon>\omega$	E	5		parallel		**RUTILE**	TiO_2
Uniaxial −								
	W	+L	1			long. sect. length fast, x-sect. length slow	**BERYL** (incl. EMERALD)	$Be_3Al_2\{Si_6O_{18}\}$
	W - S $\gamma=\beta>\alpha$	+L - +M	1 - 2 (anom.)	$0°-20°$ r>v	$0°-$ small	cleavage length slow	**CHLORITE** (Oxidised)	$(Mg,Al,Fe)_{12}\{(Si,Al)_8O_{20}\}(OH)_1$
	W - S $\gamma=\beta>\alpha$	+L - +M	1 - 2 (anom.)	$0°-20°$ r>v	$0°-$ small	cleavage length slow	**CHLORITE** (Unoxidised)	$(Mg,Al,Fe)_{12}\{(Si,Al)_8O_{20}\}(OH)_1$
	S $\gamma=\beta>\alpha$	+L - H	3 - 5	$\simeq 0°$	$\simeq$ parallel	length slow	**STILPNOMELANE**	$(K,Na,Ca)_{0-1.4}$ $(Fe^{+3},Fe^{+2},Mg,Al,Mn)_{5.9-8.2}$ $\{Si_8O_{20}\}(OH)_4(O,OH,H_2O)_{3.6-8.5}$
	S $\omega>\epsilon$	+M	2 - 3		parallel	length fast	**ELBAITE** (TOURMALINE)	$Na(Li,Al)_3Al_6B_3Si_6O_{27}(OH,F)_4$
	S $\omega>\epsilon$	+M	3		parallel	length fast	**SCHORL** (TOURMALINE)	$Na(Fe,Mn)_3Al_6B_3Si_6O_{27}(OH,F)_4$

System	Form	Cleavage	Twinning	Zoning	Alteration	Occurrence	Remarks
Tet.	variable, prismatic, fibrous, granular, radial	imperfect	sector	Z		M, I, S	cf. zoisite, clinozoisite, apatite, grossularite, melilite, andalusite
Mon., Tr.	tabular	one perfect, one imperfect, parting	polysynthetic	Z hour-glass		M, I	inclusions common cf. chlorite, clintonite, biotite, stilpnomelane
Tet.	small, prismatic					I, M, S	inclusions, pleochroic haloes around grains cf. zircon, sphene, monazite
Tet.	prismatic, acicular, reticulate network, inclusions	good	geniculate			M, I, S	often opaque cf. brookite, anatase, sphene, baddeleyite, cassiterite
Hex.	prismatic, inclusions zoned	imperfect		Z	kaolin	I, M	liquid inclusions cf. apatite, quartz
Mon. (Pseudo-Hex.)	tabular, scaly, radiating, pseudomorphs	perfect basal	simple, polysynthetic	Z		I, M, S, O	pleochroic haloes cf. bowlingite
Mon. (Pseudo-Hex.)	tabular, scaly, radiating, pseudomorphs	perfect basal	simple, polysynthetic	Z		I, S, M, O	pleochroic haloes cf. bowlingite
Mon.	plates, pseudo-hex., micaceous masses	two at 90°, perfect basal	polysynthetic	Z		I, M, O	cf. biotite
Trig.	prismatic, radiating	fractures at 90°	rare	Z		I, M	cf. tourmaline group, biotite, hornblende
Trig.	hexagonal, rounded, suns, triangular x-sect.	fractures		Z		I, M, S	cf. tourmaline group, biotite, hornblende

GREEN MINERALS	Pleochroism	Relief	δ	2V Dispersion	Extinction	Orientation	Mineral	Composition
Uniaxial −								
		+M	3		parallel		ZUSSMANITE	$K(Fe,Mg,Mn)_{13}(Si,Al)_{18}O_{42}(OH)_{14}$
	S $\gamma \geqslant \beta > \alpha$ $\beta > \gamma > \alpha$	+M	4 − 5	$0^o - 25^o$ Mg r$\lessgtr$v Fe r$>$v	$0^o - 9^o$ wavy	cleavage length slow	BIOTITE	$K_2(Mg,Fe^{+2})_{6-4}(Fe^{+3},Al,Ti)_{0-2}$ $\{Si_{6-5}Al_{2-3}O_{20}\}(OH,F)_4$
	H	1 anom.			parallel	length fast	VESUVIANITE (IDOCRASE)	$Ca_{10}(Mg,Fe)_2Al_4\{Si_2O_7\}_2$ $\{SiO_4\}_5(OH,F)_4$
	W $\omega > \epsilon$	H	1	$0^o - 30^o$	parallel	tabular length slow, prismatic length fast	CORUNDUM	$\alpha-Al_2O_3$
Biaxial +								
	M	−L − +L	1 − 2	$65^o - 90^o$ r$<$v	parallel		CORDIERITE	$Al_3(Mg,Fe^{+2})_2\{Si_5AlO_{18}\}$
	W $\alpha = \beta > \gamma$	+L	1 (anom.)	$0^o - 20^o$ r$<$v	$\simeq$ parallel	cleavage length fast	PENNINITE	$(Mg,Al,Fe)_{12}\{(Si,Al)_8O_{20}\}(OH)_{16}$
	W $\alpha = \beta > \gamma$	+L	1	$\simeq 20^o$ r$<$v	$\simeq$ parallel	cleavage length fast	SHERIDANITE	$(Mg,Al,Fe)_{12}\{(Si,Al)_8O_{20}\}(OH)_{16}$
	W $\alpha = \beta > \gamma$	+L	1 − 2	$0^o - 40^o$ r$<$v	$0^o - 9^o$	cleavage length fast	CLINOCHLORE	$(Mg,Al,Fe)_{12}\{(Si,Al)_8O_{20}\}(OH)_{16}$
	W $\alpha = \beta > \gamma$	+L − +M	1 (anom.)	$0^o - 30^o$ r$<$v	$\simeq$ parallel	cleavage length fast	RIPIDOLITE- KLEMENTITE (PROCHLORITE)	$(Mg,Al,Fe)_{12}\{(Si,Al)_8O_{20}\}(OH)_{16}$
	M − S $\alpha = \beta > \gamma$	+L − +M	1	$0^o - 60^o$ r$<$v	0° − small	cleavage length fast	Mn and Cr- CHLORITES	$(Mg,Mn,Al,Cr,Fe)_{12}$ $\{(Si,Al)_8O_{20}\}(OH)_{16}$

System	Form	Cleavage	Twinning	Zoning	Alteration	Occurrence	Remarks
Hex.	tabular	one perfect				M	
Mon. (Pseudo-Hex.)	tabular, flakes, bent plates, pseudo-hex.	perfect basal	simple	Z	chlorite, vermiculite, prehnite	M, I, S	pleochroic haloes, inclusions common, birds-eye maple structure cf. stilpnomelane, phlogopite
Tet.	variable, prismatic, fibrous, granular, radial	imperfect		Z		M, I, S	cf. zoisite, clinozoisite, apatite, grossularite, melilite, andalusite
Trig.	tabular, prismatic, six-sided x-sect.	parting	simple, lamellar seams	Z colour banding		M, I	inclusions
Orth. (Pseudo-Hex.)	pseudo-hex., anhedral	moderate to absent	simple, polysynthetic, cyclic, sector, interpenetrant		pinite, talc	M, I	pleochroic haloes, inclusions (including opaque dust) common cf. quartz, plagioclase
Mon.	tabular, vermicular, radiating, pseudomorphs	perfect basal	simple, pennine law	Z		I, M	cf. clinochlore, prochlorite
Mon.	tabular, spherulitic	perfect basal	simple			I	
Mon.	tabular, fibrous, pseudo-hex.	perfect basal	polysynthetic			M	pleochroic haloes, cf. penninite, prochlorite, leuchtenbergite, katschubeite
Mon.	tabular, scaly, vermicular, fan-shaped aggregates	perfect basal				M, I, O	cf. clinochlore, penninite
Mon.	tabular	perfect basal				O	

GREEN MINERALS	Pleochroism	Relief	δ	2V Dispersion	Extinction	Orientation	Mineral	Composition
Biaxial +								
	W – S $\alpha=\beta>\gamma$	+L – +M	1 – 2 (anom.)	$0°$ – $60°$ $r<v$	$0°$ – $9°$	cleavage length fast	CHLORITE (Unoxidised)	$(Mg,Al,Fe)_{12}\{(Si,Al)_8O_{20}\}(OH)$
	W $\alpha=\beta>\gamma$	+L – +M	2	$31°$ $r<v$	$8°$ – $10°$	cleavage length fast	CORUNDOPHILITE	$(Mg,Al,Fe)_{12}\{(Si,Al)_8O_{20}\}(OH)$
	S $\gamma\geq\beta>\alpha$	+M	2	$52°$ – $83°$ $r>v$	$17°$ – $27°$	length slow	EDENITE	$NaCa_2(Mg,Fe^{+2})_5AlSi_7O_{22}(OH)_2$
	W	+M	2	$67°$ – $70°$ $r>v$	$33°$ – $40°$	extinction nearest cleavage slow	JADEITE	$NaAl\{Si_2O_6\}$
	W $\alpha>\beta=\gamma$	+M	2	$73°$ – $86°$ $r<v$	parallel	length slow	VIRIDINE (ANDALUSITE)	$(Al,Fe^{+3})_2SiO_5$
	W	+M	2 – 3	$58°$ – $68°$ $r<v$	$22°$ – $26°$	extinction nearest cleavage slow	HIDDENITE (SPODUMENE)	$LiAl\{Si_2O_6\}$
	W $\gamma=\beta>\alpha$ $\gamma>\beta=\alpha$	+M	2 – 3	$68°$ – $90°$ $r\lesssim v$	parallel, symmetrical	length slow	ANTHOPHYLLITE (Mg-rich)	$(Mg,Fe^{+2})_7\{Si_8O_{22}\}(OH,F)_2$
	W $\gamma>\alpha$	+M	3	$21°$ – $30°$ $r>v$	parallel	length slow	SILLIMANITE	Al_2SiO_5
	W	+M	3	$50°$ – $62°$ $r>v$	$38°$ – $48°$	extinction nearest cleavage slow	DIOPSIDE (SALITE)	$Ca(Mg,Fe)\{Si_2O_6\}$
	M – S $\gamma\geq\beta>\alpha$	+M	3	$67°$ – $90°$ $r>v$	$26°$	length slow	PARGASITE	$NaCa_2Mg_4Al_3Si_6O_{22}(OH)_2$
	W $\alpha>\gamma$	+M	3	$76°$ – $87°$ $r>v$	parallel, symmetrical	length slow, long diagonal of rhombic sect. slow	LAWSONITE	$CaAl_2(OH)_2\{Si_2O_7\}H_2O$

System	Form	Cleavage	Twinning	Zoning	Alteration	Occurrence	Remarks
Mon. (Pseudo-Hex.)	tabular, scaly, radiating, pseudomorphs	perfect basal	simple, polysynthetic	Z		I, M, S, O	pleochroic haloes cf. bowlingite
Mon.	tabular	perfect basal				M	
Mon.	prismatic	two at 56°		Z		M, I	cf. hornblende
Mon.	prismatic	two at 87°	simple, polysynthetic	Z	tremolite-actinolite	M	cf. nephrite, diopside, omphacite, fassaite
Orth.	columnar, square x-sect.	two at 89°				M	cf. andalusite
Mon.	tabular	two at 87°, parting	simple		muscovite, cymatolite, kaolinite, eucryptite	I	cf. aegirine-augite, diopside, eucryptite
Orth.	bladed, prismatic, fibrous, asbestiform	two at $54\frac{1}{2}^{o}$			talc	M	cf. gedrite, tremolite, zoisite, cummingtonite, holmquistite
Orth.	needles, slender prisms, fibrous, rhombic x-sect., faserkiesel	one, // long diagonal of x-sect.				M	cf. andalusite, mullite
Mon.	short prismatic	two at 87°	simple, polysynthetic	Z	tremolite-actinolite	M, I	cf. hedenbergite, tremolite, omphacite, wollastonite, epidote
Mon.	prismatic	two at 56°, partings	simple, polysynthetic			M, I	cf. hornblende, hastingsite, cummingtonite
Orth.	rectangular, rhombic	two perfect, one imperfect	simple, polysynthetic			M	cf. zoisite, prehnite, pumpellyite, andalusite, scapolite

GREEN MINERALS	Pleochroism	Relief	δ	2V Dispersion	Extinction	Orientation	Mineral	Composition
Biaxial +								
	N – W $\gamma>\beta=\alpha$	+M	3 – 4	$65^\circ - 90^\circ$ Mg r<v Fe r>v	$15^\circ - 21^\circ$	length slow	CUMMINGTONITE	$(Mg,Fe^{+2})_7\{Si_8O_{22}\}(OH)_2$
	S $\alpha>\beta>\gamma$ $\alpha>\gamma\geqslant\beta$	+M – H	1 – 2	$40^\circ - 50^\circ$	$\beta : z$ $15^\circ - 30^\circ$	length fast	MAGNESIORIEBECKITE	$Na_2Mg_3Fe_2^{+3}\{Si_8O_{22}\}(OH)_2$
	S $\alpha>\beta>\gamma$ $\alpha>\gamma\geqslant\beta$	+M – H	1 – 2	$40^\circ - 90^\circ$ r$\leqslant$v	$3^\circ - 21^\circ$	length fast	RIEBECKITE	$Na_2Fe_3^{+2}Fe_2^{+3}\{Si_8O_{22}\}(OH)_2$
	W $\beta>\alpha=\gamma$	+M – H	2	$26^\circ - 85^\circ$ r<v	$4^\circ - 32^\circ$	length slow	PUMPELLYITE	$Ca_4(Mg,Fe^{+2})(Al,Fe^{+3})_5O(OH)_3$ $\{Si_2O_7\}_2\{SiO_4\}_2\cdot 2H_2O$
	M – S	+M – H	2 – 3 (anom.)	$25^\circ - 50^\circ$ r>v	$35^\circ - 48^\circ$	extinction nearest cleavage fast	TITANAUGITE	$(Ca,Na,Mg,Fe^{+2},Mn,Fe^{+3},Al,Ti)$ $\{(Si,Al)_2O_6\}$
	N – W	+M – H	2 – 3	$25^\circ - 83^\circ$ r>v	$35^\circ - 48^\circ$	extinction nearest cleavage fast	AUGITE-FERROAUGITE (incl. SALITE)	$(Ca,Na,Mg,Fe^{+2},Mn,Fe^{+3},Al,Ti)$ $\{(Si,Al)_2O_6\}$
	W	+M – H	2 – 3	$51^\circ - 62^\circ$	$35^\circ - 48^\circ$	extinction nearest cleavage fast	FASSAITE	$(Ca,Na,Mg,Fe^{+2},Mn,Fe^{+3},Al,Ti)$ $\{(Si,Al)_2O_6\}$
	W	+M – H	2 – 3	$60^\circ - 67^\circ$	$39^\circ - 41^\circ$	extinction nearest cleavage fast	OMPHACITE	$(Ca,Na,Mg,Fe^{+2},Mn,Fe^{+3},Al,Ti)$ $\{(Si,Al)_2O_6\}$
	W $\gamma=\beta>\alpha$ $\gamma>\beta=\alpha$	+M – H	2 – 3	$68^\circ - 90^\circ$ r$\leqslant$v	parallel, symmetrical	length slow	FERROGEDRITE	$Fe_5Al_4Si_6O_{22}(OH)_2$
	N – M $\gamma=\alpha>\beta$ $\beta>\alpha=\gamma$	+M – H	3	$0^\circ - 30^\circ$ r$\leqslant$v	$37^\circ - 44^\circ$	extinction nearest cleavage slow	PIGEONITE	$(Mg,Fe^{+2},Ca)(Mg,Fe^{+2})\{Si_2O_6\}$
	W $\gamma>\beta>\alpha$	+M – H	3	$50^\circ - 62^\circ$ r>v	$38^\circ - 48^\circ$	extinction nearest cleavage fast	HEDENBERGITE	$Ca(Fe,Mg)\{Si_2O_6\}$

System	Form	Cleavage	Twinning	Zoning	Alteration	Occurrence	Remarks
Mon.	prismatic, subradiating, fibrous, asbestiform	two at 55°	simple, polysynthetic			M, I	cf. tremolite, actinolite, grunerite, anthophyllite
Mon.	prismatic, fibrous, columnar aggregates	two at 56°	simple, polysynthetic	Z		I, M	cf. glaucophane, crossite, arfvedsonite, hastingsite
Mon.	prismatic, fibrous, columnar aggregates	two at 56°	simple, polysynthetic	Z		I, M	cf. glaucophane, crossite, arfvedsonite, hastingsite
Mon.	fibrous, needles, bladed, radial aggregates	two moderate	common	Z		M	cf. clinozoisite, zoisite, epidote, lawsonite
Mon.	prismatic	two at 93°	polysynthetic, twin seams	Z hour-glass		I	cf. hypersthene
Mon.	prismatic	two at 87°, parting	simple, polysynthetic, twin seams	Z	amphibole	I, M	herringbone structure, exsolution lamellae cf. pigeonite, diopside, epidote
Mon.	prismatic	two at 87°	simple, polysynthetic	Z	amphibole	I, M	cf. augite, diopside, omphacite, jadeite
Mon.	prismatic	two at 87°	simple, polysynthetic			M	inclusions cf. fassaite, diopside, jadeite
Orth.	bladed, prismatic, fibrous, asbestiform	two at 54$\frac{1}{2}$$^\circ$			talc	M	cf. anthophyllite, zoisite, cummingtonite, grunerite
Mon.	prismatic, anhedral, overgrowths	two at 87°, parting	simple, polysynthetic	Z		I	exsolution lamellae cf. augite, olivine
Mon.	prismatic	two at 87°	simple, polysynthetic			M, I	cf. diopside, aegirine-augite, augite

GREEN MINERALS	Pleochroism	Relief	δ	2V Dispersion	Extinction	Orientation	Mineral	Composition
Biaxial +								
	W - M $\beta>\alpha>\gamma$	H	1 - 2 anom.	$45^\circ - 68^\circ$ r>v	$2^\circ - 30^\circ$	length fast	CHLORITOID (OTTRELITE)	$(Fe^{+2},Mg,Mn)_2(Al,Fe^{+3})Al_3O_2\{SiO_4\}_2(OH)_4$
	S $\alpha>\gamma$	H	3 - 4	$70^\circ - 90^\circ$ r>v	$\gamma : z$ $70^\circ - 90^\circ$	extinction nearest cleavage fast	AEGIRINE-AUGITE	$(Na,Ca)(Fe^{+3},Fe^{+2},Mg)\{Si_2O_6\}$
	W - M $\alpha>\beta>\gamma$	VH - E	5 anom.	$17^\circ - 40^\circ$ r>v	40° symmetrical		SPHENE	$CaTi\{SiO_4\}(O,OH,F)$
	W $\gamma>\alpha$	+E	1	90° r>v	$\simeq 45^\circ$		PEROVSKITE (DYSANALYTE)	$(Ca,Na,Fe^{+2},Ce)(Ti,Nb)O_3$
Biaxial -								
	M	-L - +L	1 - 2	$65^\circ - 90^\circ$ r<v	parallel		CORDIERITE	$Al_3(Mg,Fe^{+2})_2\{Si_5AlO_{18}\}$
	W $\gamma>\beta>\alpha$	-L - +M	3	$0^\circ - 30^\circ$	small		MONTMORILLONITE-BEIDELLITE (SMECTITE)	$(\frac{1}{2}Ca,Na)_{0.7}(Al,Mg,Fe)_4\{(Si,Al)_8O_{20}\}(OH)_4 \cdot nH_2O$
	W $\gamma=\beta>\alpha$	-L - +M	3 - 4	$25^\circ - 70^\circ$	indistinct	length slow	NONTRONITE (SMECTITE)	$(\frac{1}{2}Ca,Na)_{0.7}(Al,Mg,Fe)_4\{(Si,Al)_8O_{20}\}(OH)_4 \cdot nH_2O$
		+L	1			cleavage slow	LIZARDITE (SERPENTINE)	$Mg_3\{Si_2O_5\}(OH)_4$
	W $\gamma=\beta>\alpha$	+L	1 anom.	$0^\circ - 40^\circ$ r>v	$\simeq$ parallel	cleavage length slow	PENNINITE	$(Mg,Al,Fe)_{12}\{(Si,Al)_8O_{20}\}(OH)_{16}$
	W $\gamma>\alpha$	+L	1 anom.	$37^\circ - 61^\circ$ r>v	parallel	length slow	ANTIGORITE (SERPENTINE)	$Mg_3\{Si_2O_5\}(OH)_4$

System	Form	Cleavage	Twinning	Zoning	Alteration	Occurrence	Remarks
Mon., Tr.	tabular	one perfect, one imperfect, parting	polysynthetic	Z hour-glass		M	inclusions common cf. chlorite, clintonite, biotite, stilpnomelane
Mon.	short prismatic, needles, felted aggregates	two at 87°, parting	simple, lamellar, twin seams	Z hour-glass		I	cf. aegirine, acmite
Mon.	rhombic, irregular grains	parting	simple, polysynthetic		leucoxene	I, M, S	cf. monazite, rutile
Mon? Pseudo-Cub.	small cubes, skeletal	poor to distinct	polysynthetic, complex, interpenetrant	Z	leucoxene	I, M	cf. melanite, picotite, ilmenite
Orth. (Pseudo-Hex.)	pseudo-hex., anhedral	moderate to absent	simple, polysynthetic, cyclic, sector, interpenetrant		pinite, talc	M, I	pleochroic haloes, inclusions (including opaque dust) common cf. quartz, plagioclase
Mon.	shards, scales, massive, microcrystalline	perfect basal				I, S	cf. nontronite
Mon.	shards, scales, fibrous, massive	perfect basal				I, M	cf. montmorillonite, kaolinite
Mon.	tabular, fine grained aggregates	perfect basal	occasional			I, M	mesh, hour-glass structure cf. chlorite group
Mon.	tabular, vermicular, radiating, pseudomorphs	perfect basal	simple	Z		I, M	pleochroic haloes cf. clinochlore, prochlorite
Mon.	anhedral, fibrolamellar, flakes, laths, plates	perfect basal	occasional			I, M	mesh, hour-glass structure cf. antigorite, serpophite, chlorite, amphibole

GREEN MINERALS	Pleochroism	Relief	δ	2V Dispersion	Extinction	Orientation	Mineral	Composition
Biaxial −								
		+L	2	$30° - 32°$	parallel	length slow	CHRYSOTILE (SERPENTINE)	$Mg_3\{Si_2O_5\}(OH)_4$
	W $\gamma=\beta>\alpha$	+L	3	$0° - 8°$ r$\leqslant$v	$1° - 2°$	cleavage length slow	VERMICULITE	$(Mg,Ca)_{0.7}(Mg,Fe^{+3},Al)_{6.0}$ $\{(Al,Si)_8O_{20}\}(OH)_4 \cdot 8H_2O$
	W $\gamma=\beta>\alpha$	+L	3 - 4	small	$\simeq$ parallel	cleavage length slow	MINNESOTAITE	$(Fe^{+2},Mg)_3\{(Si,Al)_4O_{10}\}OH_2$
	M $\gamma\geqslant\beta>\alpha$	+L	4	$32° - 46°$ r>v	$1° - 3°$	cleavage length slow	FUCHSITE	Cr- muscovite
	W $\gamma>\beta>\alpha$	+L - +M	1	$0° - 20°$	$0° - 7°$	cleavage length slow	DELESSITE	$(Mg,Al,Fe)_{12}\{(Si,Al)_8O_{20}\}(OH)_{16}$
	W - M $\gamma=\beta>\alpha$	+L - +M	1 anom.	$0° - 20°$ r>v	$\simeq$ parallel	cleavage length slow	DIABANTITE BRUNSVIGITE	$(Mg,Al,Fe)_{12}\{(Si,Al)_8O_{20}\}(OH)_{16}$
	M - S $\gamma=\beta>\alpha$	+L - +M	1	$0° - 20°$ r>v	$0°$ - small	cleavage length slow	Mn and Cr- CHLORITES	$(Mg,Mn,Al,Cr,Fe)_{12}$ $\{(Si,Al)_8O_{20}\}(OH)_{16}$
	W - S $\gamma=\beta>\alpha$	+L - +M	1 - 2 (anom.)	$0° - 20°$ r>v	$0°$ - small	cleavage length slow	CHLORITE (Oxidised)	$(Mg,Al,Fe)_{12}\{(Si,Al)_8O_{20}\}(OH)_{16}$
	W - S $\gamma=\beta>\alpha$	+L - +M	1 - 2 (anom.)	$0° - 20°$ r>v	$0°$ - small	cleavage length slow	CHLORITE (Unoxidised)	$(Mg,Al,Fe)_{12}\{(Si,Al)_8O_{20}\}(OH)_{16}$
	M $\gamma\geqslant\beta>\alpha$ $\alpha>\beta\ \gamma$	+L - +M	3 - 4	$0° - 15°$ r<v	$0° - 5°$	cleavage length slow	PHLOGOPITE	$K_2(Mg,Fe^{+2})_6\{Si_6Al_2O_{20}\}(OH,F)_4$
	W $\gamma=\beta>\alpha$	+M	1 anom.	$0°$ - small r>v	$\simeq$ parallel	cleavage length slow	RIPIDOLITE PROCHLORITE	$(Mg,Al,Fe)_{12}\{(Si,Al)_8O_{20}\}(OH)_{16}$
	W $\gamma=\beta>\alpha$	+M	1 anom.	$0° - 20°$	small	cleavage length slow	DAPHNITE	$(Mg,Al,Fe)_{12}\{(Si,Al)_8O_{20}\}(OH)_{16}$
	N - W	+M	1	small	small	cleavage length slow	CHAMOSITE	$(Mg,Al,Fe)_{12}\{(Si,Al)_8O_{20}\}(OH)_{16}$

System	Form	Cleavage	Twinning	Zoning	Alteration	Occurrence	Remarks
Mon.	fibrous, cross-fibre veinlets	fibrous				I, M	mesh, hour-glass structure cf. antigorite, serpophite, chlorite, amphibole
Mon.	minute particles, plates	perfect basal				S, I, M	cf. biotite, smectites
Mon.	minute plates, needles, radiating aggregates	perfect basal				M	cf. talc
Mon.	thin tablets, shreds	perfect basal	simple			M	birds-eye maple structure cf. muscovite
Mon.	spherulitic	perfect basal	simple			I	cf. thuringite, greenalite
Mon.	tabular, fibrous	perfect basal				M, I, O	cf. chlorite group
Mon.	tabular, radiating	perfect basal				O, I, M	
Mon. (Pseudo-Hex.)	tabular, scaly, radiating, pseudomorphs	perfect basal	simple, polysynthetic	Z		I, S, M, O	pleochroic haloes cf. bowlingite
Mon. (Pseudo-Hex.)	tabular, scaly, radiating, pseudomorphs	perfect basal	simple, polysynthetic	Z		I, M, S, O	pleochroic haloes cf. bowlingite
Mon.	tabular, flakes, plates, pseudo-hex.	perfect basal	inconspicuous	Z colour zoning		I, M	inclusions common, birds-eye maple structure cf. biotite, muscovite, lepidolite, rutile, tourmaline
Mon.	tabular, vermicular, fan-shaped aggregates	perfect basal				M, I, O	cf. clinochlore, penninite
Mon.	concentric aggregates, fibrous plates	perfect basal				O	
Mon.	pseudospherulitic, concentric, tabular, massive	one good, concentric parting				S, O	cf. glauconite, collophane, greenalite, thuringite

GREEN MINERALS	Pleochroism	Relief	δ	2V Dispersion	Extinction	Orientation	Mineral	Composition
Biaixial –								
	M – S $\gamma>\beta>\alpha$ $\gamma<\beta>\alpha$	+M	1 – 3	$0°$ – $50°$ r<v	$36°$ – $70°$	length fast	KATOPHORITE-MAGNESIOKATOPHORITE	$Na_2Ca(Mg,Fe^{+2})_4Fe^{+3}\{Si_7AlO_{22}\}(OH,F)_2$
	M $\gamma=\beta>\alpha$	+M	2	$0°$ – $23°$ r<v	parallel	cleavage length slow	XANTHOPHYLLITE	$Ca_2(Mg,Fe)_{4\cdot6}Al_{1\cdot4}\{Si_{2\cdot5}Al_{5\cdot5}O_{20}\}(OH)_4$
	W	+M	2	$0°$ – $20°$	small	cleavage length slow	THURINGITE	$(Mg,Al,Fe)_{12}\{(Si,Al)_8O_{20}\}(OH)_{16}$
	S $\alpha>\beta>\gamma$	+M	2	$15°$ – $80°$ r>v	$18°$ – $53°$ $(40°)$ 'flamy'	length slow	ECKERMANNITE	$Na_3(Mg,Fe^{+2})_4Al\{Si_8O_{22}\}(OH,F)_2$
	M $\gamma=\beta>\alpha$	+M	2	$32°$ r<v		parallel	cleavage length slow	CLINTONITE	$Ca_2(Mg,Fe)_{4\cdot6}Al_{1\cdot4}\{Si_{2\cdot5}Al_{5\cdot5}O_{20}\}(OH)_4$
	S $\gamma\geqslant\beta>\alpha$	+M	2	$52°$ – $83°$ r>v	$17°$ – $27°$	length slow	EDENITE	$NaCa_2(Mg,Fe^{+2})_5AlSi_7O_{22}(OH)_2$
	W patchy $\alpha>\gamma$	+M	2	$73°$ – $86°$ r<v		parallel	length fast	ANDALUSITE	Al_2SiO_5
	M $\gamma=\beta>\alpha$	+M	2 – 3	$0°$ – $20°$ r<v	$0°$ – $3°$	cleavage length slow	GLAUCONITE (incl. CELADONITE)	$(K,Na,Ca)_{1\cdot2-2\cdot0}(Fe^{+3},Al,Fe^{+2},Mg)_{4\cdot0}\{Si_{7-7\cdot6}Al_{1-0\cdot4}O_{20}\}(OH)_4\cdot nH_2O$
	W $\gamma=\beta>\alpha$ $\gamma>\beta=\alpha$	+M	2 – 3	$68°$ – $90°$ r$\lessgtr$v	parallel, symmetrical	length slow	ANTHOPHYLLITE (Mg-rich) – FERROGEDRITE	$(Mg,Fe^{+2})_7\{Si_8O_{22}\}(OH,F)_2$
	W $\gamma>\beta>\alpha$	+M	2 – 3	$73°$ – $86°$ r<v	$10°$ – $17°$	length slow	ACTINOLITE-FERROACTINOLITE	$Ca_2(Mg,Fe^{+2})_5\{Si_8O_{22}\}(OH,F)_2$
	S $\gamma\geqslant\beta>\alpha$ $\beta>\gamma>\alpha$	+M	4 – 5	$0°$ – $25°$ Mg r$\lessgtr$v Fe r>v	$0°$ – $9°$ wavy	cleavage length slow	BIOTITE	$K_2(Mg,Fe^{+2})_{6-4}(Fe^{+3},Al,Ti)_{0-2}\{Si_{6-5}Al_{2-3}O_{20}\}(OH,F)_4$
	S $\alpha>\beta>\gamma$	+M – H	1 – 2	$0°$ – $50°$ r<v	$0°$ – $30°$ anom.	length fast	ARFVEDSONITE	$Na_3(Mg,Fe^{+2})_4Al\{Si_8O_{22}\}(OH,F)_2$

System	Form	Cleavage	Twinning	Zoning	Alteration	Occurrence	Remarks
Mon.	prismatic	two at 56°, parting	simple	Z		I	cf. barkevikite, kaersutite, basaltic hornblende, arfvedsonite, cossyrite
Mon.	tabular, short prismatic	perfect basal	simple			M	cf. clintonite, chlorite, chloritoid
Mon. (Pseudo-Hex.)	tabular, radiating	perfect basal				O	
Mon.	prismatic	two at 56°, parting	simple, polysynthetic			I	cf. glaucophane, arfvedsonite, cummingtonite, tremolite, hornblende
Mon.	tabular, short prismatic	perfect basal	simple			M	cf. xanthophyllite, chlorite, chloritoid
Mon.	prismatic	two at 56°		Z		M, I	cf. hornblende
Orth.	columnar, square x-sect.	two at 89°	rare		sericite, kyanite, sillimanite	M	variety chiastolite, inclusions in shape of cross
Mon.	grains, pellets, plates, pseudomorphs	perfect basal			limonite, goethite	S	cf. chamosite, biotite, chlorite group
Orth.	bladed, prismatic, fibrous, asbestiform	two at $54\frac{1}{2}^{\circ}$			talc	M	cf. gedrite, zoisite, cummingtonite, grunerite, holmquistite
Mon.	long, prismatic, fibrous	two at 56°, parting	simple, polysynthetic	Z		M, I	cf. tremolite, orthoamphibole, hornblende
Mon. (Pseudo-Hex.)	tabular, flakes, plates, pseudo-hex.	perfect basal	simple	Z	chlorite, vermiculite, prehnite	M, I, S	pleochroic haloes, inclusions common, birds-eye maple structure cf. stilpnomelane, phlogopite
Mon.	prismatic	two at 56°, parting	simple, polysynthetic	Z		I	cf. riebeckite, katophorite, glaucophane, tourmaline

GREEN MINERALS	Pleochroism	Relief	δ	2V Dispersion	Extinction	Orientation	Mineral	Composition
Biaxial −								
	S $\alpha>\beta>\gamma$ $\alpha>\gamma\geqslant\beta$	+M − H	1 − 2	$40^{\circ} - 50^{\circ}$	$\beta : z$ $15^{\circ} - 30^{\circ}$	length fast		MAGNESIORIEBECKITE	$Na_2Mg_3Fe_2^{+3}\{Si_8O_{22}\}(OH)_2$
	S $\alpha>\beta>\gamma$ $\alpha>\gamma\geqslant\beta$	+M − H	1 − 2	$40^{\circ} - 90^{\circ}$ r$\leqslant$v	$3^{\circ} - 21^{\circ}$	length fast		RIEBECKITE	$Na_2Fe_3^{+2}Fe_2^{+3}\{Si_8O_{22}\}(OH)_2$
	N − S $\gamma>\beta>\alpha$	+M − H	1 − 2	$50^{\circ} - 90^{\circ}$ r$\leqslant$v	parallel	length slow		HYPERSTHENE (incl. BRONZITE-EULITE)	$(Mg,Fe^{+2})\{SiO_3\}$
	S $\gamma\geqslant\beta>\alpha$ $\beta>\gamma>\alpha$	+M − H	2 − 3	$66^{\circ} - 85^{\circ}$ r>v	$13^{\circ} - 34^{\circ}$	length slow		HORNBLENDE	$(Na,K)_{0-1}Ca_2(Mg,Fe^{+2},Fe^{+3},Al)_5$ $\{Si_{6-7}Al_{2-1}O_{22}\}(OH,F)_2$
	S $\beta\geqslant\gamma>\alpha$ $\gamma\geqslant\beta>\alpha$	+M − H	3	$10^{\circ} - 90^{\circ}$ r$\leqslant$v	$9^{\circ} - 40^{\circ}$	length slow		HASTINGSITE-FERROHASTINGSITE	$NaCa_2(Mg,Fe^{+2})_4$ $(Al,Fe^{+3})Al_2Si_6O_{22}(OH,F)_2$
	W $\gamma>\beta=\alpha$	+M − H	3 − 4	$84^{\circ} - 90^{\circ}$ r>v	$10^{\circ} - 15^{\circ}$	length slow		GRUNERITE	$(Fe^{+2},Mg)_7\{Si_8O_{22}\}(OH)_2$
	W $\omega>\epsilon$	H	1	$0^{\circ} - 30^{\circ}$ (uniaxial) moderate	parallel	tabular length slow, prismatic length fast		CORUNDUM	$\alpha-Al_2O_3$
	W − M $\beta>\alpha>\gamma$	H	1 − 2 anom.	$45^{\circ} - 68^{\circ}$ r>v	$2^{\circ} - 30^{\circ}$	length fast		CHLORITOID (OTTRELITE)	$(Fe^{+2},Mg,Mn)_2(Al,Fe^{+3})Al_3O_2$ $\{SiO_4\}_2(OH)_4$
	S	H	3	65° r<v				HOWIEITE	$Na(Fe,Mn)_{10}$ $(Fe,Al)_2Si_{12}O_{31}(OH)_{13}$
	S $\alpha>\gamma$	H	3 − 4	$70^{\circ} - 90^{\circ}$ r>v	$\gamma : z$ $70^{\circ} - 90^{\circ}$	extinction nearest cleavage fast		AEGIRINE-AUGITE	$(Na,Ca)(Fe^{+3},Fe^{+2},Mg)\{Si_2O_6\}$
	W $\beta>\gamma>\alpha$	H	3 − 4 anom.	$74^{\circ} - 90^{\circ}$ r>v	$0^{\circ} - 15^{\circ}$ parallel in elong. sect.	length fast/slow		EPIDOTE	$Ca_2Fe^{+3}Al_2O\cdot OH\{Si_2O_7\}\{SiO_4\}$

System	Form	Cleavage	Twinning	Zoning	Alteration	Occurrence	Remarks
Mon.	prismatic, fibrous, columnar aggregates	two at 56°	simple, polysynthetic	Z		I, M	cf. glaucophane, crossite, arfvedsonite, hastingsite
Mon.	prismatic, fibrous, columnar aggregates	two at 56°	simple, polysynthetic	Z		I, M	cf. glaucophane, crossite, arfvedsonite, hastingsite
Orth.	prismatic, anhedral - euhedral	two at 88°	polysynthetic, twin seams	Z		I, M	exsolution lamellae, schiller inclusions cf. andalusite, titanaugite
Mon.	prismatic	two at 56°, parting	simple, polysynthetic	Z	mica, chlorite	M, I	cf. edenite, biotite, aegirine, augite, actinolite, pargasite, ferrohastingsite
Mon.	prismatic	two at 56°, parting	simple, polysynthetic	Z		I, M	cf. arfvedsonite, hornblende
Mon.	prismatic, subradiating, fibrous, asbestiform	two at 55°, cross fractures	simple, polysynthetic		limonite	M	cf. tremolite, actinolite, cummingtonite, anthophyllite
Trig.	tabular, prismatic, six-sided x-sect.	parting	simple, lamellar seams	Z colour banding		M, I	inclusions
Mon., Tr.	tabular	one perfect, one imperfect, parting	polysynthetic	Z hour-glass		M, I	inclusions common cf. chlorite, clintonite, biotite, stilpnomelane
Tr.	bladed	three, one good				M	
Mon.	short, prismatic, needles, felted aggregates	two at 87°, parting	simple, polysynthetic, twin seams	Z hour-glass		I	cf. aegirine, acmite
Mon.	distinct crystals, columnar aggregates, six-sided x-sect.	one perfect	uncommon	Z		M, I, S	cf. zoisite, clinozoisite, diopside, augite, sillimanite

GREEN MINERALS	Pleochroism	Relief	δ	2V Dispersion	Extinction	Orientation	Mineral	Composition
Biaxial –								
	S $\alpha>\beta\geqslant\gamma$	H – VH	4 – 5	$60^{\circ} - 70^{\circ}$ r>v	$0^{\circ} - 10^{\circ}$	extinction nearest cleavage fast	AEGIRINE	$NaFe^{+3}\{Si_2O_6\}$
	M $\alpha>\beta=\gamma$	VH	3	$\simeq 0^{\circ}$			CRONSTEDTITE	$(Fe_4{}^{+2}Fe_2{}^{+3})(Si_2Fe_2{}^{+3})O_{10}(OH)$
	M $\gamma>\beta>\alpha$	VH	4	$44^{\circ} - 70^{\circ}$ r>v	parallel	length fast/slow	TEPHROITE	$Mn_2\{SiO_4\}$

System	Form	Cleavage	Twinning	Zoning	Alteration	Occurrence	Remarks
Mon.	prismatic, needles, felted aggregates, blunt terminations	two at 87°, parting	simple, polysynthetic, twin seams	Z hour-glass		I	borders may be black cf. aegirine-augite, acmite
Mon. (Pseudo-Hex.)	tabular	perfect				O	green to opaque cf. septochlorites
Orth.	anhedral - euhedral, rounded	moderate				O, M	pleochroic in reds and black cf. monticellite, olivine

BLUE MINERALS	Pleochroism	Relief	δ	2V Dispersion	Extinction	Orientation	Mineral	Composition
Isometric								
		-L	(weak)				SODALITE (incl. HACKMANITE)	$Na_8\{Al_6Si_6O_{24}\}Cl_2$
		-L	1 (weak)				NOSEAN	$Na_8\{Al_6Si_6O_{24}\}SO_4$
		-L	1 (weak)				HAÜYNE	$(Na,Ca)_{4-8}\{Al_6Si_6O_{24}\}(SO_4,S)_{1-}$
Uniaxial +								
	H		1 anom.		parallel	length fast	VESUVIANITE (IDOCRASE) (incl. WILUITE)	$Ca_{10}(Mg,Fe)_2Al_4\{Si_2O_7\}_2 \{SiO_4\}_5(OH,F)_4$
Uniaxial −								
	W	+L	1		parallel	long. sect. length fast, x-sect. length slow	BERYL	$Be_3Al_2\{Si_6O_{18}\}$
	W $\varepsilon>\omega$	+M	1		parallel	length fast, tabular length slow	APATITE (incl. DAHLLITE, FRANCOLITE)	$Ca_5(PO_4)_3(OH,F,Cl)$
	S $\omega>\varepsilon$	+M	3		parallel	length fast	SCHORL (TOURMALINE)	$Na(Fe,Mn)_3Al_6B_3Si_6O_{27}(OH,F)_4$
	H		1 anom.		parallel	length fast	VESUVIANITE (IDOCRASE)	$Ca_{10}(Mg,Fe)_2Al_4\{Si_2O_7\}_2 \{SiO_4\}_5(OH,F)_4$

System	Form	Cleavage	Twinning	Zoning	Alteration	Occurrence	Remarks
Cub.	hexagonal x-sect., anhedral aggregates	poor	simple		zeolites, diaspore, gibbsite	I	cf. fluorite, leucite
Cub.	hexagonal x-sect., anhedral aggregates	poor	simple	Z	zeolites, diaspore, gibbsite, limonite	I	clouded with inclusions
Cub.	hexagonal x-sect., anhedral aggregates	imperfect	simple	Z	zeolites, diaspore, gibbsite	I	
Tet.	variable, prismatic, fibrous, granular, radial	poor	sector	Z		M, I, S	cf. zoisite, clinozoisite, apatite, grossularite, melilite, andalusite
Hex.	prismatic, inclusions zoned	imperfect		Z	kaolin	I, M	liquid inclusions cf. apatite, quartz
Hex.	small, prismatic, hexagonal	poor basal		Z		I, S, M, O	cf. beryl, topaz, dahllite
Trig.	hexagonal, rounded, suns, triangular x-sect.	fractures		Z		I, M, S	cf. tourmaline group, Na-amphibole
Tet.	variable	poor		Z		M, I, S	cf. zoisite, clinozoisite, apatite, grossularite

BLUE MINERALS	Pleochroism	Relief	δ	2V Dispersion	Extinction	Orientation	Mineral	Composition
Uniaxial −								
	W $\omega>\epsilon$	H	1	$0°-30°$	parallel	tabular length slow, prismatic length fast	CORUNDUM	$\alpha\text{-}Al_2O_3$
	W $\omega>\epsilon,\ \epsilon>\omega$	E	5 anom.	$0°$ – small			ANATASE	TiO_2
Biaxial +								
	M	−L − +L	1 − 2	$65°-90°$ r<v	parallel		CORDIERITE	$Al_3(Mg,Fe^{+2})_2\{Si_5AlO_{18}\}$
	W $\gamma>\beta>\alpha$	+M	1	$50°$ r<v	parallel	length slow	CELESTINE	$SrSO_4$
	W $\gamma>\alpha$	+M	3	$21°-30°$ r>v	parallel	length slow	SILLIMANITE	Al_2SiO_5
	W $\alpha>\gamma$	+M	3	$76°-87°$ r>v	parallel, symmetrical	length slow, long diagonal of rhombic sect. slow	LAWSONITE	$CaAl_2(OH)_2\{Si_2O_7\}H_2O$
	S $\alpha>\beta>\gamma$ $\alpha>\gamma\gtrless\beta$	+M − H	1 − 2	$40°-50°$	$\beta:z$ $15°-30°$	length fast	MAGNESIORIEBECKITE	$Na_2Mg_3Fe_2^{+3}\{Si_8O_{22}\}(OH)_2$
	S $\alpha>\beta>\gamma$ $\alpha>\gamma\gtrless\beta$	+M − H	1 − 2	$40°-90°$ r$\lessgtr$v	$3°-21°$	length fast	RIEBECKITE (CROCIDOLITE)	$Na_2Fe_3^{+2}Fe_2^{+3}\{Si_8O_{22}\}(OH)_2$
		H	1 anom.	$50°-65°$ strong	parallel	length fast	VESUVIANITE (IDOCRASE) (incl. WILUITE)	$Ca_{10}(Mg,Fe)_2Al_4\{Si_2O_7\}_2\{SiO_4\}_5(OH,F)_4$
	W $\beta>\gamma>\alpha$	H	1	$50°-66°$ r<v	$6°-9°$	length slow	SAPPHIRINE	$(Mg,Fe)_2Al_4O_6\{SiO_4\}$

System	Form	Cleavage	Twinning	Zoning	Alteration	Occurrence	Remarks
Trig.	tabular, prismatic, six-sided x-sect.	parting	simple, lamellar seams	Z colour banding		M, I	inclusions cf. sapphirine
Tet.	small, prismatic, acicular	two perfect		Z	leucoxene	S, I, M	often yellow cf. rutile, brookite
Orth. (Pseudo-Hex.)	pseudo-hex., anhedral	moderate to absent	simple, polysynthetic, cyclic, sector, interpenetrant		pinite, talc	M, I	pleochroic haloes, inclusions (including opaque dust) common cf. quartz, plagioclase
Orth.	tabular, fibrous	three				S	cf. barytes
Orth.	needles, slender prisms, fibrous, rhombic x-sect., faserkiesel	one, // long diagonal x-sect.				M	cf. andalusite, mullite
Orth.	rectangular, rhombic	two perfect, one imperfect	simple, polysynthetic			M	cf. lawsonite, prehnite, pumpellyite, andalusite, scapolite
Mon.	prismatic, fibrous, columnar aggregates	two at 56°	simple, polysynthetic	Z		I, M	cf. glaucophane, crossite, arfvedsonite, hastingsite
Mon.	prismatic, fibrous, columnar aggregates	two at 56°	simple, polysynthetic	Z		I, M	cf. glaucophane, crossite, arfvedsonite, hastingsite
Tet.	variable	poor		Z		M, I, S	cf. zoisite, clinozoisite, apatite, grossularite
Mon.	tabular	poor	polysynthetic, uncommon			M	cf. corundum, cordierite, kyanite, zoisite, Na-amphibole

BLUE MINERALS	Pleochroism	Relief		2V Dispersion	Extinction	Orientation	Mineral	Composition
Biaxial +								
	W – M $\beta > \alpha > \gamma$	H	1 – 2 anom.	$45^\circ - 68^\circ$ r>v	$2^\circ - 30^\circ$	length fast	CHLORITOID (OTTRELITE)	$(Fe^{+2},Mg,Mn)_2(Al,Fe^{+3})Al_3O_2\{SiO_4\}_2(OH)_4$
	W $\gamma > \beta > \alpha$	H	4	$84^\circ - 86^\circ$ r<v	parallel	length fast	DIASPORE	$\alpha-AlO(OH)$
Biaxial –								
	M	–L – +L	1 – 2	$65^\circ - 90^\circ$ r<v	parallel		CORDIERITE	$Al_3(Mg,Fe^{+2})_2\{Si_5AlO_{18}\}$
	S $\gamma > \beta > \alpha$ $\beta > \gamma > \alpha$	+M	1 – 2 anom.	$2^\circ - 65^\circ$ r<v	$12^\circ - 30^\circ$	length slow	CROSSITE	$Na_2(Mg_3,Fe_3^{+2},Fe_2^{+3},Al_2)\{Si_8O_{22}\}(OH)_2$
	S $\gamma > \beta > \alpha$	+M	1 – 3 anom.	$0^\circ - 50^\circ$ r>v	$4^\circ - 14^\circ$	length slow	GLAUCOPHANE	$Na_2Mg_3Al_2\{Si_8O_{22}\}(OH)_2$
	S patchy $\alpha > \beta \geqslant \gamma$	+M	2	$20^\circ - 40^\circ$ r≶v	parallel	length fast	DUMORTIERITE	$HBAl_8Si_3O_{20}$
	M $\beta > \gamma > \alpha$	+M	2 – 3 anom.	$66^\circ - 87^\circ$ r<v	$15^\circ - 40^\circ$	length slow	RICHTERITE-FERRORICHTERITE	$Na_2Ca(Mg,Fe^{+3},Fe^{+2},Mn)_5\{Si_8O_{22}\}(OH,F)_2$
	S	+M	3	51° r>v		parallel	length slow	HOLMQUISTITE	$Li_2(Mg,Fe^{+2})_3(Al,Fe^{+3})_2\{Si_8O_{22}\}(OH)_2$
	S $\gamma \geqslant \beta > \alpha$	+M	4	$61^\circ - 70^\circ$ r<v	$9^\circ - 12^\circ$	long diagonal fast	LAZULITE	$Al_2(Mg,Fe)(OH)_2(PO_4)_2$
	S $\alpha > \beta > \gamma$	+M – H	1 – 2	$0^\circ - 50^\circ$ r<v	$0^\circ - 30^\circ$ anom.	length fast	ARFVEDSONITE	$Na_3(Mg,Fe^{+2})_4Al\{Si_8O_{22}\}(OH,F)_2$

System	Form	Cleavage	Twinning	Zoning	Alteration	Occurrence	Remarks
Mon., Tr.	tabular	one perfect, one imperfect, parting	polysynthetic	Z hour-glass		M, I	often green, inclusions common
Orth.	large tablets	simple				S, M, I	cf. anhydrite, boehmite, gibbsite, corundum, sillimanite
Orth. (Pseudo-Hex.)	pseudo-hex., anhedral	moderate to absent	simple, polysynthetic, cyclic, sector, interpenetrant		pinite, talc	M, I	pleochroic haloes, inclusions (including opaque dust) common cf. quartz, plagioclase
Mon.	prismatic, columnar aggregates	two at 58°	simple, polysynthetic	Z		M	cf. glaucophane, riebeckite
Mon.	prismatic, columnar aggregates	two at 58°	simple, polysynthetic	Z		M	resembles holmquistite in pleochroism cf. arfvedsonite, eckermannite, riebeckite
Orth.	prismatic, acicular	imperfect, cross fractures	penetration trillings		sericite	M, I	cf. tourmaline, sillimanite
Mon.	long prismatic, fibrous	two at 56°, parting	simple, polysynthetic	Z		M, I	
Orth.	fibrous, bladed, prismatic	two at 54$\frac{1}{2}$^{o}				M, I	cf. anthophyllite, glaucophane
Mon.	diamond-shaped, anhedral	two	polysynthetic			M	cf. tourmaline, dumortierite
Mon.	prismatic	two at 56°, parting	simple, polysynthetic	Z		I	cf. riebeckite, katophorite, glaucophane, tourmaline

BLUE MINERALS	Pleochroism	Relief	δ	2V Dispersion	Extinction	Orientation	Mineral	Composition
Biaxial −								
	S $\alpha>\beta>\gamma$ $\alpha>\gamma\geqslant\beta$	+M − H	1 − 2	$40^{\circ} - 50^{\circ}$ $\beta : z$ $15^{\circ} - 30^{\circ}$		length fast	MAGNESIORIEBECKITE	$Na_2Mg_3Fe_2^{+3}Si_8O_{22}(OH)_2$
	S $\alpha>\beta>\gamma$ $\alpha>\gamma\geqslant\beta$	+M − H	1 − 2	$40^{\circ} - 90^{\circ}$ $r\lessgtr v$	$3^{\circ} - 21^{\circ}$	length fast	RIEBECKITE (CROCIDOLITE)	$Na_2Fe_3^{+2}Fe_2^{+3}Si_8O_{22}$
	S $\beta>\gamma>\alpha$ $\gamma\geqslant\beta>\alpha$	+M − H	3	$10^{\circ} - 90^{\circ}$ $r\lessgtr v$	$9^{\circ} - 40^{\circ}$	length slow	HASTINGSITE- FERROHASTINGSITE	$NaCa_2(Mg,Fe^{+2})_4$ $(Al,Fe^{+3})Al_2Si_6O_{22}(OH,F)_2$
	W $\omega>\varepsilon$	H	1	$0^{\circ} - 30^{\circ}$ (uniaxial) moderate	parallel	tabular length slow, prismatic length fast	CORUNDUM	$\alpha-Al_2O_3$
	W $\beta>\gamma>\alpha$	H	1	$50^{\circ} - 66^{\circ}$ $r<v$	$6^{\circ} - 9^{\circ}$	length slow	SAPPHIRINE	$(Mg,Fe)_2Al_4O_6\{SiO_4\}$
	W − M $\beta>\alpha>\gamma$	H	1 − 2 anom.	$45^{\circ} - 68^{\circ}$ $r>v$	$2^{\circ} - 30^{\circ}$	length fast	CHLORITOID (OTTRELITE)	$(Fe^{+2},Mg,Mn)_2(Al,Fe^{+3})Al_3O_2$ $\{SiO_4\}_2(OH)_4$
	W $\gamma>\beta>\alpha$	H	2	$82^{\circ} - 83^{\circ}$ $r>v$	$0^{\circ} - 32^{\circ}$	length slow	KYANITE	Al_2SiO_5
	W $\gamma>\beta>\alpha$	VH	4	$44^{\circ} - 70^{\circ}$ $r>v$	parallel	length fast/slow	TEPHROITE (incl. ROEPPERITE)	$Mn_2\{SiO_4\}$
	W $\gamma>\beta>\alpha$	VH	4	$44^{\circ} - 70^{\circ}$ $r>v$	parallel	length fast/slow	KNEBELITE	$(Mn,Fe)_2\{SiO_4\}$
	W $\omega>\varepsilon, \varepsilon>\omega$	E	5 anom.	(uniaxial)			ANATASE	TiO_2

System	Form	Cleavage	Twinning	Zoning	Alteration	Occurrence	Remarks
Mon.	prismatic, fibrous, columnar aggregates	two at 56°	simple, polysynthetic	Z		I, M	cf. glaucophane, crossite, arfvedsonite, hastingsite
Mon.	prismatic, fibrous, columnar aggregates	two at 56°	simple, polysynthetic	Z		I, M	cf. glaucophane, crossite, arfvedsonite, hastingsite
Mon.	prismatic	two at 56°, parting	simple, polysynthetic	Z		I, M	cf. arfvedsonite, hornblende
Trig.	tabular, prismatic, six-sided x-sect.	parting	simple, lamellar seams	Z colour banding		M, I	inclusions cf. sapphirine
Mon.	tabular	poor	polysynthetic, uncommon			M	cf. corundum, cordierite, kyanite, zoisite, Na-amphibole
Mon., Tr.	tabular	one perfect, one imperfect, parting	polysynthetic	Z hour-glass		M, I	inclusions common
Tr.	bladed, prismatic	two, parting	simple, polysynthetic			M	cf. sillimanite, pyroxene
Orth.	anhedral - euhedral, rounded	two moderate, imperfect	uncommon			O, M	cf. knebelite, olivine
Orth.	anhedral - euhedral, rounded	two moderate, imperfect	uncommon			O, M	cf. tephroite, olivine
Tet.	small, prismatic, acicular	two perfect		Z	leucoxene	S, I, M	often yellow cf. rutile, brookite

PURPLE MINERALS	Pleochroism	Relief	δ	2V Dispersion	Extinction	Orientation	Mineral	Composition
Isometric								
		$-M$					FLUORITE	CaF_2
Uniaxial +								
	M – S $\alpha=\beta>\gamma$	$+L - +M$	1	$0^{\circ} - 60^{\circ}$ $r<v$	0° – small	cleavage length fast	Mn and Cr- CHLORITES	$(Mg,Mn,Al,Cr,Fe)_{12}$ $\{(Si,Al)_8O_{20}\}(OH)_{16}$
Uniaxial –								
	M – S $\gamma=\beta>\alpha$	$+L - +M$	1	$0^{\circ} - 20^{\circ}$ $r>v$	0° – small	cleavage length slow	Mn and Cr- CHLORITES	$(Mg,Mn,Al,Cr,Fe)_{12}$ $\{(Si,Al)_8O_{20}\}(OH)_{16}$
Biaxial +								
	M	$-L - +L$	1 – 2	$65^{\circ} - 90^{\circ}$ $r<v$	parallel		CORDIERITE	$Al_3(Mg,Fe^{+2})_2\{Si_5AlO_{18}\}$
	M – S $\alpha=\beta>\gamma$	$+L - +M$	1	$0^{\circ} - 60^{\circ}$ $r<v$	0° – small	length fast	Mn and Cr- CHLORITES	$(Mg,Mn,Al,Cr,Fe)_{12}$ $\{(Si,Al)_8O_{20}\}(OH)_{16}$
	W $\gamma>\beta>\alpha$	$+M$	1	50°	parallel	length slow	CELESTINE	$SrSO_4$
	W	$+M$	2 – 3	$58^{\circ} - 68^{\circ}$	$22^{\circ} - 26^{\circ}$	extinction nearest cleavage slow	KUNZITE (SPODUMENE)	$LiAl\{Si_2O_6\}$
	S $\alpha>\beta>\gamma$ $\alpha>\gamma\geqslant\beta$	$+M - H$	1 – 2	$40^{\circ} - 50^{\circ}$	$\beta : z$ $15^{\circ} - 30^{\circ}$	length fast	MAGNESIORIEBECKITE	$Na_2Mg_3Fe_2^{+3}\{Si_8O_{22}\}(OH)_2$

System	Form	Cleavage	Twinning	Zoning	Alteration	Occurrence	Remarks
Cub.	anhedral, hexagonal x-sect.	two or three perfect	interpenetrant	Z		I, S, M, O	colour spots cf. cryolite, halite
Mon.	tabular, radiating	perfect basal				O, I, M	
Mon.	tabular, radiating	perfect basal				O, I, M	
Orth. (Pseudo-Hex.)	pseudo-hex., anhedral	moderate to absent	simple, polysynthetic, cyclic, sector, interpenetrant			M, I	pleochroic haloes, inclusions (including opaque dust) common cf. quartz, plagioclase
Mon.	tabular, radiating	perfect basal				O, I, M	
Orth.	tabular, fibrous	three				S	cf. barytes
Mon.	tabular	two at 87°, parting	simple			I	cf. diopside
Mon.	prismatic, fibrous, columnar aggregates	two at 56°	simple, polysynthetic	Z		I, M	cf. glaucophane, crossite, arfvedsonite, hastingsite

PURPLE MINERALS	Pleochroism	Relief	δ	2V Dispersion	Extinction	Orientation	Mineral	Composition
Biaxial +								
	S $\alpha>\beta>\gamma$ $\alpha>\gamma\geq\beta$	+M – H	1 – 2	$40^\circ - 90^\circ$ $r\lesssim v$	$3^\circ - 21^\circ$	length fast	RIEBECKITE	$Na_2Fe_3^{+2}Fe_2^{+3}\{Si_8O_{22}\}(OH)_2$
	M – S	+M – H	2 – 3 (anom.)	$25^\circ - 50^\circ$ $r>v$	$35^\circ, 48^\circ$	extinction nearest cleavage fast	TITANAUGITE	$(Ca,Na,Mg,Fe^{+2},Mn,Fe^{+3},Al,Ti)_2\{(Si,Al)_2O_6\}$
		H	2	$35^\circ - 46^\circ$ $r>v$	64°		PYROXMANGITE	$(Mn,Fe)\{SiO_3\}$
	S $\gamma>\alpha>\beta$ $\gamma>\beta>\alpha$	H – VH	3 – 5	$64^\circ - 85^\circ$ $r\lesssim v$	$2^\circ - 9^\circ$	length fast/slow	PIEMONTITE	$Ca_2(Mn,Fe^{+3},Al)_2AlO\cdot OH\{Si_2O_7\}\{SiO_4\}$
Biaxial –								
	M	-L – +L	1 – 2	$65^\circ - 90^\circ$ $r<v$	parallel		CORDIERITE	$Al_3(Mg,Fe^{+2})_2\{Si_5AlO_{18}\}$
	W $\gamma>\alpha$	+L	1	small			KÄMMERERITE (KOCHUBEITE)	$(Mg,Al,Cr,Fe)_{12}\{(Si,Al)_8O_{20}\}OH_{16}$
	M – S $\gamma=\beta>\alpha$	+L – +M	1	$0^\circ - 20^\circ$ $r>v$	0° – small	length slow	Mn and Cr-CHLORITES	$(Mg,Mn,Al,Cr,Fe)_{12}\{(Si,Al)_8O_{20}\}(OH)_{16}$
	S $\gamma>\beta>\alpha$ $\beta>\gamma>\alpha$	+M	1 – 2 anom.	$12^\circ - 65^\circ$ $r<v$	$2^\circ - 30^\circ$	length slow	CROSSITE	$Na_2(Mg_3,Fe_3^{+2},Fe_2^{+3},Al_2)\{Si_8O_{22}\}(OH)_2$
	S $\gamma>\beta>\alpha$	+M	1 – 3 anom.	$0^\circ - 50^\circ$ $r>v$	$4^\circ - 14^\circ$	length slow	GLAUCOPHANE	$Na_2Mg_3Al_2\{Si_8O_{22}\}(OH)_2$
	S patchy $\alpha>\beta\geq\gamma$	+M	2	$20^\circ - 40^\circ$ $r\lesssim v$	parallel	length fast	DUMORTIERITE	$HBAl_8Si_3O_{20}$

System	Form	Cleavage	Twinning	Zoning	Alteration	Occurrence	Remarks
Mon.	prismatic, fibrous, columnar aggregates	two at 56°	simple, polysynthetic	Z		I, M	cf. glaucophane, crossite, arfvedsonite, hastingsite
Mon.	prismatic	two at 93°	polysynthetic, twin seams	Z hour-glass		I	
Tr.	prismatic	four, two at 92°	simple, polysynthetic			O, M	cf. rhodonite, bustamite
Mon.	columnar, six-sided x-sect.	one perfect	polysynthetic, uncommon			M, I	cf. thulite, titanaugite, dumortierite
Orth. (Pseudo-Hex.)	pseudo-hex., anhedral	moderate to absent	simple, polysynthetic, cyclic, sector, interpenetrant		pinite, talc	M, I	pleochroic haloes, inclusions (including opaque dust) common cf. quartz
Mon.	tabular	perfect				O	cf. chlorite
Mon.	tabular, radiating	perfect basal				O, I, M	
Mon.	prismatic, columnar aggregates	two at 58°	simple, polysynthetic	Z		M	cf. glaucophane, riebeckite
Mon.	prismatic, columnar aggregates	two at 58°	simple, polysynthetic	Z		M	resembles holmquistite in pleochroism cf. arfvedsonite, eckermannite, riebeckite
Orth.	prismatic, acicular	imperfect, cross fractures	interpenetrant, trillings		sericite	M, I	cf. tourmaline, sillimanite

PURPLE MINERALS	Pleochroism	Relief	δ	2V Dispersion	Extinction	Orientation	Mineral	Composition
Biaxial –								
	W	+M	2	$63^{\circ} - 80^{\circ}$ r<v	inclined		AXINITE	$(Ca,Mn,Fe^{+2})_3Al_2BO_3\{Si_4O_{12}\}OH$
	M β>γ>α	+M	2 – 3 anom.	$66^{\circ} - 87^{\circ}$ r<v	$15^{\circ} - 40^{\circ}$	length slow	RICHTERITE- FERRORICHTERITE	$Na_2Ca(Mg,Fe^{+3},Fe^{+2},Mn)_5$ $\{Si_8O_{22}\}(OH,F)_2$
	S	+M	3	51° r>v	parallel	length slow	HOLMQUISTITE	$Li_2(Mg,Fe^{+2})_3(Al,Fe^{+3})_2$ $\{Si_8O_{22}\}(OH)_2$
	S α>β>γ	+M – H	1 – 2	$0^{\circ} - 50^{\circ}$ r<v	$0^{\circ} - 30^{\circ}$ anom.	length fast	ARFVEDSONITE	$Na_3(Mg,Fe^{+2})_4Al\{Si_8O_{22}\}(OH,F)$
	S α>β>γ α>γ≳β	+M – H	1 – 2	$40^{\circ} - 50^{\circ}$	$15^{\circ} - 30^{\circ}$	length fast	MAGNESIORIEBECKITE	$Na_2Mg_3Fe_2^{+3}\{Si_8O_{22}\}(OH)_2$
	S α>β>γ α>γ≳β	+M – H	1 – 2	$40^{\circ} - 90^{\circ}$ r≲v	$3^{\circ} - 21^{\circ}$	length fast	RIEBECKITE	$Na_2Fe_3^{+2}Fe_2^{+3}\{Si_8O_{22}\}(OH)_2$
	N – S γ>β>α	+M – H	1 – 2	$50^{\circ} - 90^{\circ}$ r≲v	parallel	length slow	HYPERSTHENE	$(Mg,Fe^{+2})\{SiO_3\}$
	S	H	3	65° r<v			HOWIEITE	$Na(Fe,Mn)_{10}$ $(Fe,Al)_2Si_{12}O_{31}(OH)_{13}$

System	Form	Cleavage	Twinning	Zoning	Alteration	Occurrence	Remarks
Tr.	bladed, wedge-shaped, clusters	imperfect in several directions		Z		I, M	inclusions common cf. quartz
Mon.	long prismatic, fibrous	two at 56°, parting	simple, polysynthetic	Z		M, I	
Orth.	fibrous, bladed, prismatic	two at $54\frac{1}{2}^{o}$				M, I	cf. anthophyllite, glaucophane
Mon.	prismatic	two at 56°, parting	simple, polysynthetic	Z		I	cf. riebeckite, katophorite, glaucophane, tourmaline
Mon.	prismatic, fibrous, columnar aggregates	two at 56°	simple, polysynthetic	Z		I, M	cf. glaucophane, crossite, arfvedsonite
Mon.	prismatic, fibrous, columnar aggregates	two at 56°	simple, polysynthetic	Z		I, M	cf. glaucophane, crossite, arfvedsonite
Orth.	prismatic, anhedral – euhedral	two at 88°	polysynthetic, twin seams	Z		I, M	exsolution lamellae, schiller inclusions cf. andalusite
Tr.	bladed	three, one good				M	

BLACK MINERALS	Pleochroism	Relief	δ	2V Dispersion	Extinction	Orientation	Mineral	Composition
Isometric								
	H – VH	anom.					GARNET	$Ca_3(Fe^{+3},Ti)_2Si_3O_{12}$
	H – E						SPINEL (incl. PLEONASTE)	$(Mg,Fe^{+2})Al_2O_4$
Uniaxial –								
	W $\omega>\epsilon,\ \epsilon>\omega$	E	5	0° – small			ANATASE	TiO_2
Biaxial +								
	M – S $\gamma>\alpha$	VH	5	32° r<v	4° – 45°		COSSYRITE (AENIGMATITE)	$Na_2Fe_5^{+2}TiSi_6O_{20}$
	W $\gamma>\alpha$	E	(weak)	90° r>v	$\simeq 45^{o}$		PEROVSKITE (incl. LOPARITE-DYSANALYTE)	$(Ca,Na,Fe^{+2},Ce)(Ti,Nb)O_3$
Biaxial –								
	S $\gamma=\beta>\alpha$	+L – H	3 – 5	$\simeq 0^{o}$	$\simeq$ parallel	length slow	STILPNOMELANE	$(K,Na,Ca)_{0-1.4}$ $(Fe^{+3},Fe^{+2},Mg,Al,Mn)_{5.9-8.2}$ $\{Si_8O_{20}\}(OH)_4(O,OH,H_2O)_{3.6-8.5}$
	M – S $\gamma>\beta>\alpha$ $\gamma<\beta>\alpha$	+M	1 – 3	0° – 50° r<v	36° – 70°	length fast	KATOPHORITE-MAGNESIOKATOPHORITE	$Na_2Ca(Mg,Fe^{+2})_4Fe^{+3}$ $\{Si_7AlO_{22}\}(OH,F)_2$
	W – M $\beta>\alpha>\gamma$	H	1 – 2	45° – 68° r>v	2° – 30°	length fast	CHLORITOID (OTTRELITE)	$(Fe^{+2},Mg,Mn)_2(Al,Fe^{+3})Al_3O_2$ $\{SiO_4\}_2(OH)_4$
	S $\alpha>\beta\geqslant\gamma$	H – VH	4 – 5	60° – 70° r>v	0° – 10°	extinction nearest cleavage fast	AEGIRINE	$NaFe^{+3}\{Si_2O_6\}$

System	Form	Cleavage	Twinning	Zoning	Alteration	Occurrence	Remarks
Cub.	four, six, eight-sided, polygonal x-sect., aggregates	parting, irregular fractures	complex, sector	Z	chlorite	M, I, S	inclusions cf. spinel
Cub.	small grains, cubes, octahedra, rhombic x-sect.	parting				M, I	
Tet.	small, prismatic, acicular	two perfect		Z	leucoxene	S, I, M	usually yellow to blue cf. rutile, brookite
Tr.	small, prismatic, aggregates	two at 66°	simple, repeated			I	cf. katophorite, kaersutite, basaltic hornblende
Mon? Pseudo-Cub.	small cubes, skeletal	poor to distinct	polysynthetic, complex, interpenetrant	Z	leucoxene	I, M	cf. melanite, picotite, ilmenite
Mon.	plates, pseudo-hex., micaceous masses	two at 90°, perfect basal	polysynthetic	Z		I, M, O	cf. biotite, chlorite, chloritoid, clintonite
Mon.	prismatic	two at 56°, parting	simple	Z		I	cf. barkevikite, kaersutite, basaltic hornblende, arfvedsonite, cossyrite
Mon., Tr.	tabular	one perfect, one imperfect, parting	polysynthetic	Z hour-glass		M, I	inclusions common cf. chlorite, clintonite, biotite, stilpnomelane
Mon.	prismatic, needles, felted aggregates, blunt terminations	two at 87°, parting	simple, polysynthetic, twin seams	Z hour-glass		I	usually green cf. aegirine-augite, acmite

BLACK MINERALS	Pleochroism	Relief	δ	2V Dispersion	Extinction	Orientation	Mineral	Composition
Biaxial –								
	W	VH	3				DEERITE	$(Fe,Mn)_{13}(Fe,Al)_7Si_{13}O_{44}(OH)_{11}$
	W	VH	4	$44^\circ - 70^\circ$ r<v	parallel	length fast/slow	KNEBELITE	$(Mn,Fe)_2\{SiO_4\}$
	W $\omega>\varepsilon$, $\varepsilon>\omega$	E	5 anatase	(uniaxial)			ANATASE	TiO_2

The following common opaque minerals also appear black in thin sections:

	Pleochroism	Relief	δ	2V Dispersion	Extinction	Orientation	Mineral	Composition
	H – E						SPINEL GROUP	$(Mg,Mn,Zn,Ni,Al_2,Fe^{+2})Fe_2{}^{+3}O_4$
							CHALCOPYRITE	$CuFeS_2$
	E						CHROMITE (MAGNESIOCHROMITE)	$(Fe^{+2},Mg)Cr_2O_4$
	E						GALENA	PbS
							GRAPHITE	C
	E						MAGNETITE	$Fe^{+2}Fe_2{}^{+3}O_4$
			(weak)				PYRITE	FeS_2
							PYRRHOTITE	Fe_7S_8-FeS
	E		(weak)				LIMONITE	$FeO \cdot OH \cdot nH_2O$
	E		5				HAEMATITE	$\alpha-Fe_2O_3$
	E		5				ILMENITE	$FeTiO_3$

System	Form	Cleavage	Twinning	Zoning	Alteration	Occurrence	Remarks
Mon.	acicular, amphibole-like	one good	simple			M	
Orth.	anhedral - euhedral, rounded	two moderate, imperfect	uncommon			O, M	cf. tephroite
Tet.	small, prismatic, acicular	two perfect		Z	leucoxene	S, I, M	usually yellow to blue cf. rutile, brookite
Cub.	small grains, cubes, octahedra, rhombic x-sect.	parting				M, I, O	opaque
Tet.	aggregates					O, S	opaque
Cub.	subhedral grains, octahedra, aggregates					I, O	
Cub.	cubes, octahedra	perfect, parting	interpenetrant			O	opaque
Hex.	thin ragged flakes, scales	parting	perfect basal			M	
Cub.	small grains, octahedra	parting				I, M, S, O	opaque, exsolved ulvöspinel
Cub.	cubes, octahedra, irregular masses		interpenetrant			O, I, M, S	opaque amorphous = melnikovite
Hex.	grains, irregular masses	parting				O, I, M	opaque
M'loid.	stain or border to other minerals, pseudomorphs					I, M, S, O	opaque to translucent cf. goethite
Trig.	scales, flakes, grains and irregular masses	parting	polysynthetic			M, I, O, S	opaque to translucent red cf. goethite, limonite
Trig.	skeletal, grains and irregular masses				leucoxene	I, M, S	opaque

GREY MINERALS	Pleochroism	Relief	δ	2V Dispersion	Extinction	Orientation	Mineral	Composition
Isometric								
		$-M$					FLUORITE	CaF_2
		$-M - -L$	rarely weak				OPAL	SiO_2
		$-L$	(weak)				SODALITE (incl. HACKMANITE)	$Na_8\{Al_6Si_6O_{24}\}Cl_2$
		$-L$	(weak)				NOSEAN	$Na_8\{Al_6Si_6O_{24}\}SO_4$
		$-L$	(weak)				HAÜYNE	$(Na,Ca)_{4-8}\{Al_6Si_6O_{24}\}(SO_4,S)_{1-}$
		$-L - +M$	(weak)				VOLCANIC GLASS	
		E	(weak)				SPHALERITE	ZnS
	W $\gamma>\alpha$	E	(weak)				PEROVSKITE (KNOPITE-LOPARITE-DYSANALYTE)	$(Ca,Na,Fe^{+2},Ce)(Ti,Nb)O_3$
Uniaxial +								
		H	1 anom.		parallel	length fast	VESUVIANITE (IDOCRASE) (incl. WILUITE)	$Ca_{10}(Mg,Fe)_2Al_4\{Si_2O_7\}_2$ $\{SiO_4\}_5(OH,F)_4$
	W $\epsilon>\omega$	VH $-$ E	$4 - 5$		parallel	length slow	ZIRCON	$Zr\{SiO_4\}$

System	Form	Cleavage	Twinning	Zoning	Alteration	Occurrence	Remarks
Cub.	hexagonal x-sect.	two or three perfect	interpenetrant	Z		I, S, O, M	colour spots cf. cryolite, halite
M'loid.	colloform, veinlets, cavity fillings	irregular fractures				I	
Cub.	hexagonal x-sect., anhedral aggregates	poor	simple		zeolites, diaspore, gibbsite	I	cf. fluorite, leucite
Cub.	hexagonal x-sect., anhedral aggregates	imperfect	simple	Z	zeolites, diaspore, gibbsite, limonite	I	clouded with inclusions
Cub.	hexagonal x-sect., anhedral aggregates	imperfect	polysynthetic	Z	zeolites, diaspore, gibbsite	I	
M'loid.	amorphous, massive	perlitic parting			frequent, devitrification	I	often with crystallites and phenocrysts cf. tachylyte, lechatelierite
Cub.	irregular, anhedral, curved faces	six perfect	polysynthetic, lamellar intergrowths	Z		O	colour variable, (uniaxial) cf. cassiterite
Mon? Pseudo-Cub.	small cubes, skeletal	poor to distinct	polysynthetic, complex, interpenetrant	Z	leucoxene	I, M	cf. melanite, picotite, ilmenite
Tet.	variable, prismatic, fibrous, granular, radial	imperfect	sector	Z		M, I, S	cf. zoisite, clinozoisite, apatite, grossularite, melilite, andalusite
Tet.	minute prisms	poor, absent		Z	metamict	I, S, M	cf. apatite

GREY MINERALS	Pleochroism	Relief	δ	2V Dispersion	Extinction	Orientation	Mineral	Composition
Uniaxial +								
	W $\varepsilon > \omega$	VH - E	5		parallel, oblique to twin plane	length slow	CASSITERITE	SnO_2
Uniaxial -								
		-L - +M	5		symmetrical to cleavage		CALCITE	$CaCO_3$
		-L - +M	5		symmetrical to cleavage		DOLOMITE	$CaMg(CO_3)_2$
		+L - VH	5		symmetrical to cleavage		SIDERITE	$FeCO_3$
	W $\varepsilon > \omega$	+M	1		parallel	length fast, tabular length slow	APATITE (incl. DAHLLITE, FRANCOLITE)	$Ca_5(PO_4)_3(OH,F,Cl)$
	S $\omega > \varepsilon$	+M	3		parallel	length fast	SCHORL (TOURMALINE)	$Na(Fe,Mn)_3Al_6B_3Si_6O_{27}(OH,F)_4$
		+H	1 anom.		parallel	length fast	VESUVIANITE (IDOCRASE)	$Ca_{10}(Mg,Fe)_2Al_4\{Si_2O_7\}_2\{SiO_4\}_5(OH,F)_4$
	W $\omega > \varepsilon$	H	1	$0° - 30°$	parallel, symmetrical	tabular length slow, prismatic length fast	CORUNDUM	$\alpha\text{-}Al_2O_3$
	W $\omega > \varepsilon,\ \varepsilon > \omega$	E	5 anom.	$0° - small$			ANATASE	TiO_2

System	Form	Cleavage	Twinning	Zoning	Alteration	Occurrence	Remarks
Tet.	subhedral, veinlets, diamond-shaped x-sect.	prismatic	geniculate, cyclic, common	Z		O, I, S	cf. sphalerite, rutile
Trig.	anhedral, oolitic, spherulitic	rhombohedral	polysynthetic, // long diagonal			S, M, I, O	twinkling cf. rhombohedral carbonates
Trig.	rhombohedral	rhombohedral	polysynthetic, // long and short diagonals	Z	huntite	S, M, I	twinkling cf. rhombohedral carbonates
Trig.	rhombohedral	rhombohedral	polysynthetic, // long diagonal, uncommon			S, O, I, M	twinkling, brown stain around borders and along cleavage cracks cf. rhombohedral carbonates
Hex.	small, prismatic, hexagonal	poor basal				I, S, M, O	cf. beryl, topaz, dahllite
Trig.	hexagonal, rounded, suns, triangular x-sect.	fractures		Z		I, M, S	cf. tourmaline group
Tet.	variable, prismatic, fibrous, granular, radial	imperfect	sector	Z		M, I, S	cf. zoisite, clinozoisite, apatite, grossularite, melilite, andalusite
Trig.	tabular, prismatic, six-sided x-sect.	parting	simple, lamellar seams	Z colour banding		M, I	inclusions cf. sapphirine
Tet.	small, prismatic, acicular	simple	rare			S, I, M	usually yellow to blue cf. rutile, brookite

GREY MINERALS	Pleochroism	Relief	δ	2V Dispersion	Extinction	Orientation	Mineral	Composition
Biaxial +								
	W	+M	3	$50°$ – $62°$ r>v	$38°$ – $48°$	extinction nearest cleavage slow	DIOPSIDE (SALITE)	$Ca(Mg,Fe)\{Si_2O_6\}$
		+M	4	$82°$ – $90°$ r>v	parallel	cleavage length slow	FORSTERITE	Mg_2SiO_4
	W – M $\beta>\alpha>\gamma$	H anom.	1 – 2	$45°$ – $68°$ r>v	$2°$ – $30°$	length fast	CHLORITOID (OTTRELITE)	$(Fe^{+2},Mg,Mn)_2(Al,Fe^{+3})Al_3O_2\{SiO_4\}_2(OH)_4$
Biaxial –								
		–L	2 – 3	$40°$ – $60°$	$\simeq$ parallel	length slow	SEPIOLITE	$H_4Mg_2\{Si_3O_{10}\}$
		–L – +M	5	$4°$ – $14°$ (uniaxial)	symmetrical to cleavage		CALCITE	$CaCO_3$
	N – W	+M	1	small	small	cleavage length slow	CHAMOSITE	$(Mg,Al,Fe)_{12}\{(Si,Al)_8O_{20}\}(OH)_1$
		+M	2	$38°$ – $60°$ r>v	$\alpha{:}z\ 30°\text{-}44°$ $\simeq$ parallel	length fast/slow	WOLLASTONITE	$Ca\{SiO_3\}$
	S variable $\alpha>\beta>\gamma$	+M – H	1 – 2	$0°$ – $50°$ r<v	$0°$ – $30°$ anom.	length slow	ARFVEDSONITE	$Na_3(Mg,Fe^{+2})_4Al\{Si_8O_{22}\}(OH,F)_2$
	N – S $\gamma>\beta>\alpha$	+M – H	1 – 2	$50°$ – $90°$ r$\lesssim$v	parallel	length slow	HYPERSTHENE	$(Mg,Fe^{+2})\{SiO_3\}$
	W	+M – VH	4	$48°$ – $90°$	parallel	cleavage length slow	OLIVINE	$(Mg,Fe)_2\{SiO_4\}$

System	Form	Cleavage	Twinning	Zoning	Alteration	Occurrence	Remarks
Mon.	short prismatic	two at 87°	simple, polysynthetic	Z	tremolite-actinolite	M, I	cf. hedenbergite, tremolite, omphacite, wollastonite, epidote
Orth.	anhedral – euhedral, rounded	uncommon, irregular fractures	uncommon	Z	chlorite, antigorite, serpentine, iddingsite, bowlingite	I, M	deformation lamellae cf. diopside, augite, pigeonite, humite group, epidote
Mon., Tr.	tabular	perfect basal, one imperfect, parting	polysynthetic	Z hour-glass		M, I	inclusions common cf. clintonite, biotite, stilpnomelane
Mon. (Orth.)	fibrous aggregates, curved, matted					I, S	
Trig.	anhedral	rhombohedral	polysynthetic, // long diagonal			M	twinkling cf. rhombohedral carbonates
Mon.	pseudo-spherulitic, concentric, tabular, massive	concentric parting				S, O	cf. glauconite, collophane, greenalite, thuringite
Tr.	subhedral – euhedral, columnar, fibrous	three	polysynthetic	Z	pectolite, calcite	M, I	cf. tremolite, pectolite
Mon.	prismatic	two at 56°, parting	simple, polysynthetic	Z		I	cf. riebeckite, katophorite, glaucophane, tourmaline
Orth.	prismatic, anhedral – subhedral	two at 88°	polysynthetic, twin seams	Z		I, M	exsolution lamellae, schiller inclusions cf. andalusite
Orth.	anhedral – euhedral, rounded	uncommon, one moderate, irregular fractures	uncommon, vicinal	Z	chlorite, antigorite, serpentine, iddingsite, bowlingite	I, M	deformation lamellae cf. diopside, augite, pigeonite, humite group, epidote

GREY MINERALS	Pleochroism	Relief	δ	2V Dispersion	Extinction	Orientation	Mineral	Composition
Biaxial –								
	W – M $\beta>\alpha>\gamma$	H	1 – 2	$45^\circ - 68^\circ$ r>v	$2^\circ - 30^\circ$	length fast	CHLORITOID (OTTRELITE)	$(Fe^{+2},Mg,Mn)_2(Al,Fe^{+3})Al_3O_2\{SiO_4\}_2(OH)_4$
	W $\gamma>\beta>\alpha$	H	2	$82^\circ - 83^\circ$ r>v	$0^\circ - 32^\circ$	length slow	KYANITE	Al_2SiO_5
	S	H	3	65° r<v			HOWIEITE	$Na(Fe,Mn)_{10}(Fe,Al)_2Si_{12}O_{31}(OH)_{13}$
	W $\beta>\gamma>\alpha$	H anom.	3 – 4	$74^\circ - 90^\circ$ r>v	$0^\circ - 15^\circ$ parallel in elong. sect.	length fast/slow	EPIDOTE	$Ca_2Fe^{+3}Al_2O\cdot OH\{Si_2O_7\}\{SiO_4\}$
	W – M $\alpha>\beta>\gamma$	VH – E	5 anom.	$17^\circ - 40^\circ$ r>v	40° symmetrical		SPHENE	$CaTi\{SiO_4\}(O,OH,F)$

System	Form	Cleavage	Twinning	Zoning	Alteration	Occurrence	Remarks
Mon., Tr.	tabular	one perfect, one imperfect, parting	polysynthetic	Z hour-glass		M, I	inclusions common cf. chlorite, clintonite, biotite, stilpnomelane
Tr.	bladed, prismatic	two, parting	simple, polysynthetic			M	cf. sillimanite, pyroxene
Tr.	bladed	three, one good				M	
Mon.	distinct crystals, columnar aggregates	one perfect	uncommon			M, I, S	cf. zoisite, clinozoisite, diopside, augite, sillimanite
Mon.	rhombic, irregular grains	parting	simple, polysynthetic		leucoxene	I, M, S	cf. monazite, calcite

WHITE MINERALS	Pleochroism	Relief	δ	2V Dispersion	Extinction	Orientation	Mineral	Composition
							LEUCOXENE	$TiO_2 n \cdot H_2O$
		H	(anom.)				GROSSULARITE	$Ca_3Al_2Si_3O_{12}$

System	Form	Cleavage	Twinning	Zoning	Alteration	Occurrence	Remarks
Amor-phous	finely crystalline, pseudomorphs, alteration of ilmenite					I, M, O	opaque, finely crystalline, rutile or brookite
Cub.	four, six, eight-sided, polygonal x-sect.	parting, irregular fractures	sector	Z		M	cf. garnet group, periclase, vesuvianite

Tables 3–11

Mineral groups

TABLE 3 AMPHIBOLE GROUP

Mineral	Composition	System	Colour	Pleochroism	Relief	δ	2V Dispersion	Extinction
Orthorhombic								
Biaxial +								
ANTHOPHYLLITE (Fe-rich)	$(Mg,Fe^{+2})_7$ $\{Si_8O_{22}\}(OH,F)_2$	Orth.	colourless, brown, yellow, green	W $\gamma=\beta>\alpha$ $\gamma>\beta=\alpha$	+M	2 – 3	$68^\circ - 90^\circ$ $r\lessgtr v$	parallel, symmetrical
GEDRITE	$(Mg,Fe^{+2})_5Al_2$ $\{Si_6Al_2O_{22}\}(OH,F)_2$	Orth.	colourless, brown, yellow	W $\gamma=\beta>\alpha$ $\gamma>\beta=\alpha$	+M – H	2 – 3	$68^\circ - 90^\circ$ $r\lessgtr v$	parallel, symmetrical
Biaxial –								
ANTHOPHYLLITE (Mg-rich)	$(Mg,Fe^{+2})_7$ $\{Si_8O_{22}\}(OH,F)_2$	Orth.	colourless, brown, yellow, green	W $\gamma>\beta>\alpha$	+M	2 – 3	$68^\circ - 90^\circ$ $r\lessgtr v$	parallel, symmetrical
HOLMQUISTITE	$Li_2(Mg,Fe^{+2})_3(Al,Fe^{+3})_2$ $\{Si_8O_{22}\}(OH)_2$	Orth.	yellow, blue, violet	S	+M	3	51° $r>v$	parallel
FERROGEDRITE	$Fe_5Al_4Si_6O_{22}(OH)_2$	Orth.	colourless, yellow, green	W $\gamma=\beta>\alpha$ $\gamma>\beta=\alpha$	+M – H	2 – 3	$68^\circ - 90^\circ$ $r\lessgtr v$	parallel, symmetrical
Monoclinic								
Biaxial +								
EDENITE	$NaCa_2(Mg,Fe^{+2})_5$ $AlSi_7O_{22}(OH)_2$	Mon.	green	$\gamma\gtrless\beta>\alpha$	+M	2	$58^\circ - 64^\circ$ $r>v$	$22^\circ - 27^\circ$
PARGASITE	$NaCa_2Mg_4Al_3Si_6O_{22}(OH)_2$	Mon.	colourless, brown, yellow, green	M – S $\gamma\gtrless\beta>\alpha$	+M	3	$67^\circ - 90^\circ$ $r>v$	26°
CUMMINGTONITE	$(Mg,Fe^{+2})_7\{Si_8O_{22}\}(OH)_2$	Mon.	colourless, green	N – W $\gamma>\beta>\alpha$	+M	3 – 4	$65^\circ - 90^\circ$ Mg $r<v$ Fe $r>v$	$15^\circ - 21^\circ$

Orientation	Form	Cleavage	Twinning	Zoning	Alteration	Occurrence	Remarks
length slow	bladed, prismatic, fibrous, asbestiform	two at $54\frac{1}{2}^{\circ}$			talc	M	cf. gedrite, tremolite, cummingtonite, holmquistite
length slow	bladed, prismatic, fibrous, asbestiform	two at $54\frac{1}{2}^{\circ}$			talc	M	cf. Fe-anthophyllite, tremolite, holmquistite, cummingtonite, grunerite, zoisite
length slow	bladed, prismatic, fibrous, asbestiform	two at $54\frac{1}{2}^{\circ}$			talc	M	cf. gedrite, tremolite, zoisite, cummingtonite, grunerite, holmquistite
length slow	fibrous, bladed, prismatic	two at $54\frac{1}{2}^{\circ}$				M, I	cf. anthophyllite, glaucophane
length slow	bladed, prismatic, fibrous, asbestiform	two at $54\frac{1}{2}^{\circ}$				M	cf. zoisite, cummingtonite, grunerite, anthophyllite
length slow	prismatic	two at 56°		Z		M, I	cf. hornblende
length slow	prismatic	two at 56°, parting	simple, polysynthetic			M, I	cf. hornblende, hastingsite, cummingtonite
length slow	prismatic, sub-radiating, fibrous, asbestiform	two at 55°	simple, polysynthetic			M, I	cf. tremolite, actinolite, grunerite, anthophyllite

TABLE 3 AMPHIBOLE GROUP	Mineral	Composition	System	Colour	Pleochroism	Relief	δ	2V Dispersion	Extinction
Monoclinic									
Biaxial +									
	MAGNESIORIEBECKITE	$Na_2Mg_3Fe_2^{+3}$ $\{Si_8O_{22}\}(OH)_2$	Mon.	green, blue, lavender	S $\alpha>\beta>\gamma$ $\alpha>\gamma\geqslant\beta$	+M - H	1 - 2	$40^\circ - 50^\circ$	β:z 15°-30°
	RIEBECKITE	$Na_2Fe_3^{+2}Fe_2^{+3}$ $\{Si_8O_{22}\}(OH)_2$	Mon.	green, blue, lavender	S $\alpha>\beta>\gamma$ $\alpha>\gamma\geqslant\beta$	+M - H	1 - 2	50° r>v	α:z 3°-10° β:x 17°-24°
	RIEBECKITE	$Na_2Fe_3^{+2}Fe_2^{+3}$ $\{Si_8O_{22}\}(OH)_2$	Mon	green, blue, lavender	S $\alpha>\beta>\gamma$ $\alpha>\gamma\geqslant\beta$	+M - H	1 - 2	$68^\circ - 85^\circ$ r<v	α:z 0°-8° γ:x 14°-22°
Biaxial -									
	CROSSITE	$Na_2(Mg_3,Fe_3^{+2},$ $Fe_2^{+3},Al_2)\{Si_8O_{22}\}(OH)_2$	Mon.	colourless, yellow, blue, lavender	S $\gamma>\beta>\alpha$ $\beta>\gamma>\alpha$	+M	1 - 2 anom.	$12^\circ - 65^\circ$ r<v	$2^\circ - 30^\circ$
	GLAUCOPHANE	$Na_2Mg_3Al_2\{Si_8O_{22}\}(OH)_2$	Mon.	colourless, blue, lavender	S $\gamma>\beta>\alpha$	+M	1 - 3 anom.	$0^\circ - 50^\circ$ r>v	$4^\circ - 14^\circ$
	KATOPHORITE-MAGNESIO-KATOPHORITE	$Na_2Ca(Mg,Fe^{+2})_4Fe^{+3}$ $\{Si_7AlO_{22}\}(OH,F)_2$	Mon.	brown, yellow, green, black	M - S $\gamma>\beta>\alpha$ $\gamma<\beta>\alpha$	+M	1 - 3	$0^\circ - 50^\circ$ r<v	$36^\circ - 70^\circ$
	BARKEVIKITE	$Ca_2(Na,K)$ $(Fe^{+2},Mg,Fe^{+3},Mn)_5$ $\{Si_{6.5}Al_{1.5}O_{22}\}(OH)_2$	Mon.	brown, orange, yellow	S $\gamma>\beta>\alpha$	+M	2	$40^\circ - 50^\circ$ r>v	$11^\circ - 18^\circ$
	EDENITE	$NaCa_2(Mg,Fe^{+2})_5$ $AlSi_7O_{22}(OH)_2$	Mon.	green	S $\gamma\geqslant\beta>\alpha$	+M	2	62° r>v	21°
	ECKERMANNITE	$Na_3(Mg,Fe^{+2})_4Al$ $\{Si_8O_{22}\}(OH,F)_2$	Mon.	green	S $\alpha>\beta>\gamma$	+M	2 - 3	$15^\circ - 80^\circ$ r>v	$18^\circ - 53^\circ$ (> 40°) flamy

Orientation	Form	Cleavage	Twinning	Zoning	Alteration	Occurrence	Remarks
length fast	prismatic, fibrous, columnar aggregates	two at 56°	simple, polysynthetic	Z		I, M	cf. glaucophane, crossite, arfvedsonite, hastingsite
OAP ⊥ (010) length fast	prismatic, fibrous, columnar aggregates	two at 56°	simple, polysynthetic	Z		I, M	cf. glaucophane, crossite, arfvedsonite, hastingsite
OAP // (010) length fast	prismatic, fibrous, columnar aggregates	two at 56°	simple, polysynthetic	Z		I, M	cf. glaucophane, crossite, arfvedsonite, hastingsite
length slow	prismatic, columnar aggregates	two at 58°	simple, polysynthetic	Z		M	cf. glaucophane, riebeckite
length slow	prismatic, columnar aggregates	two at 58°	simple, polysynthetic	Z		M	resembles holmquistite in pleochroism cf. arfvedsonite, eckermannite, riebeckite
length fast	prismatic	two at 56°, parting	simple	Z		I	cf. barkevikite, kaersutite, basaltic hornblende, arfvedsonite, cossyrite
length slow	prismatic	two at 56°, parting	simple, polysynthetic	Z		I	cf. hastingsite, kaersutite, basaltic hornblende
length slow	prismatic	two at 56°		Z		M, I	cf. hornblende
length slow	prismatic	two at 56°, parting	simple, polysynthetic			I	cf. arfvedsonite, glaucophane, cummingtonite, tremolite, hornblende

TABLE 3
AMPHIBOLE
GROUP

Mineral	Composition	System	Colour	Pleochroism	Relief	δ	2V Dispersion	Extinction

Monoclinic

Biaxial −

Mineral	Composition	System	Colour	Pleochroism	Relief	δ	2V Dispersion	Extinction
RICHTERITE-FERRORICHTERITE (incl. WINCHITE, CHIKLITE)	$Na_2Ca(Mg,Fe^{+3},Fe^{+2},Mn)_5$ $\{Si_8O_{22}\}(OH,F)_2$	Mon.	colourless, red, orange, yellow, blue, lavender	M $\beta>\gamma>\alpha$	+M	2 − 3 anom.	$66^{\circ} - 87^{\circ}$ r<v	$15^{\circ} - 40^{\circ}$
ACTINOLITE-FERROACTINOLITE	$Ca_2(Mg,Fe^{+2})_5$ $\{Si_8O_{22}\}(OH,F)_2$	Mon.	colourless, yellow, green	W $\gamma>\beta>\alpha$	+M	2 − 3	$73^{\circ} - 86^{\circ}$ r<v	$10^{\circ} - 17^{\circ}$
TREMOLITE	$Ca_2(Mg,Fe^{+2})_5$ $\{Si_8O_{22}\}(OH,F)_2$	Mon.	colourless		+M	3	$65^{\circ} - 86^{\circ}$ r<v	$15^{\circ} - 21^{\circ}$
ARFVEDSONITE	$Na_3(Mg,Fe^{+2})_4Al$ $\{Si_8O_{22}\}(OH,F)_2$	Mon.	yellow, green, blue, lavender, violet, grey	S variable $\alpha>\beta>\gamma$	+M − H	1 − 2	$0^{\circ} - 50^{\circ}$ r<v	$0^{\circ} - 30^{\circ}$ anom.
RIEBECKITE	$Na_2Fe_3^{+2}Fe_2^{+3}$ $\{Si_8O_{22}\}(OH)_2$	Mon.	blue, green, lavender	S $\alpha>\beta>\gamma$ $\alpha>\gamma>\beta$	+M − H	1 − 2	$70^{\circ} - 87^{\circ}$ r>v	$\alpha{:}z$ $0^{\circ}- 5$ $\gamma{:}x$ $14^{\circ}{-}19$
HORNBLENDE	$(Na,K)_{0-1}Ca_2$ $(Mg,Fe^{+2},Fe^{+3},Al)_5$ $\{Si_{6-7}Al_{2-1}O_{22}\}(OH,F)_2$	Mon.	red, brown, green	S $\gamma>\beta>\alpha$ $\beta>\gamma>\alpha$	+M − H	2 − 3	$60^{\circ} - 85^{\circ}$ r>v	$13^{\circ} - 34^{\circ}$
BASALTIC HORNBLENDE	$(Ca,Na)_{2-3}(Mg,Fe^{+2})_{3-2}$ $(Fe^{+3},Al)_{2-3}O_2$ $\{Si_6Al_2O_{22}\}$	Mon.	brown, yellow	S $\gamma>\beta>\alpha$	+M − H	2 − 5	$60^{\circ} - 82^{\circ}$ r<v	$0^{\circ} - 18^{\circ}$
HASTINGSITE-FERROHASTINGSITE	$NaCa_2(Mg,Fe^{+2})_4$ $(Al,Fe^{+3})Al_2Si_6O_{22}$ $(OH,F)_2$	Mon.	brown, yellow, green, blue	S $\beta>\gamma>\alpha$ $\gamma>\beta>\alpha$	+M − H	3	$10^{\circ} - 90^{\circ}$ r$\lessgtr$v	$9^{\circ} - 40^{\circ}$
GRUNERITE	$(Fe^{+2},Mg)_7\{Si_8O_{22}\}(OH)_2$	Mon.	colourless, brown, yellow, green	W $\gamma>\beta=\alpha$	+M − H	3 − 4	$84^{\circ} - 90^{\circ}$ r>v	$10^{\circ} - 15^{\circ}$
KAERSUTITE	$Ca_2(Na,K)$ $(Mg,Fe^{+2},Fe^{+3})_4Ti$ $\{Si_6Al_2O_{22}\}(O,OH,F)_2$	Mon.	brown, orange, yellow	S $\gamma>\beta>\alpha$	+M − H	3 − 5	$66^{\circ} - 82^{\circ}$ r>v	$0^{\circ} - 19^{\circ}$

Orientation	Form	Cleavage	Twinning	Zoning	Alteration	Occurrence	Remarks
length slow	long prismatic, fibrous	two at 56°, parting	simple, polysynthetic	Z		M, I	
length slow	long prismatic, fibrous	two at 56°, parting	simple, polysynthetic	Z		M, I	cf. tremolite, orthoamphibole, hornblende
length slow	long prismatic, fibrous	two at 56°, parting	simple, polysynthetic			M, I	cf. actinolite, orthoamphiboles, cummingtonite, wollastonite
length fast	prismatic	two at 56°, parting	simple, polysynthetic	Z		I	cf. riebeckite, katophorite, ferrohastingsite, glaucophane, tourmaline
OAP // (010) length fast	prismatic, fibrous, columnar aggregates	two at 56°	simple, polysynthetic	Z		I, M	cf. glaucophane, crossite, arfvedsonite
length slow	prismatic	two at 56°	simple, polysynthetic		mica, chlorite	M, I	cf. edenite, aegirine-augite, biotite, basaltic hornblende, augite, actinolite, pargasite, ferrohastingsite
length slow	short prismatic	two at 56°	simple, polysynthetic	Z		I	cf. kaersutite, barkevikite, katophorite, cossyrite
length slow	prismatic	two at 56°, parting	simple, polysynthetic	Z		I, M	cf. arfvedsonite, hornblende
length slow	prismatic, subradiating, fibrous, asbestiform	two at 55°, cross fractures	simple, polysynthetic		limonite	M	cf. tremolite, actinolite, cummingtonite, anthophyllite
length slow	prismatic	two at 56°, parting	simple, polysynthetic	Z		I	cf. cossyrite, barkevikite, katophorite, basaltic hornblende, titanaugite

TABLE 4 CHLORITE GROUP

Mineral	Composition	System	Colour	Pleochroism	Relief	δ	2V Dispersion	Extinction
Monoclinic								
Biaxial +								
PENNINITE	$(Mg,Al,Fe)_{12}$ $\{(Si,Al)_8O_{20}\}(OH)_{16}$	Mon.	colourless, yellow, green	W $\alpha=\beta>\gamma$	+L	1 anom.	0° – 20° r<v	$\simeq$ parallel
SHERIDANITE	$(Mg,Al,Fe_{12}$ $\{(Si,Al)_8O_{20}\}(OH)_{16}$	Mon.	colourless, green	W $\alpha=\beta>\gamma$	+L	1	$\simeq 20^{\circ}$ r<v	$\simeq$ parallel
CLINOCHLORE	$(Mg,Al,Fe)_{12}$ $\{(Si,Al)_8O_{20}\}(OH)_{16}$	Mon.	colourless, yellow, green	W $\alpha=\beta>\gamma$	+L	1 – 2	0° – 40° r<v	0° – 9°
RIPIDOLITE (PROCHLORITE) KLEMENTITE	$(Mg,Al,Fe)_{12}$ $\{(Si,Al)_8O_{20}\}(OH)_{16}$	Mon.	colourless, green	W $\alpha=\beta>\gamma$	+L – +M	1 anom.	0° – 30° r<v	$\simeq$ parallel
CORUNDOPHILITE	$(Mg,Al,Fe)_{12}$ $\{(Si,Al)_8O_{20}\}(OH)_{16}$	Mon.	colourless, green	W $\alpha=\beta>\gamma$	+L – +M	2	31° r<v	8° – 10°
Mn and Cr- CHLORITES	$(Mg,Mn,Al,Cr,Fe)_{12}$ $\{(Si,Al)_8O_{20}\}(OH)_{16}$	Mon.	pink, orange-red, green, violet	M – S $\alpha=\beta>\gamma$	+L – +M	1	0° – 60° r<v	0° – small
Biaxial –								
PENNINITE	$(Mg,Al,Fe)_{12}$ $\{(Si,Al)_8O_{20}\}(OH)_{16}$	Mon.	colourless, yellow, green	W $\gamma=\beta>\alpha$	+L	1 anom.	0° – 40° r>v	$\simeq$ parallel
DIABANTITE- BRUNSVIGITE	$(Mg,Al,Fe)_{12}$ $\{(Si,Al)_8O_{20}\}(OH)_{16}$	Mon.	green	W – M $\gamma=\beta>\alpha$	+L – +M	1 anom.	0° – 20°	$\simeq$ parallel
DELESSITE	$(Mg,Al,Fe)_{12}$ $\{(Si,Al)_8O_{20}\}(OH)_{16}$	Mon.	pink, green	W $\gamma=\beta>\alpha$	+L – +M	1	0° – 20°	0° – 7°
RIPIDOLITE (PROCHLORITE)	$(Mg,Al,Fe)_{12}$ $\{(Si,Al)_8O_{20}\}(OH)_{16}$	Mon.	colourless, green	W $\gamma=\beta>\alpha$	+M	1 anom.	0° – small r>v	$\simeq$ parallel
DAPHNITE	$(Mg,Al,Fe)_{12}$ $\{(Si,Al)_8O_{20}\}(OH)_{16}$	Mon.	colourless, green	W $\gamma=\beta>\alpha$	+M	1	0° – 20°	small

Orientation	Form	Cleavage	Twinning	Zoning	Alteration	Occurrence	Remarks
cleavage length fast	tabular, vermicular, radiating, pseudomorphs	perfect basal	simple, pennine law	Z		I, M	pleochroic haloes cf. clinochlore, prochlorite
cleavage length fast	tabular, spherulitic	perfect basal	simple			I	
cleavage length fast	tabular, fibrous, pseudo-hex.	perfect basal	polysynthetic			M	pleochroic haloes cf. penninite, prochlorite, leuchtenbergite, katschubeite
cleavage length fast	tabular, scaly, vermicular, fan-shaped aggregates	perfect basal				M, I, O	cf. clinochlore, penninite
cleavage length fast	tabular, radiating	perfect basal				M	
cleavage length fast	tabular	perfect basal				O	
cleavage length slow	tabular, vermicular, radiating, pseudomorphs	perfect basal	simple, pennine law	Z		I, M	pleochroic haloes cf. clinochlore, prochlorite
cleavage length slow	tabular, fibrous	perfect basal				M, I, O	
cleavage length slow	spherulites	perfect basal				I	
cleavage length slow	tabular, scaly, vermicular, fan-shaped aggregates	perfect basal				M, I, O	negative ripidolite, uncommon
cleavage length slow	concentric aggregates, fibrous, plates	perfect basal				O	

TABLE 4 CHLORITE GROUP	Mineral	Composition	System	Colour	Pleochroism	Relief	δ	2V Dispersion	Extinction
Monoclinic									
Biaxial −									
	CHAMOSITE	$(Mg,Al,Fe)_{12}$ $\{(Si,Al)_8O_{20}\}(OH)_{16}$	Mon.	green, brown	N − W	+M	1 anom.	small	small
	THURINGITE	$(Mg,Al,Fe)_{12}$ $\{(Si,Al)_8O_{20}\}(OH)_{16}$	Mon.	colourless, green	W	+M	2	$0^{o} - 20^{o}$	small
	Mn and Cr- CHLORITES (PENNANTITE- GONYERITE- KÄMMERERITE- KOCHUBEITE)	$(Mg,Mn,Al,Cr,Fe)_{12}$ $\{(Si,Al)_8O_{20}\}(OH)_{16}$	Mon.	pink, orange-red, violet	M − S $\gamma=\beta>\alpha$	+L − +M	1	$0^{o} - 20^{o}$ r>v	0° − small

Orientation	Form	Cleavage	Twinning	Zoning	Alteration	Occurrence	Remarks
cleavage length slow	pseudo-spherulitic, concentric, tabular, massive	one good, concentric parting				S, O	cf. glauconite, collophane, greenalite, thuringite
cleavage length slow	tabular, radiating	perfect basal				O	
cleavage length slow	tabular, radiating	perfect basal				O, I, M	Cr-Chlorites usually positive

TABLE 5 EPIDOTE GROUP

Mineral	Composition	System	Colour	Pleochroism	Relief	δ	2V Dispersion	Extinction
Orthorhombic								
Biaxial +								
α - ZOISITE	$Ca_2Al \cdot Al_2O \cdot OH$ $\{Si_2O_7\}\{SiO_4\}$	Orth.	colourless		+M – H	1 anom.	0° – 30° r<v	parallel
β - ZOISITE (Fe-rich)	$Ca_2Al \cdot Al_2O \cdot OH$ $\{Si_2O_7\}\{SiO_4\}$	Orth.	colourless		+M – H	1 normal	0° – 60° r>v	parallel
THULITE	$Ca_2Mn^{+3}Al_2O \cdot OH$ $\{Si_2O_7\}\{SiO_4\}$	Orth.	pink, yellow	M	+M – H	1 – 3	0° – 60° r>v	parallel
Monoclinic								
Biaxial +								
CLINOZOISITE	$Ca_2Al \cdot Al_2O \cdot OH$ $\{Si_2O_7\}\{SiO_4\}$	Mon.	colourless		+M – H	1 – 2 anom.	14° – 90° r$\lessgtr$v	0° – 7°
ALLANITE (ORTHITE)	$(Ca,Ce)_2(Fe^{+2},Fe^{+3})$ $Al_2O \cdot OH\{Si_2O_7\}\{SiO_4\}$	Mon.	red, brown, yellow	W	H – VH	2 – 3	40° – 90° r$\lessgtr$v	1° – 42° parallel in elong. sect
PIEMONTITE	$Ca_2(Mn,Fe^{+3},Al)_2AlO \cdot OH$ $\{Si_2O_7\}\{SiO_4\}$	Mon.	pink, red, yellow, violet	S $\gamma>\alpha>\beta$ $\gamma>\beta>\alpha$	H – VH	3 – 5	64° – 85° r$\lessgtr$v	2° – 9°
Biaxial –								
EPIDOTE	$Ca_2Fe^{+3}Al_2O \cdot OH$ $\{Si_2O_7\}\{SiO_4\}$	Mon.	colourless, yellow, green, grey	W $\beta>\gamma>\alpha$	H	3 – 4 anom.	74° – 90° r>v	0° – 15° parallel in elong. sect
ALLANITE (ORTHITE)	$(Ca,Ce)_2(Fe^{+2},Fe^{+3})$ $Al_2O \cdot OH\{Si_2O_7\}\{SiO_4\}$	Mon.	red, brown, yellow	W	H – VH	2 – 3	40° – 90° r$\lessgtr$v	1° – 42° parallel in elong. sect

Orientation	Form	Cleavage	Twinning	Zoning	Alteration	Occurrence	Remarks
length fast	columnar aggregates, euhedral	one perfect, one imperfect	polysynthetic, rare	Z		M	cf. clinozoisite, epidote
length fast/slow	columnar aggregates, euhedral	one perfect, one imperfect	polysynthetic, rare	Z		M	cf. clinozoisite, epidote, diopside, augite
length fast/slow	columnar aggregates	one perfect, one imperfect	polysynthetic, rare			M	cf. zoisite, clinozoisite, vesuvianite, sillimanite
length fast/slow	columnar, six-sided, x-sect.	one perfect	polysynthetic, uncommon	Z		M	cf. epidote, zoisite, diopside, augite
length fast/slow	distinct crystals, columnar, six-sided x-sect., irregular	two imperfect	uncommon	Z	metamict	I, M	anastomosing cracks, isotropic in metamict state cf. melanite
length fast/slow	columnar, six-sided x-sect.	one perfect	polysynthetic, uncommon			M, I	cf. thulite, titanaugite, dumortierite
length fast/slow	distinct crystals, columnar aggregates, six-sided x-sect.	one perfect	uncommon	Z		M, I, S	cf. zoisite, clinozoisite, diopside, augite, sillimanite
length fast/slow	distinct crystals, columnar, six-sided x-sect., irregular	two imperfect	uncommon	Z	metamict	I, M	anastomosing cracks, isotropic in metamict state cf. melanite

TABLE 6
FELDSPAR
GROUP

Mineral	Composition	System	Colour	Pleochroism	Relief	δ	2V Dispersion	Extinction
Alkali Feldspars								
Biaxial +								
ISOSANIDINE	$(K,Na)\{AlSi_3O_8\}$	Mon.	colourless		$-L$	1	$0^{\circ} - 63^{\circ}$	
ISOORTHOCLASE	$(K,Na)\{AlSi_3O_8\}$	Mon.	colourless		$-L$	1	$\simeq 90^{\circ}$ variable	
ISOMICROCLINE	$(K,Na)\{AlSi_3O_8\}$	Tr.	colourless		$-L$	$1 - 2$	$66^{\circ} - 70^{\circ}$	$15^{\circ} - 20^{\circ}$
CELSIAN	$Ba\{Al_2Si_2O_8\}$	Mon.	colourless		$+L$	2	$83^{\circ} - 90^{\circ}$	α:z $3^{\circ} - 5^{\circ}$
Biaxial −								
ADULARIA	$(K,Na)\{AlSi_3O_8\}$	Mon.	colourless		$-L$	1	small-large	α:(001) $5^{\circ} - 8^{\circ}$
SANIDINE	$(K,Na)\{AlSi_3O_8\}$	Mon.	colourless		$-L$	1 (anom.)	$0^{\circ} - 25^{\circ}$ r>v	α:(001) $5^{\circ} - 8^{\circ}$
HIGH SANIDINE	$(K,Na)\{AlSi_3O_8\}$	Mon.	colourless		$-L$	1	$0^{\circ} - 63^{\circ}$ r<v	α:(001) $5^{\circ} - 11^{\circ}$
ORTHOCLASE	$(K,Na)\{AlSi_3O_8\}$	Mon.	colourless		$-L$	1	$33^{\circ} - 85^{\circ}$ r>v	α:(001) $5^{\circ} - 19^{\circ}$
ANORTHOCLASE	$(K,Na)\{AlSi_3O_8\}$	Mon.	colourless		$-L$	1	$43^{\circ} - 60^{\circ}$ r>v	α:(001) $1^{\circ} - 6^{\circ}$

Orientation	Form	Cleavage	Twinning	Zoning	Alteration	Occurrence	Remarks
	clear, distinct, crystals, tabular	two	Carlsbad			I	cf. sanidine
	anhedral – subhedral, phenocrysts	three	Carlsbad			I	cf. orthoclase
cleavage length fast	anhedral – subhedral	two, parting	albite, Carlsbad, pericline, tartan	Z	cloudy, sericite, kaolinite	I, M, S	cf. microcline. orthoclase, albite, plagioclase
	prismatic	three	Carlsbad, Baveno, Manebach			O, I	cf. orthoclase, hyalophane
	minute crystals, rhombic x-sect.	three	albite, pericline	Z sector		O, S, M	cf. alkali feldspars
optic plane ⊥ (010)	clear, distinct crystals, tabular, microlites	two, parting	Carlsbad, Baveno, Manebach	Z		I, M	perthitic cf. orthoclase, nepheline
optic plane // (010)	clear, distinct crystals, tabular, microlites	two, parting	Carlsbad, Baveno, Manebach	Z		I, M	perthitic cf. orthoclase, nepheline
(010) cleavage length fast	phenocrysts, subhedral – anhedral, spherulitic aggregates	three	Carlsbad, Baveno, Manebach	Z	sericite, kaolinite	I, M	perthitic, inclusions cf. sanidine, nepheline
	phenocrysts, microlites	two	albite, Carlsbad, pericline, fine gridiron			I	perthitic cf. sanidine, orthoclase

TABLE 6 FELDSPAR GROUP	Mineral	Composition	System	Colour	Pleochroism	Relief	δ	2V Dispersion	Extinction

Alkali Feldspars

Biaxial −

| | MICROCLINE | $(K,Na)\{AlSi_3O_8\}$ | Tr. | colourless | | −L | 1 − 2 | $66^{\circ} - 90^{\circ}$ r>v | $\alpha:(001)$ $15^{\circ} - 20^{\circ}$ |
| | HYALOPHANE | $(K,Na,Ba)\{(Al,Si)_4O_8\}$ | Mon. | colourless | | −L − N | 1 − 2 | $48^{\circ} - 79^{\circ}$ r>v | $\alpha:x$ $0^{\circ} - 20^{\circ}$ |

Plagioclase Feldspars

Biaxial +

	ALBITE (An_{0-10})	$Na\{AlSi_3O_8\}-Ca\{AlSi_2O_8\}$	Tr.	colourless		−L	1 − 2	$77^{\circ} - 82^{\circ}$ r<v	$12^{\circ} - 19^{\circ}$ (in albite twins)
	OLIGOCLASE (An_{10-30})	$Na\{AlSi_3O_8\}-Ca\{AlSi_2O_8\}$	Tr.	colourless		−L − +N	1	$82^{\circ} - 90^{\circ}$ r>v	$0^{\circ} - 12^{\circ}$ (in albite twins)
	ANDESINE (An_{30-50})	$Na\{AlSi_3O_8\}-Ca\{AlSi_2O_8\}$	Tr.	colourless		+L	1	$76^{\circ} - 90^{\circ}$ r<v	$13^{\circ} - 27\frac{1}{2}^{\circ}$ (in albite twins)
	LABRADORITE (An_{50-70})	$Na\{AlSi_3O_8\}-Ca\{AlSi_2O_8\}$	Tr.	colourless		+L	1	$76^{\circ} - 86^{\circ}$ r<v	$27\frac{1}{2}^{\circ} - 39^{\circ}$ (in albite twins)

Orientation	Form	Cleavage	Twinning	Zoning	Alteration	Occurrence	Remarks
cleavage length fast	anhedral – subhedral, phenocrysts	two, parting	albite Carlsbad, pericline, tartan	Z	cloudy, sericite, kaolinite	I, M, S	perthitic cf. isomicrocline, orthoclase, albite, plagioclase
	tabular, prismatic	two	Carlsbad, Baveno, Manebach			O, M	cf. orthoclase
	anhedral – euhedral, laths	three	albite, Carlsbad, pericline, complex	Z	sericite, calcite, kaolinite, zeolites	I, S, M	peristerite, perthitic, antiperthitic cf. cordierite
	anhedral – euhedral, laths, perthite	three	albite, Carlsbad, pericline, complex	Z	sericite, calcite, kaolinite, zeolites	I, M	peristerite, antiperthitic cf. quartz, cordierite
	anhedral – euhedral, laths	three	albite, Carlsbad, pericline, complex	Z	sericite, calcite, kaolinite, albite, saussurite	I, M	antiperthitic cf. cordierite
	anhedral – euhedral, laths	three	albite, Carlsbad, pericline, complex	Z	sericite, calcite, kaolinite, zeolites, albite, saussurite	I, M	schiller, antiperthitic cf. cordierite

345

TABLE 6
FELDSPAR
GROUP

Plagioclase Feldspars

Biaxial –

Mineral	Composition	System	Colour	Pleochroism	Relief	δ	2V Dispersion	Extinction
HIGH ALBITE	$Na\{AlSi_3O_8\}$	Tr.	colourless		–L	1	$45°$ r>v	inclined to twin lamellae
HIGH OLIGOCLASE	$Na\{AlSi_3O_8\}–Ca\{Al_2Si_2O_8\}$	Tr.	colourless		–L	1	$52° – 73°$ r>v	$0° – 12°$ (in albite twins)
OLIGOCLASE (An_{10-30})	$Na\{AlSi_3O_8\}–Ca\{Al_2Si_2O_8\}$	Tr.	colourless		N – +L	1	$86° – 90°$ r>v	$0° – 12°$ (in albite twins)
ANDESINE (An_{30-50})	$Na\{AlSi_3O_8\}–Ca\{Al_2Si_2O_8\}$	Tr.	colourless		+L	1	$76° – 90°$ r<v	$13° – 27\frac{1}{2}°$ (in albite twins)
BYTOWNITE (An_{70-90})	$Na\{AlSi_3O_8\}–Ca\{Al_2Si_2O_8\}$	Tr.	colourless		+L	2	$79° – 88°$ r>v	$39° – 51°$ (in albite twins)
ANORTHITE (An_{90-100})	$Na\{AlSi_3O_8\}–Ca\{Al_2Si_2O_8\}$	Tr.	colourless		+L	2	$77° – 79°$ r>v	$51° – 70°$ (in albite twins)

Orientation	Form	Cleavage	Twinning	Zoning	Alteration	Occurrence	Remarks
	anhedral – euhedral, plates, laths	three	albite, Carlsbad, pericline, complex	Z	sericite, calcite, kaolinite, zeolites	I	perthitic cf. cordierite
	anhedral – euhedral, laths	three	albite, Carlsbad, pericline, complex	Z	sericite, calcite, kaolinite, albite	I	
	anhedral – euhedral, laths	three	albite, Carlsbad, pericline, complex	Z	sericite, calcite, kaolinite, saussurite	I, M	peristerite, antiperthitic cf. cordierite
	anhedral – euhedral, laths	three	albite, Carlsbad, pericline, complex	Z	sericite, calcite, kaolinite, albite, saussurite	I, M	antiperthitic cf. cordierite
	anhedral – euhedral, laths	three	albite, Carlsbad, pericline, complex	Z	calcite, albite, saussurite	I, M	antiperthitic cf. cordierite
	anhedral – euhedral, laths	three	albite, Carlsbad, pericline, complex	Z	calcite, albite, saussurite	I, M	antiperthitic cf. cordierite

TABLE 7 FELDSPATHOID GROUP	Mineral	Composition	System	Colour	Pleochroism	Relief	δ	2V Dispersion	Extinction
Isometric									
	SODALITE (incl. HACKMANITE)	$Na_8\{Al_6Si_6O_{24}\}Cl_2$	Cub.	colourless, pink, red, yellow, blue, grey		$-L$	(weak)		
	NOSEAN	$Na_8\{Al_6Si_6O_{24}\}SO_4$	Cub.	colourless, brown, grey		$-L$	(weak)		
	HAÜYNE	$(Na,Ca)_{4-8}\{Al_6Si_6O_{24}\}(SO_4,S)_{1-2}$	Cub.	colourless, grey		$-L$	(weak)		
	ANALCITE	$Na\{AlSi_2O_6\}\cdot H_2O$	Cub.	colourless, brown		$-L$	1		
	WAIRAKITE	$CaAl_2Si_4O_{12}\cdot 2H_2O$	Pseudo-Cub.	colourless		$-L$	1	$70^\circ - 90^\circ$	parallel
	LEUCITE	$K\{AlSi_2O_6\}$	Tet. (Pseudo-Cub.)	colourless		$-L$	(aniso-tropic)		wavy
	MELILITE	$(Ca,Na)_2\{(Mg,Fe^{+2},Al,Si)_3O_7\}$	Tet.	colourless, brown, yellow		$+M$	1 anom.		parallel
Uniaxial +									
	LEUCITE	$K\{AlSi_2O_6\}$	Tet. (Pseudo-Cub.)	colourless		$-L$	1	small anom.	parallel
	DAVYNE-NATRODAVYNE	K-cancrinite	Hex.	colourless		$-L$	1		parallel
	MICROSOMMITE	$K,NaAlSiO_4\cdot Ca(Cl_2,SO_4)$	Hex.	colourless		$-L$	1 - 3 (iso-tropic)	anom.	parallel

Orientation	Form	Cleavage	Twinning	Zoning	Alteration	Occurrence	Remarks
	hexagonal x-sect., anhedral aggregates	poor	simple		zeolites, diaspore, gibbsite	I	cf. fluorite, leucite
	hexagonal x-sect., anhedral aggregates	imperfect	simple	Z	zeolites, diaspore, gibbsite, limonite	I	clouded with inclusions
	hexagonal x-sect., anhedral aggregates	imperfect	polysynthetic	Z	zeolites, diaspore, gibbsite	I	
	trapezohedral, rounded, radiating, irregular	poor	polysynthetic, complex, interpenetrant			I, S, M	cf. leucite, sodalite, wairakite
	trapezohedral, anhedral	imperfect	simple			I	cf. analcite
	always euhedral, octagonal	poor	polysynthetic, complex	Z		I	inclusions common cf. analcite, microcline
length slow	tabular, peg structure	moderate, single crack		Z	zeolites, carbonate	I	cf. zoisite, vesuvianite, apatite, nepheline
	always euhedral, octagonal	poor	polysynthetic, complex	Z		I	inclusions common cf. analcite, microcline
	anhedral	two perfect	rare			I	cf. cancrinite, microsommite
length slow	anhedral, prismatic	one perfect, one poor	polysynthetic, rare	Z		I	some zones isotropic cf. nepheline, quartz, cancrinite, vishnevite

TABLE 7 FELDSPATHOID GROUP	Mineral	Composition	System	Colour	Pleochroism	Relief	δ	2V Dispersion	Extinction
Uniaxial +									
	MELILITE	$(Ca,Na)_2${$(Mg,Fe^{+2},Al,Si)_3O_7$}	Tet.	colourless, brown, yellow	W $\varepsilon>\omega$	+M	1 anom.		parallel
	ÅKERMANITE	Ca_2{$MgSi_2O_7$}	Tet.	colourless		M	1		parallel
Uniaxial −									
	KALSILITE-KALIOPHILITE	K{$AlSiO_4$}	Hex.	colourless		−L	1		parallel
	DAVYNE-NATRODAVYNE	K-cancrinite	Hex.	colourless		−L	1		parallel
	VISHNEVITE	$(Na,Ca,K)_{6-8}${$Al_6Si_6O_{24}$} $(CO_3,SO_4,Cl)_{1-2}\cdot 1\text{-}5H_2O$	Hex.	colourless			1	anom.	parallel
	CANCRINITE	$(Na,Ca,K)_{6-8}${$Al_6Si_6O_{24}$} $(CO_3,SO_4,Cl)_{1-2}\cdot 1\text{-}5H_2O$	Hex.	colourless, yellow		−L	3	anom.	parallel
	NEPHELINE	$Na_3(Na,K)${$Al_4Si_4O_{16}$}	Hex.	colourless		−L − N	1		parallel
	EUCRYPTITE	$LiAlSiO_4$	Hex.	colourless		N	1		parallel
	MELILITE	$(Ca,Na)_2${$(Mg,Fe^{+2},Al,Si)_3O_7$}	Tet.	colourless, brown, yellow	W $\omega>\varepsilon$	+M	1 anom.		parallel
	GEHLENITE	Ca_2{Al_2SiO_7}	Tet.	colourless		+M	2		parallel

Orientation	Form	Cleavage	Twinning	Zoning	Alteration	Occurrence	Remarks
length fast	tabular, peg structure	moderate, single crack		Z	zeolites, carbonate	I	cf. zoisite, vesuvianite, apatite, nepheline
length fast	tabular, peg structure	moderate, single crack		Z	zeolites, carbonate	I	cf. zoisite, vesuvianite, apatite, nepheline
	prismatic, hexagonal	two poor	rare	Z		I	cf. nepheline
	anhedral	two perfect	rare			I	davyne rarely negative cf. cancrinite, microsommite
length fast	trapezohedral, anhedral	one perfect, one poor	polysynthetic, rare			I	cf. cancrinite, microsommite, nepheline
length fast	anhedral	one perfect, one poor	polysynthetic, rare			I	cf. cancrinite, microsommite, muscovite
rect. sect., length fast	prismatic, hexagonal	two poor	rare	Z	zeolites, cancrinite, muscovite	I	inclusions cf. alkali feldspars, analcite, sodalite, leucite, scapolite
	anhedral	one distinct				I	cf. spodumene
length slow	tabular, peg structure	moderate, single crack		Z		I	cf. zoisite, vesuvianite, apatite, nepheline
length slow	tabular, peg structure	moderate, single crack		Z		I	cf. zoisite, vesuvianite, apatite, nepheline

TABLE 7 FELDSPATHOID GROUP	Mineral	Composition	System	Colour	Pleochroism	Relief	δ	2V Dispersion	Extinction
Biaxial –									
	ANALCITE	$Na\{AlSi_2O_6\}\cdot H_2O$	Cub.	colourless, brown		–L	1	small–large	
	WAIRAKITE	$CaAl_2Si_4O_{12}\cdot 2H_2O$	Pseudo-Cub.	colourless		–L	1	$70^\circ - 90^\circ$	parallel

Orientation	Form	Cleavage	Twinning	Zoning	Alteration	Occurrence	Remarks
	trapezohedra, rounded, radiating, irregular	poor	polysynthetic, complex, interpenetrant			I, S, M	cf. leucite, sodalite, wairakite
	trapezohedral, anhedral	imperfect	simple			I	cf. analcite

TABLE 8
MICA
GROUP

Monoclinic

Biaxial −

Mineral	Composition	System	Colour	Pleochroism	Relief	δ	2V Dispersion	Extinction
LEPIDOLITE	$K_2(Li,Al)_{5-6}$ $\{Si_{6-7}Al_{2-1}O_{20}\}(OH,F)_4$	Mon. (or Trig.)	colourless	W $\gamma=\beta>\alpha$	+L	2 − 4	$0°$ − $58°$ $(30°$ − $50°)$ r>v	$0°$ − $7°$
ZINNWALDITE	$K_2(Fe^{+2}_{2-1},Li_{2-3},Al_2)$ $\{Si_{6-7}Al_{2-1}O_{20}\}(F,OH)_4$	Mon.	colourless, brown	W $\gamma=\beta>\alpha$	+L	3	$0°$ − $40°$ r>v	$0°$ − $2°$
FUCHSITE	Cr- muscovite	Mon.	colourless, green	M $\gamma\geqslant\beta>\alpha$	+L	4	$32°$ − $46°$ r>v	$1°$ − $3°$
PHLOGOPITE	$K_2(Mg,Fe^{+2})_6$ $\{Si_6Al_2O_{20}\}(OH,F)_4$	Mon.	colourless, red, orange, yellow	M $\gamma\geqslant\beta>\alpha$ $\alpha>\beta=\gamma$	+L − +M	3 − 4	$0°$ − $15°$ r<v	$0°$ − $5°$
PARAGONITE	$Na_2Al_4\{Si_6Al_2O_{20}\}(OH)_4$	Mon.	colourless		+L − +M	3 − 4	$0°$ − $40°$ r>v	$\simeq$ parallel
MUSCOVITE (incl. PHENGITE, SERICITE)	$K_2Al_4\{Si_6Al_2O_{20}\}(OH,F)_4$	Mon.	colourless	W	+L − +M	4	$30°$ − $47°$ r>v	$1°$ − $3°$
GLAUCONITE (incl. CELADONITE)	$(K,Na,Ca)_{1.2-2.0}$ $(Fe^{+3},Al,Fe^{+2},Mg)_{4.0}$ $\{Si_{7-7.6}Al_{1-0.4}O_{20}\}$ $(OH)_4 \cdot nH_2O$	Mon.	green	M $\gamma=\beta>\alpha$	+M	2 − 3	$0°$ − $20°$ r<v	$0°$ − $3°$
BIOTITE	$K_2(Mg,Fe^{+2})_{6-4}$ $(Fe^{+3},Al,Ti)_{0-2}$ $\{Si_{6-5}Al_{2-3}O_{20}\}(OH,F)_4$	Mon.	red, brown, orange, yellow, green	S $\gamma\geqslant\beta>\alpha$ $\beta>\gamma>\alpha$	+M	4 − 5	$0°$ − $25°$ Mg r$\lessgtr$v Fe r>v	$0°$ − $9°$

Orientation	Form	Cleavage	Twinning	Zoning	Alteration	Occurrence	Remarks
cleavage length slow	tabular, short, prismatic, flakes, pseudo-hex.	perfect basal	simple			I	pleochroic haloes, inclusions cf. muscovite, zinnwaldite, phlogopite
cleavage length slow	tabular, short, prismatic, flakes, pseudo-hex.	perfect basal	simple			I	cf. lepidolite, biotite
cleavage length slow	thin tablets, shreds	perfect basal	simple			M	birds-eye maple structure cf. muscovite
cleavage length slow	tabular, flakes, plates, pseudo-hex.	perfect basal	inconspicuous	Z colour zoning		I, M	inclusions common, birds-eye maple structure cf. biotite, muscovite, lepidolite, rutile, tourmaline
cleavage length slow	scaly, aggregates	perfect basal				M, S	cf. muscovite
cleavage length slow	tabular, flakes, scales, aggregates	perfect basal	simple			M, I, S	birds-eye maple structure cf. talc, pyrophyllite
cleavage length slow	grains, pellets, plates, pseudomorphs	perfect basal			limonite, goethite	S	cf. chamosite, biotite, chlorite
cleavage length slow	tabular, flakes, plates, pseudo-hex.	perfect basal	simple	Z	chlorite, vermiculite, prehnite	M, I	birds-eye maple structure pleochroic haloes, inclusions common cf. stilpnomelane, phlogopite

TABLE 8 MICA GROUP	Mineral	Composition	System	Colour	Pleochroism	Relief	δ	2V Dispersion	Extinction
Brittle Micas									
Biaxial +									
	CHLORITOID (OTTRELITE)	$(Fe^{+2},Mg,Mn)_2(Al,Fe^{+3})$ $Al_3O_2\{SiO_4\}_2(OH)_4$	Mon., Tr.	colourless, yellow, green, blue, grey	W – M $\beta>\alpha>\gamma$	H	1 – 2 anom.	$45^o – 68^o$ r>v	$2^o – 30^o$
Biaxial –									
	STILPNOMELANE	$(K,Na,Ca)_{0-1\cdot4}(Fe^{+3},$ $Fe^{+2},Mg,Al,Mn)_{5\cdot9-8\cdot2}$ $\{Si_8O_{20}\}(OH)_4$ $(O,OH,H_2O)_{3\cdot6-8\cdot5}$	Mon.	brown, yellow, green, black	S $\gamma=\beta>\alpha$	+L – H	3 – 5	$\simeq 0^o$	$\simeq$ parallel
	XANTHOPHYLLITE	$Ca_2(Mg,Fe)_{4\cdot6}Al_{1\cdot4}$ $\{Si_{2\cdot5}Al_{5\cdot5}O_{20}\}(OH)_4$	Mon.	colourless, red, brown, orange, yellow	M $\gamma=\beta>\alpha$	+M	2	$0^o – 23^o$ r<v	parallel
	CLINTONITE	$Ca_2(Mg,Fe)_{4\cdot6}Al_{1\cdot4}$ $\{Si_{2\cdot5}Al_{5\cdot5}O_{20}\}(OH)_4$	Mon.	colourless, brown, orange, yellow	M $\gamma=\beta>\alpha$	+M	2	32° r<v	parallel
	MARGARITE	$Ca_2Al_4\{Si_4Al_4O_{20}\}(OH)_4$	Mon.	colourless	N – W	+M	2	$40^o – 67^o$ r<v	$6^o – 8^o$
	CHLORITOID (OTTRELITE)	$(Fe^{+2},Mg,Mn)_2(Al,Fe^{+3})$ $Al_3O_2\{SiO_4\}_2(OH)_4$	Mon., Tr.	colourless, yellow, green, blue, grey	W – M $\beta>\alpha>\gamma$	H	1 – 2 anom.	$45^o – 68^o$ r>v	$2^o – 30^o$

Orientation	Form	Cleavage	Twinning	Zoning	Alteration	Occurrence	Remarks
length fast	tabular	one perfect, one imperfect, parting	polysynthetic	Z hour-glass		M, I	inclusions common cf. chlorite, clintonite, biotite, stilpnomelane
length slow	plates, pseudo-hex., micaceous masses	two at 90°, perfect basal	polysynthetic	Z		I, M, O	cf. biotite, chlorite, chloritoid, clintonite
cleavage length slow	tabular, short prismatic	perfect basal	simple			M	cf. clintonite, chlorite, chloritoid
cleavage length slow	tabular, short prismatic	perfect basal	simple			M	cf. xanthophyllite, chlorite, chloritoid
cleavage length slow	tabular, flakes, pseudo-hex.	perfect basal	polysynthetic		vermiculite	M	cf. muscovite, talc, chlorite, chloritoid
length fast	tabular	one perfect, one imperfect, parting	polysynthetic	Z hour-glass		M, I	inclusions common cf. chlorite, clintonite, biotite, stilpnomelane

TABLE 9 OLIVINE GROUP	Mineral	Composition	System	Colour	Pleochroism	Relief	δ	2V Dispersion	Extinction
Orthorhombic									
Biaxial +									
	FORSTERITE	Mg_2SiO_4	Orth.	colourless		+M	4	$82^{\circ} - 90^{\circ}$ r<v	parallel
Biaxial –									
	MONTICELLITE	$CaMg\{SiO_4\}$	Orth.	colourless		+M	2	$72^{\circ} - 82^{\circ}$ r>v	parallel
	CHRYSOLITE- HYALOSIDERITE- HORTONOLITE- FERROHORTONOLITE	$(Mg,Fe)_2\{SiO_4\}$	Orth.	colourless		+M – VH	4	$52^{\circ} - 90^{\circ}$ r>v	parallel
	FAYALITE	Fe_2SiO_4	Orth.	colourless, brown, yellow	W	VH	4	$48^{\circ} - 52^{\circ}$ r>v	parallel
	TEPHROITE (incl. ROEPPERITE)	$Mn_2\{SiO_4\}$	Orth.	colourless, red, yellow, blue	W $\gamma>\beta>\alpha$	VH	4	$44^{\circ} - 70^{\circ}$ r>v	parallel
	KNEBELITE	$(Mn,Fe)_2\{SiO_4\}$	Orth.	colourless, yellow, blue	W $\gamma>\beta>\alpha$	VH	4	$44^{\circ} - 70^{\circ}$ r>v	parallel

Orientation	Form	Cleavage	Twinning	Zoning	Alteration	Occurrence	Remarks
cleavage length slow	anhedral – euhedral, rounded	uncommon, irregular fractures	uncommon	Z	chlorite, antigorite, serpentine, iddingsite, bowlingite, magnetite	I, M	deformation lamellae cf. diopside, augite, humite group, epidote
cleavage length slow	aggregates, euhedral, rounded	poor			idocrase, fassaite, serpentine	M, I	
cleavage length slow	anhedral	moderate, irregular fractures		Z	chlorite, antigorite, serpentine, iddingsite, bowlingite, magnetite	I, M	members of olivine isomorphous series
cleavage length slow	anhedral – euhedral, rounded	one moderate, irregular fractures	uncommon, vicinal		grunerite, serpentine, magnetite	I, M	cf. knebelite, pyroxene
length fast/slow	anhedral – euhedral, rounded	two moderate, imperfect	uncommon			O, M	cf. fayalite
length fast/slow	anhedral – euhedral, rounded	two moderate, imperfect	uncommon			O, M	cf. fayalite

TABLE 10 PYROXENE GROUP	Mineral	Composition	System	Colour	Pleochroism	Relief	δ	2V Dispersion	Extinction
Orthorhombic									
Biaxial +									
	ENSTATITE	$(Mg,Fe^{+2})\{SiO_3\}$	Orth.	colourless		+M – H	2	$55^{\circ} - 90^{\circ}$ r<v	parallel
	ORTHOFERROSILITE	$Fe^{+2}\{SiO_3\}$	Orth.	colourless		H	3	$55^{\circ} - 90^{\circ}$ r<v	parallel
Biaxial –									
	HYPERSTHENE (incl. BRONZITE-EULITE)	$(Mg,Fe^{+2})\{SiO_3\}$	Orth.	colourless, pink, red, brown, green, grey	N – S $\gamma>\beta>\alpha$	+M – H	1 – 2	$50^{\circ} - 90^{\circ}$ r$\lessgtr$v	parallel
Monoclinic									
Biaxial +									
	JADEITE	$NaAl\{Si_2O_6\}$	Mon.	colourless, green	W	+M	2	$67^{\circ} - 70^{\circ}$ r>v	$33^{\circ} - 40^{\circ}$
	SPODUMENE (HIDDENITE, KUNZITE)	$LiAl\{Si_2O_6\}$	Mon.	colourless	W	+M	2 – 3	$58^{\circ} - 68^{8}$ r<v	$22^{\circ} - 26^{\circ}$
	DIOPSIDE (incl. SALITE)	$Ca(Mg,Fe)\{Si_2O_6\}$	Mon.	colourless, green, grey	W	+M	3	$50^{\circ} - 62^{\circ}$ r>v	$38^{\circ} - 48^{8}$
	FASSAITE	$(Ca,Na,Mg,Fe^{+2},Mn,Fe^{+3},$ $Al,Ti)_2\{(Si,Al)_2O_6\}$	Mon.	colourless, green	W	+M – H	2 – 3	$51^{\circ} - 62^{\circ}$	$35^{\circ} - 48^{\circ}$

Orientation	Form	Cleavage	Twinning	Zoning	Alteration	Occurrence	Remarks
length slow	prismatic, anhedral - euhedral, kelyphitic, borders	two at 88°, parting	rare			I, M	often schillerized, Fe-enstatite = bronzite
length slow	prismatic, anhedral	two at 88°	rare			M	
length slow	prismatic, anhedral - euhedral	two at 88°	polysynthetic, twin seams	Z		I, M	exsolution lamellae, schiller inclusions cf. andalusite, titanaugite
extinction nearest cleavage slow	prismatic	two at 87°	simple, polysynthetic	Z	tremolite-actinolite	M	cf. nephrite, diopside, omphacite, fassaite
extinction nearest cleavage slow	tabular	two at 87°, parting	simple		muscovite, cymatolite, kaolinite, eucryptite	I	cf. aegirine-augite, diopside
extinction nearest cleavage slow	short, prismatic	two at 87°	simple, polysynthetic	Z	tremolite-actinolite	M, I	cf. hedenbergite, tremolite, omphacite, wollastonite, epidote
extinction nearest cleavage fast	prismatic	two at 87°	simple, polysynthetic	Z	amphibole	I, M	cf. augite, diopside, omphacite, jadeite

TABLE 10
PYROXENE
GROUP

Mineral	Composition	System	Colour	Pleochroism	Relief	δ	2V Dispersion	Extinction
Monoclinic								
Biaxial +								
TITANAUGITE	$(Ca,Na,Mg,Fe^{+2},Mn,Fe^{+3}, Al,Ti)_2\{(Si,Al)_2O_6\}$	Mon.	pink, red, brown, green, lilac	M – S	+M – H	2 – 3 (anom.)	$25^{\circ} - 50^{\circ}$ r>v	$35^{\circ} - 48^{\circ}$
AUGITE-FERROAUGITE (incl. SALITE)	$(Ca,Na,Mg,Fe^{+2},Mn,Fe^{+3}, Al,Ti)_2\{(Si,Al)_2O_6\}$	Mon.	colourless, brown, green	N – W	+M – H	2 – 3	$25^{\circ} - 83^{\circ}$ r>v	$35^{\circ} - 48^{\circ}$
OMPHACITE	$(Ca,Na,Mg,Fe^{+2},Mn,Fe^{+3}, Al,Ti)_2\{(Si,Al)_2O_6\}$	Mon.	colourless, green	W	+M – H	2 – 3	$60^{\circ} - 67^{\circ}$	$39^{\circ} - 41^{\circ}$
PIGEONITE	$(Mg,Fe^{+2},Ca)(Mg,Fe^{+2}) \{Si_2O_6\}$	Mon.	colourless, pink, brown, yellow, green	N – M $\gamma=\alpha>\beta$ $\beta>\alpha=\gamma$	+M – H	3	$0^{\circ} - 30^{\circ}$ r$\lessgtr$v	$37^{\circ} - 44^{\circ}$
JOHANNSENITE	$Ca(Mn,Fe)\{Si_2O_6\}$	Mon.	colourless		H	3	$68^{\circ} - 70^{\circ}$ r>v	$46^{\circ} - 48^{\circ}$
HEDENBERGITE-FERROSALITE	$Ca(Fe,Mg)\{Si_2O_6\}$	Mon.	brown, green	W $\gamma>\beta>\alpha$	+M – H	3	$50^{\circ} - 62^{\circ}$ r>v	$38^{\circ} - 48^{\circ}$
AEGIRINE-AUGITE	$(Na,Ca)(Fe^{+3},Fe^{+2},Mg) \{Si_2O_6\}$	Mon.	brown, green	S $\alpha>\gamma$	H	3 – 4	$70^{\circ} - 90^{\circ}$ r>v	γ:z $70^{\circ}-90^{\circ}$
Biaxial –								
ACMITE	$NaFe^{+3}\{Si_2O_6\}$	Mon.	brown, yellow	W $\alpha>\beta\geqslant\gamma$	H – VH	4 – 5	$60^{\circ} - 70^{\circ}$ r>v	$0^{\circ} - 10^{\circ}$
AEGIRINE	$NaFe^{+3}\{Si_2O_6\}$	Mon.	green, black	S $\alpha>\beta\geqslant\gamma$	H – VH	4 – 5	$60^{\circ} - 70^{\circ}$ r>v	$0^{\circ} - 10^{\circ}$

Orientation	Form	Cleavage	Twinning	Zoning	Alteration	Occurrence	Remarks
extinction nearest cleavage fast	prismatic	two at 93°	polysynthetic, twin seams	Z hour-glass		I	cf. hypersthene
extinction nearest cleavage fast	prismatic	two at 87°, parting	simple, polysynthetic, twin seams	Z	amphibole	I, M	herringbone structure, exsolution lamellae cf. pigeonite, diopside, epidote
extinction nearest cleavage fast	prismatic	two at 87°	simple, polysynthetic			M	inclusions cf. fassaite, diopside, jadeite
extinction nearest cleavage slow	prismatic, anhedral, overgrowths	two at 87°, parting	simple, polysynthetic	Z		I	exsolution lamellae cf. augite, olivine
extinction nearest cleavage slow	prismatic	two at 87°	simple, polysynthetic			O	
extinction nearest cleavage fast	prismatic	two at 87°	simple, polysynthetic			M, I	cf. diopside, aegirine-augite
extinction nearest cleavage fast	short prismatic, needles, felted aggregates	two at 87°, parting	simple, polysynthetic, twin seams	Z hour-glass		I	cf. aegirine, acmite
extinction nearest cleavage fast	short prismatic, needles, pointed terminations	two at 87°, parting	simple, polysynthetic, twin seams	Z		I	cf. aegirine, aegirine-augite, Na-amphibole
extinction nearest cleavage fast	prismatic, needles, felted aggregates, blunt terminations	two at 87°, parting	simple, polysynthetic, twin seams	Z hour-glass		I	borders may be black cf. aegirine-augite, acmite

TABLE 11 ZEOLITE GROUP	Mineral	Composition	System	Colour	Pleochroism	Relief	δ	2V Dispersion	Extinction
Isometric									
	ANALCITE	$Na\{AlSi_2O_6\}\cdot H_2O$	Cub.	colourless, brown		-L	1 (weak)		
	FAUJASITE	$(Na_2,Ca)\{Al_2Si_4O_{12}\}\cdot 8H_2O$	Cub.	colourless		-L	1 (weak)		
Uniaxial +									
	ERIONITE	$(Na_2,K_2,Ca,Mg)_{4.5}\{Al_9Si_{27}O_{72}\}\cdot 27H_2O$	Hex.	colourless		-L	1		parallel
	CHABAZITE	$Ca\{Al_2Si_4O_{12}\}\cdot 6H_2O$	Trig.	colourless		-L	1		symmetrical
	ASHCROFTINE	$KNaCa\{Al_4Si_5O_{18}\}\cdot 8H_2O$	Tet.	colourless		-L - N	1		parallel
Uniaxial −									
	GMELINITE	$(Na_2,Ca)\{Al_2Si_4O_{12}\}\cdot 6H_2O$	Trig.	colourless		-L	1		symmetrical
	CHABAZITE	$Ca\{Al_2Si_4O_{12}\}\cdot 6H_2O$	Tet.	colourless		-L	1		symmetrical
	LEVYNE	$Ca\{Al_2Si_4O_{12}\}\cdot 6H_2O$	Trig.	colourless		-L	1		parallel
	GARRONITE	$NaCa_{2.5}\{Al_6Si_{10}O_{32}\}13\cdot 5H_2O$	Tet.	colourless		-L	1		parallel

Orientation	Form	Cleavage	Twinning	Zoning	Alteration	Occurrence	Remarks
	trapezohedral, rounded, radiating, irregular	poor	polysynthetic, complex, interpenetrant			I, S, M	cf. leucite, sodalite, wairakite
	octahedral, rounded	one distinct				I	
length slow	fibrous, radiating					I	
	rhombohedral, approaching cube, anhedral, granular	one poor	interpenetrant	Z		I	basal section in six segments cf. gmelinite, analcite
length slow	prismatic, granular, anhedral - euhedral, needles	two at 90°, one perfect				I, O	
	tabular, prismatic, rhombohedral, approaching cube, radiating	one good, one imperfect, parting	interpenetrant			I	cf. chabazite
	rhombohedral, approaching cube, anhedral, granular	one poor	interpenetrant	Z		I	basal section in six segments cf. gmelinite, analcite
	tabular, sheaf-like, aggregates, rhombohedral	indistinct, rhombohedral	interpenetrant			I	
						I	

TABLE 11 ZEOLITE GROUP	Mineral	Composition	System	Colour	Pleochroism	Relief	δ	2V Dispersion	Extinction
Orthorhombic									
Biaxial +									
	FERRIERITE	$(Na,K)_4Mg_2\{Al_6Si_{30}O_{72}\}(OH)_2 18H_2O$	Orth.	colourless		$-L$	1	50°	parallel
	MORDENITE	$(Na_2,K_2,Ca)\{Al_2Si_{10}O_{24}\}\cdot 7H_2O$	Orth.	colourless		$-L$	1	$76^{\circ} - 90^{\circ}$	parallel
	THOMSONITE	$NaCa_2\{(Al,Si)_5O_{10}\}_2\cdot 6H_2O$	Orth. (Pseudo-Tet.)	colourless		$-L$	1 - 2	$42^{\circ} - 75^{\circ}$ r>v	parallel
	NATROLITE	$Na_2\{Al_2Si_3O_{10}\}\cdot 2H_2O$	Orth. (Pseudo-Tet.)	colourless		$-L$	2	$58^{\circ} - 64^{\circ}$ r<v	parallel, symmetrical
	METANATROLITE	$Na_2\{Al_2Si_3O_{10}\}$	Orth. (Pseudo-Tet.)	colourless		$-L$	2	$58^{\circ} - 64^{\circ}$	$7\frac{1}{2}^{\circ}$
Biaxial −									
	ANALCITE	$Na\{AlSi_2O_6\}\cdot H_2O$	Cub.	colourless, brown		$-L$	1 (weak)	$0^{\circ} - 85^{\circ}$ anom.	parallel
	GONNARDITE	$Na_2Ca\{(Al,Si)_5O_{10}\}_2\cdot 6H_2O$	Orth. (Pseudo-Tet.)	colourless		$-L$	1	50°	parallel
	WAIRAKITE	$CaAl_2Si_4O_{12}\cdot 2H_2O$	Pseudo-Cub.	colourless		$-L$	1	$70^{\circ} - 90^{\circ}$	parallel
	MORDENITE	$(Na_2,K_2,Ca)\{Al_2Si_{10}O_{24}\}\cdot 7H_2O$	Orth.	colourless		$-L$	1	$76^{\circ} - 90^{\circ}$	parallel
	EDINGTONITE	$Ba\{Al_2Si_3O_{10}\}\cdot 4H_2O$	Orth. (Pseudo-Tet.)	colourless		$+L$	2	54° r<v	parallel

Orientation	Form	Cleavage	Twinning	Zoning	Alteration	Occurrence	Remarks
length slow	tabular, laths, radiating	one perfect				I	
length slow	tabular, acicular, fibrous	one perfect				I	
length fast/slow	fibrous, columnar, radiating	two at 90°				I	cf. natrolite, scolecite, mesolite, cancrinite
length slow	prismatic, fibrous, needles, radiating	parallel to length				I	cf. scolecite, thomsonite
length slow	prismatic, fibrous, needles, radiating	parallel to length	sector			I	cf. natrolite
	trapezohedral, rounded, radiating, irregular	poor	polysynthetic, complex, interpenetrant			I, S, M	cf. leucite, sodalite, wairakite
length fast	fibrous, spherulitic					I	
	trapezohedral, anhedral	imperfect	simple			I	cf. analcite
length slow	tabular, acicular, fibrous	one perfect				I	
length fast	minute, fibrous	two at 90°				I	

TABLE 11
ZEOLITE
GROUP

Mineral	Composition	System	Colour	Pleochroism	Relief	δ	2V Dispersion	Extinction

Monoclinic

Biaxial +

Mineral	Composition	System	Colour	Pleochroism	Relief	δ	2V Dispersion	Extinction
GMELINITE	$(Na_2,Ca)\{Al_2Si_4O_{12}\}\cdot 6H_2O$	Trig.	colourless		$-L$	1	0° – moderate	symmetrical
CHABAZITE	$Ca\{Al_2Si_4O_{12}\}\cdot 6H_2O$	Trig.	colourless		$-L$	1	0° – 32°	symmetrical
HEULANDITE	$(Ca,Na_2)\{Al_2Si_7O_{18}\}\cdot 6H_2O$	Pseudo-Mon.	colourless		$-L$	1	0° – 48° variable r>v	6° variable
DACHIARDITE	$(\frac{1}{2}Ca,Na,K)_5\{Al_5Si_{19}O_{48}\}\cdot 12H_2O$	Mon.	colourless		$-L$	1	65° – 73°	38°
YUGAWARALITE	$Ca\{Al_2Si_5O_{14}\}\cdot 4H_2O$	Mon.	colourless		$-L$	1	70°	
MESOLITE	$Na_2Ca_2\{Al_2Si_3O_{10}\}_3\cdot 8H_2O$	Mon. (Pseudo-Orth.)	colourless		$-L$	1 nearly iso-tropic	80° r>v	8°
HARMOTONE	$Ba\{Al_2Si_6O_{16}\}\cdot 6H_2O$	Mon. (or Orth.)	colourless		$-L$	1	80° weak x-dispersion	63° – 67°
PHILLIPSITE	$(\frac{1}{2}Ca,Na,K)_3\{Al_3Si_5O_{16}\}\cdot 6H_2O$	Mon. (or Orth.)	colourless		$-L$	1 – 2	60° – 80° r<v	46° – 85°
BREWSTERITE	$(Sr,Ba,Ca)\{Al_2Si_6O_{16}\}5H_2O$	Mon.	colourless		$-L$	2	47° r>v	22°

Orientation	Form	Cleavage	Twinning	Zoning	Alteration	Occurrence	Remarks
	tabular, prismatic, rhombohedral, approaching, cube, radiating	one good, one imperfect, parting	interpenetrant			I	chabazite
	rhombohedral, approaching cube, anhedral, granular	one poor	interpenetrant	Z		I	basal section in six segments cf. gmelinite, analcite
cleavage length fast	tabular, aggregates	one perfect				I, M	cf. stilbite, clinoptilite, epistilbite, mordenite, brewsterite
	prismatic	two perfect	cyclic, sector			I	
						I	
length fast/slow	fibrous, aggregates, needles	two perfect at 90°	universal but inconspicuous			I	cf. natrolite, thomsonite, scolecite
	groups, radiating	two, one good	interpenetrant, complex, sector, cruciform			I, O	
length slow	groups, radiating	two at 90°	interpenetrant, cyclic, sector			I, S	
length slow	prismatic	one perfect		Z		I	cf. heulandite, epistilbite

TABLE 11 ZEOLITE GROUP	Mineral	Composition	System	Colour	Pleochroism	Relief	2V Dispersion	Extinction
Monoclinic Biaxial −								
	ANALCITE	$Na\{AlSi_2O_6\}\cdot H_2O$	Cub.	colourless, brown		−L	1	$0°$ − $85°$ anom.	parallel
	GMELINITE	$(Na_2Ca)\{Al_2Si_4O_{12}\}\cdot 6H_2O$	Mon. (Pseudo-Trig.)	colourless		−L	1	$0°$ − moderate	symmetrical
	CHABAZITE	$Ca\{Al_2Si_4O_{12}\}\cdot 6H_2O$	Mon. (Pseudo-Trig.)	colourless		−L	1	$0°$ − $32°$	symmetrical
	SCOLECITE	$Ca\{Al_2Si_3O_{10}\}\cdot 3H_2O$	Mon. (Pseudo-Tet.)	colourless		−L	1	$36°$ − $56°$ r<v	$18°$
	GISMONDINE	$Ca\{Al_2Si_2O_8\}\cdot 4H_2O$	Mon. (Pseudo-Tet.)	colourless		N	1 − 2	$15°$ − $90°$ r<v	small
	LEONHARDITE	$Ca\{Al_2Si_4O_{12}\}nH_2O$	Mon.	colourless		−L	2	$26°$ − $44°$ r<v	$8°$ − $33°$
	LAUMONTITE	$Ca\{Al_2Si_4O_{12}\}\cdot 4H_2O$	Mon.	colourless		−L	2	$26°$ − $47°$ r<v	$8°$ − $11°$
	STILBITE	$(Ca,Na_2,K_2)\{Al_2Si_7O_{18}\}\cdot 7H_2O$	Mon. (Pseudo-Hex.)	colourless		−L	2	$30°$ − $49°$ r<v	$0°$ − $5°$ wavy
	EPISTILBITE	$Ca\{Al_2Si_6O_{16}\}\cdot 5H_2O$	Mon.	colourless		−L	2	$44°$ r<v	$10°$

Orientation	Form	Cleavage	Twinning	Zoning	Alteration	Occurrence	Remarks
	trapezohedral, rounded, radiating, irregular	poor	polysynthetic, complex, interpenetrant			I, S, M	cf. leucite, sodalite, wairakite
	tabular, prismatic, rhombohedral, approaching cube, radiating	one good, one imperfect, parting	interpenetrant			I	cf. chabazite
	rhombohedral, approaching cube, anhedral, granular	one poor	interpenetrant	Z		I, M	basal section in six segments cf. gmelinite, analcite
length fast	columnar, fibrous	two at 88°	interpenetrant, common			I, M	cf. metascolecite, natrolite
	euhedral	distinct	complex, segmented			I, O	basal section shows four segments, opposite parts alike and extinction 5°
length fast/slow	prismatic, fibrous	two good				I	cf. laumontite
length slow	prismatic, fibrous	two good			leonhardite	I, M	cf. leonhardite, phillipsite
cleavage length fast/slow	sheaf-like aggregates, spherulitic	one good	sector, interpenetrant, cruciform			I	cf. heulandite, phillipsite
length slow	prismatic, sheaf-like aggregates, spherulitic	one good	sector, interpenetrant, cruciform			I	

Index of Mineral Species